Advances in Intelligent Systems and Computing

Volume 361

About this Series

The series "Advances in Intelligent Systems and Computing" contains publications on theory, applications, and design methods of Intelligent Systems and Intelligent Computing. Virtually all disciplines such as engineering, natural sciences, computer and information science, ICT, economics, business, e-commerce, environment, healthcare, life science are covered. The list of topics spans all the areas of modern intelligent systems and computing.

The publications within "Advances in Intelligent Systems and Computing" are primarily textbooks and proceedings of important conferences, symposia and congresses. They cover significant recent developments in the field, both of a foundational and applicable character. An important characteristic feature of the series is the short publication time and world-wide distribution. This permits a rapid and broad dissemination of research results.

More information about this series at http://www.springer.com/series/11156

Herwig Unger · Phayung Meesad
Sirapat Boonkrong
Editors

Recent Advances in Information and Communication Technology 2015

Springer

Editors
Herwig Unger
FernUniversität Hagen
Lehrgebiet Informationstechnik
Hagen
Germany

Phayung Meesad
King Mongkut's University
of Technology North Bangkok
Faculty of Information Technology
Bangkok
Thailand

Sirapat Boonkrong
King Mongkut's University
of Technology North Bangkok
Faculty of Information Technology
Bangkok
Thailand

ISSN 2194-5357 ISSN 2194-5365 (electronic)
Advances in Intelligent Systems and Computing
ISBN 978-3-319-19023-5 ISBN 978-3-319-19024-2 (eBook)
DOI 10.1007/978-3-319-19024-2

Library of Congress Control Number: 2015938582

Springer Cham Heidelberg New York Dordrecht London

Printed on acid-free paper

Springer International Publishing AG Switzerland is part of Springer Science+Business Media
(www.springer.com)

Preface

Communication and data mining are the two major, rapidly developing areas of computer science in our today's networked society. A manifold set of devices starting from mobile phones and computers until sensors and usual household appliances are able to run programs as well as send and receive data to satisfy the needs of their owners in an efficient manner. More and more, the network becomes an intelligent and context sensitive environment, which can meet own decisions in a limited area and adapt itself to permanently changing conditions.

Our published book is a try to present some recent research work and results in the area of communication and information technologies. For the eleventh time, researchers from over twenty countries contribute with their chapters to the success of this book, well guided from seventy members of our program and editorial board, which asked the thirty-two authors to present their view on current developments.

Definitely, determining the focus of this book is not an easy task over the years. Although data mining, machine learning and data networking have been attracting interests in the research community for decades, there are still many aspects that need to be explored. Many researchers, research students and the people on the streets and in the shops are waiting to see what new technologies and applications can bring to the society. On the one hand, new services, the latest hardware as well as software are sold in large quantities. On the other hand, also some fear may arise to live in a world of permanent monitoring and control by even unknown third authorities.

To present all aspects of the described developments, the book is divided into two main parts. The first is dedicated to Data Mining and Machine Learning. The second is related to topics from Data Network and Communications. After discussing new algorithms and methods of data mining, the reader will be curious to get to know about their innovative applications in the second section of our volume. Security is the topic of our third chapter, which is related to all aspects of secure data processing also including ethical guidelines for data mining as well as safety and fault tolerance of data storage and transfers in our today's computer systems. Hereby, the user and the respective person's behaviour stand in the center of interests: humans decide which solution will be accepted and applied in their daily life, but on the other hand, the exploration of the users and their habits are in the focus of most companies, governments and therefore applications. The 'unknown' user plays such an important role, that we dedicated another section of our book to this topical field. Last but not least, we will give a view on today's and next generation Internet platforms and systems, which are the fundamental for all developments cited before.

Going through this book, we hope that readers, especially researchers, will be able to see state-of-the-art technologies on data mining, machine learning and data networking. Beginners or research students should be able to grasp basic ideas, background knowledge and related theories to their interests. We hope to inspire all our readers to generate a plenty of new ideas but also think about their impact on our societies.

The actual process of publishing this book took months. It all started from gathering ideas and papers from researchers and research students that have been evaluated by experts in the areas. We are thankful to Springer for their support and for agreeing to publish this book.

We would also like to thank all authors for their submissions. A lot of technical and organisational work has been done by the staff of the Information Technology Faculty at King Mongkut's University of Technology North Bangkok. Without the meticulous work of Ms. Watchareewan Jitsakul and Ms. Areena Ruangprach, the book could not have been completed on time.

March 24, 2015
Bangkok

Herwig Unger
Phayung Meesad
Sirapat Boonkrong
(Editors)

Organization

Program Committee

M. Aiello	Uni.Groningen, The Netherlands
T. Anwar	UTM, Malaysia
T. Arreerard	RMU, Thailand
W. Arreerard	RMU, Thailand
S. Auwatanamongkol	NIDA, Thailand
G. Azzopardi	Uni.Groningen, The Netherlands
T. Bernard	Syscom CReSTIC, France
T. Boehme	TU Ilmenau, Germany
M. Caspar	TU Chemnitz, Germany
J. Chartree	SRU, Thailand
H. K. Dai	OSU, USA
N. Ditcharoen	UBU, Thailand
T. Eggendorfer	HS Weingarten, Germany
J. Grover	MMUS, India
C. Haruechaiyasak	NECTEC, Thailand
S. Hengpraprohm	NPRU, Thailand
K. Hengproprohm	NPRU, Thailand
P. Hiranvanichakorn	NIDA, Thailand
A. Intrapairoth	RMUTT, Thailand
U. Inyaem	RMUTT, Thailand
W. Jirachiefpattana	NIDA, Thailand
T. Joochim	UBU, Thailand
M. Kaenampornpan	MSU, Thailand
P. Khonchoho	PBRU, Thailand
V. Khonchoho	PBRU, Thailand
P. Kropf	UNINE, Switzerland

Mario Kubek	FU Hagen, Germany
P. Kucharoen	NIDA, Thailand
K. Kyamakya	Uni.Klagenfurt, Austria
U. Lechner	UniBwM, Germany
P. Longpradit	PBRU, Thailand
P. Luenam	NIDA, Thailand
C. Namman	UBU, Thailand
S. Nuanmeesri	RSU, Thailand
P. Palawisut	NPRU, Thailand
J. Phuboonob	MSU, Thailand
R. Poonsuph	NIDA, Thailand
P. Prathombutr	NECTEC, Thailand
S. Puangpronpitag	MSU, Thailand
A. Ruangprach	KMUTNB, Thailand
A. Runvat	RMU, Thailand
P. Saengsiri	TISTR, Thailand
T. Sahapong	RMU, Thailand
P. Sanguansat	PIM, Thailand
T. Siriborvornratanakul	NIDA, Thailand
S. Smanchat	KMUTNB, Thailand
M. Sodanil	KMUTNB, Thailand
S. Sodsee	KMUTNB, Thailand
G. Somprasertsri	MSU, Thailand
P. Songram	MSU, Thailand
O. Sornil	NIDA, Thailand
T. Srikhacha	TOT, Thailand
N. Srirajun	NPRU, Thailand
W. Sriurai	UBU, Thailand
M. Srivirat	UBU, Thailand
T. Sucontphunt	NIDA, Thailand
S. Suranauwarat	NIDA, Thailand
S. Sutanthavibul	NIDA, Thailand
P. Tanawongsuwan	NIDA, Thailand
N. Thamakornnonta	NIDA, Thailand
D. Thammasiri	NPRU, Thailand
S. Tongngam	NIDA, Thailand
D. H. Tran	HNUE, Vietnam
K. Treeprapin	UBU, Thailand
D. Tutsch	Uni.Wuppertal, Germany
S. Valuvanathorn	UBU, Thailand
N. Wisitpongphan	KMUTNB, Thailand
K. Woraratpanya	KMITL, Thailand
P. Wuttidittachotti	KMUTNB, Thailand
B. Yoosuk	RMUTT, Thailand

Organizing Partners

In Cooperation with

King Monkut's University of Technology North Bangkok (KMUTNB)
FernUniversitaet in Hagen, Germany (FernUni)
Chemnitz University, Germany (CUT)
Oklahoma State University, USA (OSU)
Edith Cowan University, Western Australia (ECU)
Hanoi National University of Education, Vietnam (HNUE)
Gesellschaft für Informatik (GI)
Mahasarakham University (MSU)
Ubon Ratchathani University (UBU)
Kanchanaburi Rajabhat University (KRU)
Nakhon Pathom Rajabhat University (NPRU)
Mahasarakham Rajabhat University (RMU)
Phetchaburi Rajabhat University (PBRU)
Rajamangala University of Technology Lanna (RMUTL)
Rajamangala University of Technology Krungthep (RMUTK)
Rajamangala University of Technology Thanyaburi (RMUTT)
Prince of Songkla University, Phuket Campus (PSU)
National Institute of Development Administration (NIDA)

Contents

Invited Papers

Data Mining

Data Mining Methods

Data Mining Applications

Networks

Security

User Behaviour

Network Systems and Platforms

An Uncomfortable Change: Shifting Perceptions to Establish Pragmatic Cyber Security

Andrew Woodward and Patricia A.H. Williams

School of Computer and Security Science, Edith Cowan University, Western Australia
{a.woodward,trish.williams@ecu.edu.au}

Abstract. The challenges that a lack of conventionally conceptualized borders in Cyberspace create are increasing in scale and form. This position paper evaluates through the myriad of reasons for this situation, from the absence of cyber security standards, an industry which values training over education for short term gains, resulting in a long term de-skilled workforce, to a solutions space that has an excessive focus on technological control. This demands a necessary change in approach to cyber security to meet the increasingly intelligent and diverse threats. As a specialist field, cyber security requires a collective proactive approach incorporating technology, government support, policy and education. Indeed, it is possible that a reversal of currently accepted perceptions, where organizations manage their security in isolation, will result in a paradigm shift. This will demand acceptance of a shift in power and influence as nation states, crime and hacktivist groups with high levels of motivation, attempt to control and exploit Cyberspace.

Keywords: Cyber security, cyberspace, security, standards, education.

1 Introduction

We are approaching a rapidly uncontrollable Cyberspace environment, one in which traditional boundaries no longer apply. The challenges revolve around a lack of conventional borders. With cyber security a specialist field, and not a mere extension of network and information security, it requires a cooperative and uncommon approach of collaboration between desperate and sometimes competing stakeholders. To be effective against cyber threats it also requires government support, policy and education. The unbounded Cyberspace environment will drive this paradigm shift in security practice.

Evidence shows that most organizations, irrespective of their size and resource availability, are simply not prepared for the breadth and depth of the threats they face. Statistics on breaches provide some insight into the significance of the problem. For example, the majority of data breaches are discovered by a third party (Figure 1), and in some instances, a significant period of time will pass before a data breach is even discovered [1]. Conversely, the time it takes to penetrate an information system is usually measured in hours, not days. This situation is affirmed by the anti-virus (AV) industry, which have effectively admitted defeat, and is confirmed by research which

H. Unger et al. (eds.), *Recent Advances in Information and Communication Technology 2015*,
Advances in Intelligent Systems and Computing 361, DOI: 10.1007/978-3-319-19024-2_1

shows that rate of creation of new malware exceeds the ability of AV software to deal with the threat [2]. Malicious activity caused through online attack includes denial of service, cyber terrorism, hacking, credential theft, distribution of malware, and phishing to name but a few [3]. In an industry that is costing Australia $1.65b per annum and affecting over 5 million people [4], there is no doubt as to the breadth and depth of the protection issues. This is also reflected in the increasing number of cyber security incidents of a significant magnitude will that require response by entities such government based cyber security response centers.

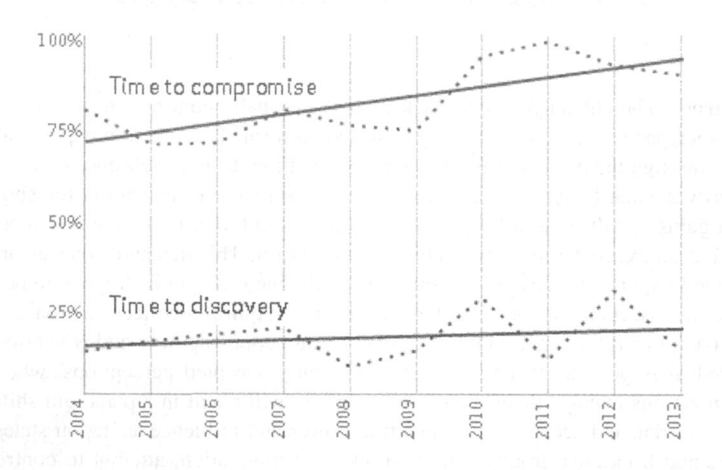

Fig. 1. A longitudinal representation of the number of compromises and discovery of compromises, where time to compromise or discover was less than one day. Source: [1].

This position paper evaluates the myriad of reasons for these challenges, the drivers for necessary change to meet the increasingly intelligent and diverse threats, and provides a discourse on the potential solutions. The result is a possible reversal of currently accepted perceptions, where organizations will be the lesser influential party. This will also demand acceptance of a shift in power and influence as nation states, crime and hacktivist groups with high levels of motivation, attempt to control and exploit Cyberspace. The impact of this discourse will be in opening up new conversations and perspectives on the conflict space of the future, and why there is a need to shift conceptual understanding if pragmatic cyber security is to be achieved.

2 Drivers for Change

2.1 Legislation

There can be little argument that legislation drives change in whichever sector or market that legislation is targeted at. Whilst some countries, such as the United States, have mandatory data breach notification, and others like Australia are in the process of legislation, many do not have such a requirement. In addition to data breach notification, there is also legislation and guidelines in Australia and other countries in relation to privacy requirements and minimum standards. The European Union is currently drafting its own data breach framework which will likely form the basis of any legislation [5]. It can be argued that in the absence of legislation, and in the absence of any other adverse consequence e.g. human lives or reputational damage, there is little incentive to make organizational change. Such legislation is vital in addressing the cyber security challenge, as it brings cyber security to the attention of corporate executives, and it will then form an integral part of a consequence framework in an organizations risk management process. Without such regulatory drivers, the organizational resources allocated to cyber security initiatives will likely not be sufficient to address the threats, as currently there are limited consequences for an adverse cyber security outcome. One recent report suggests that the actual financial loss from high profile breaches of Sony, Target and other institutions were so small that they are not sufficient to drive cyber security change [6]. However, the legislation is clearly having an effect in the US, with reports that chief financial officers are increasing their organizational spend on cyber security [7]. Additionally, a recent Gartner report stated that global information security spending would increase by 8.2% in 2015 to 76.9 billion dollars annually [8]. This suggests that until such legislation becomes widespread internationally, we are not likely to see significant change in the risk profile of most organizations in relation to cyber security.

2.2 Lack of Targeted Standards

Few standards apply specifically to cyber security. It is unfortunate that in an area but lacks standardization the word 'standard' is used to describe certification and best practice, rather than established international standardization. Similarly, there is confusion between the definitions of certification versus education, which is discussed below.

Whilst there exist standards and guidelines related to risk assessment and information security management, which can be applied to cyber security, until 2012 there was no specific cyber security standard. The *ISO/IEC 27000 Information technology -- Security techniques – Information security management systems* series comprises a suite of 37 Standards ranging from information security management, governance, specific controls and applications such as health. Similarly, *ISO 31000 Risk Management – Principles and guidelines*, and *ISO/IEC31010 Risk management – Risk assessment techniques* provide a framework and associated processes for risk assessment and management. Whilst application of these standards to cyber security may be extrapolated from the content, they do not view the increasingly complex environment of Cyberspace as a distinct problem space. In 2012, this issue was recognized with the publication of *ISO/IEC 27032:2012 Information technology -- Security techniques -- Guidelines for cybersecurity*.

ISO/IEC 27032 is the first standard to distinguish that the complex and difficult to delineate environment of Cyberspace, needed a unique approach to fill the gaps between the traditional security domains such as networks security, information security, Internet security and so on. Interestingly, this standard also identifies the fragmentation of security that exists between organizations in Cyberspace [9]. In this respect, it is a bridging standard in that it focuses on the techniques and controls to protect against threats that do not fall within the scope of more traditional security standards, such as social engineering, hacking, and malware [10]. Further, the Standard discusses how collaboration between desperate entities should occur for effective incident handling. This is the first standard to navigate the logical rather than physical boundaries of competing regulatory jurisdictions. The significance of the standard is in the perspective taken which is not solely technical but includes a framework for developing trust between entities. It is a necessarily high-level standard, yet it establishes the groundwork for further development of context specific detailed specifications. As with most standards, it also provides an educational component on Cybersecurity and its relationship to other types of security.

There are also numerous guides and frameworks produced by the US National Institute of Standards and Technology (NIST), with a recently released Cyber Security Framework. Although targeted at critical infrastructure, this framework could be applied to cyber security in general. Organizations rely on such frameworks in attempting to address the cyber security threat, as they seek some kind of reassurance that what they are doing aligns with best practice. Conversely, these same organizations are also averse to undertaking or engaging any process, practice or device, which does not seem to conform to a standard, guideline, or established best practice. This is not to necessarily suggest that there should be one, but rather, to start a discussion about the role of standards, and how they are used as part of a cyber-security strategy, or whether the lack of a specific standard for cyber security is even an issue.

2.3 Technical Issues

For many organizations, there is an over reliance on technical solutions to the cyber security threat. Many vendors of technical solutions will readily inform customers that their device is the magic bullet, and will address their cyber security challenges. The reality is that whilst technical controls have a role to play in protecting organizations information assets, technical controls alone are not able to fill the entire role. The reason for this is that the problem is multi-faceted, and ultimately, people are involved in many parts of the process of using, protecting and disseminating information, both internally and externally. As such, people themselves contain vulnerabilities, in that they are human, and can be exploited in the same way that vulnerabilities in computer systems can also be exploited. Also, humans form the link between technical controls and process implementation. Many of those occupying the role of information security officer or manager in organizations are frequently doing so as part of another role, or are simply under resourced to tackle the cyber security threats. As such, it is easy to understand why such people would consider such a solution: it claims to address the issues, it is usually from a reputable vendor so there is less risk involved, and it means that they can address at least some of the threat without having to spend time on implementing any further solution. The counterpoint to that is the high volume of data breaches which still occur in organizations, despite the significant resources and dedicated information security personnel and mangers that these organizations employ [1].

In Australia, the Australian Signals Directorate produces a Top 35 list of mitigation strategies [11]. Whilst these are sound suggestions , and the effectiveness of each is rated, the authors have experienced situations where people are implementing only the top four on the list and not implementing any of the other recommendations. The issue with this is that there are no recommendations relating to the human factor until you reach number 28 on the list. This is problematic in that most cyber security breaches involve an element of human failure, with policy, governance, and security education awareness and training all having important roles to play. It is also worth noting that these recommendations are based on firstly, what worked, and secondly what worked for Australian Government agencies and departments. As such, it is reflective of the past, and may not necessarily deal with the medium to long-term threat horizon.

2.4 Education Issues

The IT industry has traditionally relied upon certifications, both vendor and professional body based, to up-skill workers to undertake IT work. Whilst this has a place, and is relevant to train workers to operate particular software and hardware, these skills are often not transferrable. It can be argued that the knowledge and expertise required to adequately defend an organizations' data from cyber threats goes beyond that which can be offered through certifications. Further exacerbating the problem is that these certifications are usually well marketed and understood by the market, and as such organizations seeking to address their cyber security issues will often engage organizations who have people certified in particular skills. A case in point is certification based on penetration testing. People with this certification may have at least the knowledge required to undertake a penetration test, but unfortunately, penetration testing is intended to address only a particular small subset of cyber security, namely that of testing the configuration of specific systems. It is not a substitute for a thorough vulnerability analysis or risk assessment.

The discrepancy between education and certification is caused by a lack of understanding, brought about through lack of education, about the difference of the two, what they offer, and what the common causes of cyber security vulnerabilities are. Another example of an education gap is related to the concept of defense in depth, which is often quoted as being vital in cyber security, and yet is poorly understood by practitioners. This is borrowed from the physical security domain, and when implemented as intended refers to a security strategy which incorporates all elements of deter, detect, delay and respond, known as 3DR. However, reference to any IT based information about defense in depth only talks about delaying an attacker, through layering network defenses which do nothing more than delay an attack. For example, Lippman et al write "Defense in depth is a common strategy that uses layers of firewalls ..." [12]. This should be more correctly referred to as delay in depth, not defense in depth, as firewalls are consistent with a delay tactic. This was published by a reputable journal, and has since been cited in excess of 100 times. This highlights an education gap and a failure to implement a strategy incorporating all elements of 3DR, and as such will result in a flawed solution.

This lack of education in those managing an organization's information security will likely initiate a penetration test using an external consultant. This decision is driven by limited knowledge or inaccurate advice by those underneath them who are aware of these testing through promotion of certifications in this area.

This discourse does not suggests that certifications do not have their place. They are useful to address a skills gap in an organization at short notice. The reality is that whilst a Higher Education qualification provides greater benefit to an individual and their adaptability in an organization, it demands significantly greater time and resources to acquire this education. Used more correctly, certifications would be a valuable addition for an employee who holds a computer science or information technology degree, as they would have the fundamental underpinnings and knowledge to better utilize the specific training outcomes.

Another aspect of the certification vs. education argument is concerned with the lack of coherence or consistency within the profession itself. The IT security sector has such a varied employment market, that as such, it can be challenging to define a cyber-security role, and the corresponding qualifications or experience that is required. There have been attempts to address this issue, through schemes such as the skills framework for the information age (SFIA) [13]. However, the reality is that whilst a security guard outside of a pharmacy is required to be licensed and meet certain minimum standards, those proclaiming themselves as information security professionals do not have to substantiate their claim. Frequently, their evidence to support such a claim is a selection of industry and vendor certifications. From an educational perspective, these equate to Certificate or Diploma level qualifications, and at best are equivalent to the first year of a university Bachelor degree when viewed from a learning outcomes perspective. Similar professions such as Accounting and Engineering do not allow such activity to occur, and even professions which were traditionally vocational, such as Nursing, now require a tertiary education qualification.

3 Demand for a New Paradigm

The paper to this point has discussed some of the many issues and challenges that make up the current cyber security landscape. Further, the discussion substantiated these with examples of what constitute the problem. This section will highlight some key areas that we believe need to be implemented if the current issues are to be addressed.

3.1 Legislation

The need for the legal fraternity to understand better, and enact legislation which has adverse consequences for those failing to maintain adequate cyber security in their organization is required. Legislation to enforce cyber security would be counterproductive, as it would likely lead to an environment where every organization had the same level of security, and with the same potential vulnerabilities. Currently, data breach notification legislation is one part and this should be coupled with financial penalties for organizations that experience a data breach over and above that of the actual loss. It seems that having to notify those affected by a breach, which is meant to be a form of reputational consequence, is not adequate, as breaches are still too frequent.

3.2 Standards

Adherence to, and guidance from relevant standards is crucial in the gamut of cyber security protection. Standards provide requirements and specifications that can be used over and over again, consistently and across organisations. Standards are produced and published by globally recognized entities (Standards Development Organisations SDOs) to ensure reliable, quality, balanced, consensus driven expert advice, "to meet technical, safety regulatory, societal and market needs" [14]. The purpose of using standards is to guarantee fit-for-purpose whilst not constraining innovation nor limiting market potential. In this space however, more input is required from experts in the field to contribute to the development of security in Cyberspace and to provide guidance in an absence of regulatory compliance mechanisms.

3.3 Education

There are several vectors which need to be addressed under the banner of education. Firstly, a universally agreed upon and available set of established skills and foundation knowledge that would need to be demonstrated by anyone claiming to be a cyber security professional. By extension, any program, degree or course offered by an institution would need to be measured against this cyber security core body of knowledge. This could also include specializations at a later point, but an initial level of agreed upon knowledge is essential to begin to tackle the problem. Furthermore, industry itself needs to have a very clear understanding about the role of training and that of education in tackling cyber security

In addition, education of the Executive and 'C' Level Officers in organizations about the real level of risk and threat presented by cyber security, or a lack thereof is important. Critical to this is a thorough explanation of what tools and processes are valuable in increasing cyber security, and which offer very little value. Such education alongside privacy and data breach legislation would increase the visibility of the problem within the governance structure of an organization.

4 Conclusion

It is clear that the current approach to cyber security is not working, and this new conflict space needs a new approach to tackling security. The media is littered with daily attacks against individuals, small business, and major multinational corporations. The problem space is complex, multifaceted, and it will take considerable combined will to enact change. Legislation clearly has a role to play as a driving factor to ensure that cyber security is taken seriously by anyone collecting, analyzing, storing or transacting data. However, without cohesion from the IT profession around cyber security, and clear standardization of minimum requirements, this will be challenging. It will take considerable effort to create change in this space, and until such time as this occurs, the number of adverse cyber security incidents is only likely to increase. Therefore, as a specialist field, cyber security requires a collective proactive approach incorporating technology, government support, policy and education. This will introduce a reversal of currently accepted perceptions, where organizations manage their security in isolation, and will result in a paradigm shift. This will demand acceptance of a shift in power and influence as nation states, crime and hacktivist groups with high levels of motivation, attempt to control and exploit Cyberspace. Unless organisations embrace this shift they will be left more and more vulnerable.

References

1. Verizon: 2014 Data breach investigations report, Verizon: 60 (2014)
2. Haffejee, J., Irwin, B.: Testing antivirus engines to determine their effectiveness as a security layer. In: Information Security for South Africa (ISSA). IEEE (2014)
3. Zhang, Y., Xiao, Y., Ghaboosi, K., Zhang, J., Deng, H.: A survey of cyber crimes. Security and Communication Networks 5(4), 422–437 (2012)
4. Australian Government: Australian cyber security centre to be established (2013), http://www.defence.gov.au/defencenews/stories/2013/jan/0124.htm
5. Pearson, N.: A larger problem: financial and reputational risks. Computer Fraud & Security 4, 11–13 (2014)
6. Dean, B.: Why companies aren't investing in cyber security. itnews (2015)
7. Norton, S.: Tech CFOs Boost Spending on Cybersecurity: Report. CIO Journal, The Wall Street Journal (2015)
8. Gartner: Gartner Says Worldwide Information Security Spending Will Grow Almost 8 Percent in 2014 as Organizations Become More Threat-Aware (2014), http://www.gartner.com/newsroom/id/2828722
9. ISO: ISO/IEC 27032:2012 Information technology – Security techniques – Guidelines for cybersecurity (2015), http://www.iso.org/iso/catalogue_detail?csnumber=44375
10. ISO: Are you safe online? New ISO standard for cybersecurity (2012), http://www.iso.org/iso/home/news_index/news_archive/news.htm?refid=Ref1667
11. ASD: Strategies to Mitigate Targeted Cyber Intrusions (2014), http://www.asd.gov.au/publications/Mitigation_Strategies_2014.pdf
12. Lippmann, R., Ingols, K., et al.: Validating and restoring defense in depth using attack graphs. In: Military Communications Conference, MILCOM 2006. IEEE (2006)
13. Leahy, D., Wilson, D.: Digital skills for employment. In: Passey, D., Tatnall, A. (eds.) KCICTP/ITEM 2014. IFIP AICT, vol. 444, pp. 178–189. Springer, Heidelberg (2014)
14. IEEE: Purpose of Standards Education (2015), http://www.ieee.org/education_careers/education/standards/why.html (retrieved)

Feature Point Matching with Matching Distribution

San Ratanasanya, Jumpol Polvichai, and Boonchareon Sirinaovakul

Department of Computer Engineering,
King Mongkut's University of Technology Thonburi, Bangkok, Thailand
hameroon@hotmail.com, jumpol@cpe.kmutt.ac.th,
boon@kmutt.ac.th

Abstract. Most of the feature point matching techniques considers only the number of matches. The higher number of matches is, the better results are. However, reliability and quality of the matching is addressed in a few techniques. So, finding the good matches of the pairs of points from the two given point sets is one of the main issue of feature point matching. This paper presents new approach to obtain reliable and good matches. The high quality of matching can be achieved when the matches are spread all over the entire point set. Therefore, we proposed to use the distribution of the matches to verify the quality of the matching. The preliminary results show that our proposed algorithm significantly outperform SIFT even when the results of SIFT are enhanced using RANSAC.

Keywords: Feature Point Matching, Image Registration, Image Retrieval, SIFT, RANSAC.

1 Introduction

Feature point matching is a very important process for applications that need to locate objects in images or databases such as robot navigation, tele-surgery, and image retrieval [1, 2, 3]. The feature points are derived from the images, called the reference and input image, of the same scene which may have been acquired at different times, from different viewpoints, and by different sensors. The goal of feature point matching is to geometrically align these two point sets. Automatic feature point matching algorithms are now used to initiate the better input for various complex tasks such as navigating robot, finding stereo correspondences, motion tracking, and recognizing object and scene.

There are many techniques to match feature points. All of them rely on location of points and/or other information. For instance, the techniques based on SIFT [4] or SIFT-based techniques such as SURF [5], ORB, BRIEF, and FREAK [6] use descriptor, a vector describes the distinctiveness of each feature computed by using neighborhood intensities of that feature points, to match the feature points. However, the distinctiveness of SIFT-based descriptors might not be distinct enough in some cases and results in ambiguous match. That is, one feature point can be matched with more than one feature points. Moreover, the geometric structures are used in more

H. Unger et al. (eds.), *Recent Advances in Information and Communication Technology 2015*,
Advances in Intelligent Systems and Computing 361, DOI: 10.1007/978-3-319-19024-2_2

recent techniques. The geometric structures are used to select a subset of points and match them to get better results. Additionally, a subset of matches can be selected according to some conditions, such as angular [7] and pairwise constraints [8], to get better matching results. A random selection can also be applied to improve the results [9]. Unfortunately, this approach cannot provide good matching results since the same pattern geometric structure might be found in any part of the images. It is hard to find the correct match of the same structure especially in the cases that the images are partially overlapped or do not have overlapped regions.

In this paper, we proposed a new algorithm to solve feature point matching problem. We treat feature point matching problem as a searching for the closet pair of points. The one-to-one relationship is used to find a reliable match by considering a distance between the pair of points. Moreover, we take distribution of the matches into account in order to make sure that the number of overall matches is reliable.

2 Literature Reviews

From the literatures, the techniques to tackle feature point matching can be classified into three groups [10]. The first category is to use only location of the points to find the best correspondences. The techniques in this group are simple but take high computational time since it needs exhaustive search throughout the search space to find the solution. However, they are sensitive to noise and cannot provide good results in some cases. The second group is to incorporate neighborhood information of each point to get the best correspondence. The techniques in this group seem to have more robust to noises than those of the first group since they use more information to find the solution. The state of the art technique in this category is SIFT. Although SIFT variations are also widely used with the comparable performance to SIFT, but they cannot outperform SIFT in every aspect [11]. However, their results are comparable. Some techniques get rid wrong match still need to be used in post processing to improve the solution. The techniques in third group are usually based on graph matching algorithms and use structural information of the point set to find the correspondences between the two point sets. These techniques seem to be better than those of the first two categories, but they require the most similar structure of the two point set for the best results [12].The principles of the techniques in each category are presented in the following section.

2.1 Location Information

The techniques in this group use only one piece of information that is location of the points. After the feature points are extracted from the images, the location of each point is used to determine the correspondences with the other point set. The framework is divided into four sub-tasks, namely, feature extraction, transformation space, search strategy, and similarity measure [13] as can be seen from the block diagram shown in Fig. 1.

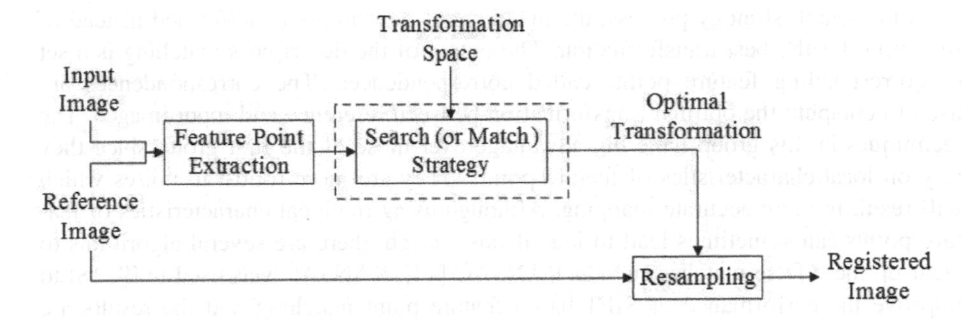

Fig. 1. Image Registration Block Diagram

A pair or a set of images is input to Feature Extraction sub-process. This sub-process will extract feature points from each image. These feature points represent the image itself with much smaller information, to cut off unnecessary computation and speed up the whole registration process, and are invariant to a specific class of transformation. Then, we decide which class of transformation is used to transform the input image. This assumption is set up in the Transformation Space sub-process. Also, we set up transformation width or range of transformation that the optimal transformation is in. Next, we search for or estimate the transformation within that range. There are several strategies for searching or estimation such as geometric branch and bound framework as proposed by [13]. So, we call this sub-process as Search Strategy. Finally, we can check whether the transformation is optimal by measuring the similarity between the reference and input image, after applying that transformation. The transformation that gives the best similarity will then be optimal transformation. This will be measured in the Similarity Measure sub-process using distance function such as Partial Hausdorff Distance.

Moreover, there is several works that apply optimization algorithm with the basic search. That is, searching for the best combination of transformation parameters such as rotation, scaling, and translation under the specific transformation. The optimization algorithm will then be applied to accelerate the search. For instance, Genetic algorithm and PSO are applied in [14, 15, 16].

Simplicity is the main advantage of the techniques in this group. Although it is computational intensive, any optimization algorithms could be used to accelerate the search. However, it may not robust to noises according to the definition of matched point.

2.2 Neighborhood Information

The techniques in this group are based on SIFT [4]. The framework of the techniques in this group is quite similar to that shown in Fig. 1 except the feature extraction and search strategy processes. In the feature extraction process, the feature points are extracted and their neighborhood information are used to build descriptor for each feature point. The details of how to extract feature point and create descriptor can be found in [4].

In the search strategy process, the matching of descriptors is performed instead of searching for the best transformation. The output of the descriptors matching is a set of corresponding feature points called correspondences. The correspondences are used to compute the optimal transformation between reference and input images. The techniques in this group have big advantage over those of the first group since they rely on local characteristics of feature points. They are more robust to noises which will result in more accurate mapping. Although using the local characteristics of feature points can sometimes lead to lots of false match, there are several algorithms to clear up these false matches such as RANSAC [17]. RANSAC was used in [9, 18] to improve the performance of SIFT-based feature point matching and the results are impressive.

Moreover, the feature points extracted by SIFT-based detector are high redundancy meaning that there are multiple features at the same location but not scale. The SIFT descriptors are build regarding to scaled local information of each feature point results in distinctive characteristic of each feature. The redundancy allows testing for best correlation of points more accurate. Unfortunately, it has a drawback. The redundancy makes it computationally intensive especially when there is enormous number of features. And the descriptors might lose their distinctiveness since it is based on intensity contrast which cannot be distinct if there are repetitive patterns in the image or there are areas that have similar contrast in the local neighborhood in the images.

2.3 Structural Information

From different perspective, feature point matching can be formulated in terms of graph matching. The feature extraction and search strategy processes in the feature point matching framework are also adopted for the techniques in this category.

After the feature points are extracted, they are treated like nodes in the graph. Then the structural information in terms of edges is added to represent the relation between each node in a graph. The search strategy can be done by graph matching which however is an NP-hard problem. The exact solution may not be found in reasonable time so approximation solution has to be found. Conventional graph matching approaches [19, 20] mainly focus on pairwise similarity between two correspondences such as distances among feature points. Most of them use Iterative Closet Point to minimize distance of set of points. Pairwise relations, however, are not enough to incorporate the information about the entire geometrical structure of features. To overcome the limitation of pairwise similarity, several researchers proposed techniques that applied graph theory, called graph matching, [21, 22] to get good matching results. However, graph matching still has two main issues. The first issue is that geometric constraints are required. Therefore, simulation of connection between edges and nodes must be tested iteratively to meet geometric constraints. Moreover, many techniques do not robust to outlier nodes. The second issue is that optimizing graph matching is difficult due to the nature of non-convex objective function of feature matching problem.

In summary, the main differences among the three groups are geometric constraints and local to global matching. In terms of overall performance, the SIFT-based techniques outperform the others because its density in both feature point and distinctive local descriptor. However, the local descriptors may lead to uncertain matching. On the other hand, if we can introduce local to global matching methods to the simplest group of feature matching, we can therefore get a low complexity technique that robust to noises and outlier so that it results in good feature matching performance.

3 Proposed Method

Considering a point set that has enormous number of points, the typical matching algorithm will be inefficient. The chances of mismatching results are high because there are too many ambiguous points to match. Not only ambiguity is introduced with algorithms in point matching but matching with neighborhood or structural information also ineffective in this case because there will be a lot of similar neighborhood or structural information if there are too many points. Moreover, in the case of symmetry object, the existing search strategy might not be able to tell the difference between the object under 0 and 180 degree of rotation. Therefore, we proposed another search algorithm that takes the distribution of the matches into account to tackle the aforementioned problem. This can be done by visualizing specific zone in the point set. The idea is that each point in the point set will be assigned to a zone or cluster, which is a quadrant in this work. The algorithm will do a search in quadrant-to-quadrant manner. Therefore, there will be 16 combination pairs of subset to be considered.

Considering the particular pair of cluster to be search, all the points from two point sets will be paired up and set as a center of transformation. Then the search aiming to get the maximum matching value will be processed by applying combinations of transformation parameters within the search interval. Matching value can be determined from the number of match pairs and the number of cluster that has at least one match as stated in the following equations.

$$\text{MatchingValue} = \sum_{C} w_c * \sum_{C} \text{match}_c \quad (1)$$

$$w_c = \begin{cases} 1, & \text{if there is at leat a match in cluster c} \\ 0, & \text{otherwise} \end{cases} \quad (2)$$

Where C = {1,2,3, …, n}, n is number of cluster which equals to 4 in this work. match_c means number of matches in cluster c. w_c is a match coefficient of each cluster. The flowchart of this algorithm is shown in Fig. 2.

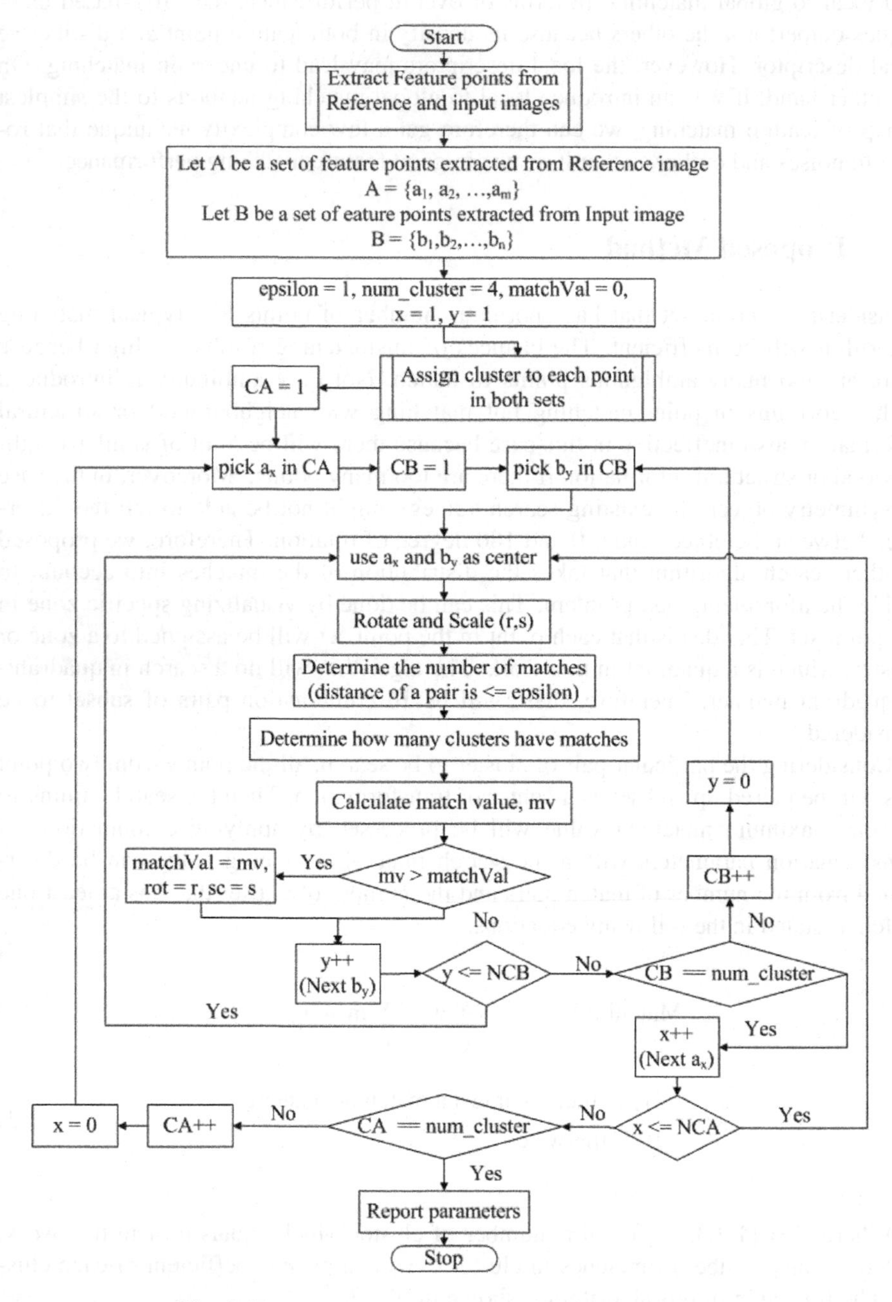

Fig. 2. The Flowchart of the Proposed Search Algorithm

4 Experiments and Discussions

We investigate the performance of our algorithm in terms of accuracy. The experiment is done using simple images representing geometric shapes. The following subsections give details of the preliminary experiment.

4.1 Point Sets

The point sets used in our experiments are derived from the images having geometrical objects. The input images are the reference images that are transformed by known similar transformation. The reference and input images are shown in Table 1 and 2.

Table 1 showed the synthesis images that have only one object and multiple objects in the image. These images represent simple case of feature point matching problem. The transformations applied to the images are to simulate various ranges of similar transformation.

Table 1. Single Object Synthesis Images with the Transformation Applied

Case	Reference	Input	Transformation ([Rotation, Scale])	Case	Reference	Input	Transformation ([Rotation, Scale])
A1			[-90, 1.0]	B1			[18, 1.0]
A2			[180, 1.0]	B2			[0, 0.8]
S1			[180, 1.0]	B3			[0, 1.5]
S2			[-45, 0.7]	O1			[-90, 1.0]
S3			[-45, 1.0]	O2			[180, 1.0]

4.2 Experimental Settings and Measurements

In this work, we assumed similarity transformation and represent it by a 4-element vector, whose entries are the rotation r, the translation vector (tx,ty), and the scaling factor s. However, the translation vector can be omitted since our approaches have already cancelled this vector. Also, we assumed that the feature points are provided. In this paper, we compare our proposed algorithm with SIFT so the SIFT detector is used extract features to be fair comparison.

To measure which set of transformation parameters is the best, we use Euclidean distance to measure shortest distance among neighbor pairs and sum up to get an overall distance. On the other hand, we can compare the transformation parameters directly since we know the exact solution or ground truth. Moreover, we can use other similarity measure tool such as normal correlation to measure similarity of the point sets. For preliminary study, we simply measure the distance of the match pairs. The extensive review of performance evaluation for feature point matching based on local descriptors is presented in [11]. In this paper, we compare our proposed algorithm with SIFT since SIFT provides better performance in terms of matching feature point and robustness to transformations.

Considering high redundancy property of SIFT, there are high chances that there will be a lot of false matches. That is, many matched pairs of points are actually mismatched. The reliable match should have one-to-one relationship meaning that any particular point in one set should be matched to exactly one point from the other set. Therefore, RANSAC would be used to select the best reliable matches for SIFT. We called it SIFT+RANSAC. Moreover, the distance between a matched pair of points directly relates to the quality of that match. The longer distance the poorer match. Therefore, in this paper, we did not measure only number of reliable matches but also the distance of each match. The particular match will be considered as mismatch and will not be counted if the pair of points is far from each other more than a tolerance distance, which is set to 3 pixels. That is, we measure only the matches that are good and reliable.

4.3 Results on the Proposed Algorithm Search

The results are shown in Table 3. It shows that both techniques can find the matches in all cases. However, the number of reliable matches, the number of good matches, and the matching precision, of the solutions obtained from each technique are also presented.

From Table 3, it is obviously seen that the number of reliable match or one-to-one relationship of points obtained from both techniques are not much different exclude the last two cases, "O1" and "O2". It is a nature of SIFT that has high redundancy of feature so the higher number of reliable match could be possible. Although the number of reliable matches obtained from both techniques are quite equal it can be obviously seen that the number of good matches are different. The proposed algorithm provides more good matches than SIFT+RANSAC. Additionally, our proposed algorithm has higher matching precision than SIFT+RANSAC. The value of matching precision also implies that SIFT+RANSAC has more false positive matches than our proposed algorithm. The higher number false positive matches the more ineffective performance to the applications that require feature point matching.

Table 2. The matching accuracy results of the preliminary experiments

Case	Number of Reliable Match Pairs		Number of Good Matches		Matching Precision	
	SIFT+ RANSAC	Proposed Algorithm	SIFT+ RANSAC	Proposed Algorithm	SIFT+ RANSAC	Proposed Algorithm
A1	26	19	6	11	23.08%	**57.89%**
A2	13	19	6	13	46.15%	**68.42%**
S1	22	19	7	17	31.82%	**89.47%**
S2	9	21	1	14	11.11%	**66.67%**
S3	12	21	3	12	25.00%	**57.14%**
B1	18	35	1	16	5.56%	**45.71%**
B2	28	35	5	14	17.86%	**40.00%**
B3	27	35	3	6	11.11%	**17.14%**
O1	131	43	39	31	29.77%	**72.09%**
O2	111	36	4	28	3.60%	**77.78%**

5 Conclusions

We have proposed feature point matching algorithm with three different perspectives from the most of the existing feature point matching algorithms. First, we omitted translation parameter from the search by setting a pair of points in consideration as origin points. Second, we not only measured the number of the matches based on Euclidean distance but also measured the spread of the matches by dividing the point sets into 4 areas and account for the matching distribution. Finally, neither neighborhood nor structural information was needed to provide more accurate mapping in our algorithm. We have done the preliminary experiment with synthesis images and compare the performance and accuracy with SIFT+RANSAC. The experimental results shows that the proposed algorithm outperform SIFT+RANSAC in terms of precision. In addition to investigate the performance of the proposed algorithm, we plan to test our algorithm with standard benchmark image data in the near future.

References

1. Tang, J., Shao, L., Zhen, X.: Robust point pattern matching based on spectral context. Pattern Recognition 47(3), 1469–1484 (2014)
2. Yongfang, G., Ming, Y., Yicai, S.: Study on an Improved Robust Algorithm for Feature Point Matching. Physics Procedia 33, 1810–1816 (2012)
3. Myronenko, A., Song, X.: Point set registration: Coherent Point drift. IEEE Transactions on Pattern Analysis and Machine Intelligence 32(2), 2262–2275 (2010)

4. Lowe, D.G.: Distinctive image features from scale-invariant keypoints. International Journal of Computer Vision 60(2), 91–110 (2004)
5. Bay, H., Ess, A., Tuytelaars, T., Van Gool, L.: SURF: Speeded Up Robust Features. Computer Vision and Image Understanding (CVIU) 110(3), 346–359 (2008)
6. Bekele, D., Teutsch, M., Schuchert, T.: Evaluation of binary keypoint descriptors. In: Proceeding of the 20th IEEE International Conference on Image Processing, pp. 3652–3656 (2013)
7. Kim, J., Choi, O., Kweon, I.S.: Efficient feature tracking for scene recognition using angular and scale constraints. In: Proceeding of the 2008 IEEE/RSJ International Conference on Intelligent Robots and Systems, pp. 4086–4091 (2008)
8. Choi, O., Kweon, I.S.: Robust feature point matching by preserving local geometric consistency. Computer Vision and Image Understanding 113(6), 726–742 (2009)
9. Wang, Y., Huang, H., Dong, Z., Wu, M.: Modified RANSAC for sift-based in SAR image registration. Progress in Electromagnetics Research M 37, 73–82 (2014)
10. Yongfang, G., Ming, Y., Yicai, S.: Study on an Improved Robust Algorithm for Feature Point Matching. Physics Procedia 33, 1810–1816 (2012)
11. Mikolajczyk, K., Schmid, C.: A performance evaluation of local descriptors. IEEE Transactions on Pattern Analysis and Machine Intelligence 27(10), 1615–1630 (2005)
12. Tang, J., Shao, L., Jones, S.: Point Pattern Matching Based on Line Graph Spectral Context and Descriptor Embedding. In: IEEE Winter Conference on Applications of Computer Vision, pp. 17–22 (2014)
13. Mount, D.M., Netanyahu, N.S., Ratanasanya, S.: New approaches to robust, point-based image registration. In: Image Registration for Remote Sensing. Cambridge Publishing (March 2011)
14. Zhang, L., Xu, W., Chang, C.: Genetic algorithm for affine point pattern matching. Pattern Recognition Letters 24(1-3), 9–19 (2003)
15. Senthilnath, J., Omkar, S.N., Mani, V., Kalro, N.P., Diwakar, P.G.: Multi-objective Genetic Algorithm for efficient point matching in multi-sensor satellite image. In: Proceeding of the IEEE Geoscience and Remote Sensing Society, pp. 1761–1764 (2012)
16. Yin, P.Y.: Particle swarm optimization for point pattern matching. Journal of Visual Communication and Image Representation 17(1), 143–162 (2006)
17. Fischler, M.A., Bolles, R.C.: Random Sample Consensus: A Paradigm for Model Fitting with Applications to Image Analysis and Automated Cartography. Communication of the ACM 24(6), 381–395 (1981)
18. Shi, G., Xu, X., Dai, Y.: SIFT Feature Point Matching Based on Improved RANSAC Algorithm. IHMSC 1, 474–477 (2013)
19. Cho, M., Lee, J., Lee, K.M.: Reweighted random walks for graph matching. In: Daniilidis, K., Maragos, P., Paragios, N. (eds.) ECCV 2010, Part V. LNCS, vol. 6315, pp. 492–505. Springer, Heidelberg (2010)
20. Leordeanu, M., Hebert, M., Sukthankar, R.: An integer projected fixed point method for graph matching and map inference. In: Proceeding of the 23rd Annual Conference on Neural Information Processing Systems (2009)
21. Chertok, M., Keller, Y.: Efficient high order matching. IEEE Transactions on Pattern Analysis and Machine Intelligence 32(12), 2205–2215 (2010)
22. Lee, J., Cho, M., Lee, K.M.: Hyper-graph Matching via Reweighted Random Walks. In: Proceeding of the 2011 IEEE Conference on Computer Vision and Pattern Recognition (2011)

Dimensionality Reduction Algorithms for Improving Efficiency of PromoRank: A Comparison Study

Metawat Kavilkrue[1] and Pruet Boonma[2]

[1] Faculty of Engineering, North-Chiang Mai University, Chiang Mai, Thailand
[2] Faculty of Engineering, Chiang Mai University, Chiang Mai, Thailand
comengi49@gmail.com, pruet@eng.cmu.ac.th

Abstract. It is often desirable to find markets or sale channels where an object, e.g., a product, person or service, can be recommended efficiently. Since the object may not be highly ranked in the global property space, PromoRank algorithm promotes a given object by discovering promotive subspace in which the target is top rank. However, the computation complexity of PromoRank is exponential to the dimension of the space. This paper studies the impact of dimensionality reduction algorithms, such as PCA or FA, in order to reduce the dimension size and, as a consequence, improve the performance of PromoRank. This paper evaluate multiple dimensionality reduction algorithms to obtains the understanding about the relationship between properties of data sets and algorithms such that an appropriate algorithm can be selected for a particular data set. The evaluation results show that dimensionality reduction algorithms can improve the performance of PromoRank while maintain an acceptable ranking accuracy.

1 Introduction

Online marketing becomes an important tool for business and organization [7]. Google[1] and Amazon[2], for instances, relies heavily on online marketing operations, e.g., online advertisement, recommendation and fraud detection. For example, when a user search for a specific term, Google will shows related products in GoogleAds. This can promote the sell of the products because it reflects user's interest [9]. This technique has been used widely not only in business but also public sectors. For instance, when a client borrows a book from a library, the library might want to suggest another related books to them based on their interest. This approach can help boost service satisfiable of clients.

Ranking is a technique to carry out recommendation. It is used widely, for instance, in many bookstores, where top selling books are shown on the front of the stores. This can increase those books selling because people are tend to believe that, because so many other customers already bought these books, they should be good. This is also applied to many other fields in business as well, e.g., American Top Forty, Fortune 500, or NASDAQ-100. Because the number of top ranking is limited, only those who are the best on every dimensions can be in the list. Nevertheless, there are many cases that when consider only a subset of the dimensions, some interesting objects can be found. Consider the following example:

[1] http://www.google.com
[2] http://www.amazon.com

H. Unger et al. (eds.), *Recent Advances in Information and Communication Technology 2015*,
Advances in Intelligent Systems and Computing 361, DOI: 10.1007/978-3-319-19024-2_3

Table 1. Example of multidimensional data

Genre	Year	Object	Score
Science	2012	O_1	0.9
Fiction	2012	O_1	0.2
Fiction	2012	O_2	0.8
Fiction	2011	O_2	0.7
Science	2011	O_2	0.5
Science	2012	O_3	0.4
Fiction	2012	O_3	0.8

Table 2. Target object O_1's subspaces and its ranks

Subspace	Rank	Object Count
{*}	3	3
{Genre=Science}	1	3
{Genre=Fiction}	3	3
{Year=2012}	2	3
{Genre=Science, Year=2012}	1	2
{Genre=Fiction, Year=2012}	3	3

Example 1. (Product Recommendation) It is impossible that Donald Knuth's The Art of Computer Programming series can be in the top list of all books in Amazon store. However, when consider only computer science books with readership toward college students, this book will be ranked on the top list. So, this book series should be recommended only in that particular category and readership.

Example 2. (Online Advertisement) a company wants to promote its product through online advertisement channel, e.g., Google's AdSense. However, the company does not have enough budget to promote the product in all market. However, when consider the sale statistics, the company observe that such product is well accepted by New England market.Therefore, the company can buy advertisement specifically to such market.

From the example, the data space is breakdown into subspaces, e.g., instead of all categories, only New York is considered; hence, the target object, e.g, the salesman, can be the top rank, i.e., top-R, in only some of the subspace. This subspace where the target object is the top rank is called *promotive subspace*.

Table 1 shows a concrete example of a multi-dimensional data set. From the table, there are one object dimension, *Object*, with three target objects, O_1, O_2, O_3. There are two subspace dimensions, *Genre and Year*, and a score dimension, *Score*. Consider O_1 as the target object to promote, Table 2 lists O_1's 6 subspaces and the corresponding rank and object count in each subspace. The rank is derived from the sum-aggregate score of all objects in the subspace. For example, in {Science, 2012}, O_1 ranks 1st because the score of $O_1 : 0.9 > O_2 : 0.5 > O_3 : 0.4$. Object count is the number of objects in that subspace. Thus {Science, 2012} is a promotive subspace of O_1.

Thus, given a target object, the goal is to find subspace with large promotiveness, i.e., subspace where the target object is top-R. For example, observe that O_1, which is ranked third in all dimensions ({*}), should be promote in subspace {Science} because it is the first rank. In other words, {Science} is a promotive subspace of O_1. The problem of finding large promotiveness subspace is formally defined in section 2.

PromoRank [10] proposes to use subspace ranking for promoting a target object by finding a subspace where the target object is in Top-R ranking. Section 3 discusses PromoRank in detail. However, the computation complexity of PromoRank is exponential to the dimension size, this paper proposes to use dimensionality reduction algorithms, such as principal component analysis (PCA) and factor analysis (FA), to reduce the number of dimensions before performing PromoRank. This approach is explained in Section 4. Dimensionality reduction in recommendation system, in generally, has been studied for many years, e.g., in [1, 5, 8]. In particular, a dimensionality reduction algorithm, i.e., PCA, is used successfully to reduce the subspace ranking used in PromoRank [6]. This paper further investigate this approach by comparing the impact of three well known dimensionality reduction algorithms on the performance of PromoRank on different datasets. The evaluation in Section 5 shows that different algorithms are suitable to different datasets, based on the datasets' properties.

2 Problem Definition

Consider a d-dimensional data set $\mathcal{D}$ with the size of n, each tuple in $\mathcal{D}$ has d **subspace dimension** $\mathcal{A} = \{A_1, A_2, ..., A_d\}$, **object dimension** I_o and **score dimension** I_s. Let $\text{dom}(I_o) = O$ is the complete set of objects and $\text{dom}(I_s) = \mathbb{R}^+$. Let $S = \{a_1, a_2, ..., a_d\}$, where $a_i \in A_i$ or $a_i = *$ ($*$ refers to any value) is a **subspace** of $\mathcal{A}$. In Table 1, $O = \{O_1, O_2, O_3\}$, $\mathcal{A} = \{A_{\text{genre}}, A_{\text{year}}\}$. An example of S is {Genre=Science, Year=2012}

As a consequence, S induces a projection of $\mathcal{D}_S \subseteq \mathcal{D}$ and a subspace of object $O_S \subseteq O$. For example, when $S = \{Genre = Science, Year = 2012\}$, $O_S = \{O_1, O_3\}$.

For a d-dimensional data, all subspaces can be group into 2^d cuboids. Thus, S belongs to a d'-dimensional cuboid $\mathcal{A}'$ denoted by $A'_1 A'_2 ... A'_{d'}$, iff S has non-star values in these d' dimensions and star values in the other $d - d'$ dimensions.

Then, for a given **target object** $t_q \in O$, $S_q = \{S_q | t_q \in O_{S_q}\}$ is the set of **target subspace** where t_q occurs. For example, O_1 in Table 2 has 6 target subspaces, as in Table 1, subspace {2011} is not a target subspace because O_1 does not occur in it.

There are many ways to measure the promotiveness of objects in each subspace. One way to measure the promotiveness is percentile-rank, calculated from inverse of the rank of the target object in the subspace times distinct object count, i.e.,

$$P = Rank^{-1} \cdot ObjCount. \quad (1)$$

For example, in Table 2, promotiveness of subspace {Genre=Fiction} is $\frac{1}{3} \cdot 3 = 1$ while the promotiveness of subspace {Genre=Science, Year=2012} is $\frac{1}{1} \cdot 2 = 2$.

Finally, the definition of the promotion query problem is, *given data set $\mathcal{D}$, target object t_q, and promotiveness measure* P, *find the top-R subspaces with the largest promotiveness values.*

This promotion query is a challenging problem because it has a combinatorial nature, the number of combination of subspace with multiple dimensions can increase exponentially. The brute-force approach that enumerates all subspaces and compute the promotiveness in each subspace is prohibitive.

3 PromoRank Algorithm

To address the aforementioned challenge, **promotion analysis through ranking (PromoRank)** [10] utilizes the concept of subspace ranking, i.e., ranking in only selected dimensions. PromoRank consists of two phases: aggregation and partition. Algorithm 1, shows the pseudo-code of PromoRank [10].

Algorithm 1. PromoRank($t_q, S, \mathcal{D}, O_S, d_0$)

Input: target object t_q, subspace S, data set $\mathcal{D}$, object set in current subspace O_S, current partition dimension d_0
Output: Top-R promotive subspaces Results

1: Results $\leftarrow \emptyset$
2: **if** $|\mathcal{D}| <$ minsup $\vee\, t_q \notin O$ **then**
3:    **return**
4: **end if**
5: Compute Rank and P
6: Enqueue (S, P) to Results
7: **for** $d' \leftarrow d_0 + 1$ **to** d **do**
8:    Sort $\mathcal{D}$ based on d'-th dimension
9:    **for all** value v in d'-th dimension **do**
10:       $S' \leftarrow S \cup \{d' : v\}$
11:       PromoRank(S', $\mathcal{D}_{S'}$, $O_{S'}$, d')
12:    **end for**
13: **end for**

In aggregation phase, if the size of data set is not less than a threshold (*minsup*) and t_q is in the given current subspace, then, promotiveness P of a subspace S is computed and kept in *Results* priority queue. From Algorithm 1, Rank and P of the target object are computed for the input subspace S (Line 5). In particular, Rank can be measured from the rank of the target object in the subspace and P is calculated using Equation 1. Then, S and P are inserted into the priority queue, where P is the key (Line 6). This priority queue maintains the top-R results.

In partition phase, the input data is iteratively processed for an addition dimension (d'). Then, for each distinct value on the d'-th dimension, a new subspace is defined and processed recursively. In particular, the input data $\mathcal{D}$ is sorted according to the d'-dimension (Line 8). Then, $\mathcal{D}$ can be projected into multiple partition, corresponding to the distinct values on the d'-t dimension. A new subspace S' is defined for each partition (Line 10). Then, PromoRank recursively computes over subspace S' (Line 9).

At each recursion, the aggregation phase runs in $O(|\mathcal{D}| + |O|)$ and the partition phase runs in $O(|\mathcal{D}|)$. Given that there are d dimension, the number of recursion will be 2^d. Thus, the computational complexity of this algorithm is $O(2^d(|\mathcal{D}| + |O|))$.

4 Dimensionality Reduction for PromoRank

In order to further improve PromoRank, this paper proposes to reduce the number of dimensions (d) of the data set. From the computational complexity of PromoRank, reduce dimensions should impact the performance greatly [2,3]. Moreover, this approach can be performed as a pre-processing for PromoRank; thus, it can be combined with the pruning approaches. However, not all reduction algorithms can be applied to all data set, this paper further investigate this approach by comparing multiple algorithms to find suitable algorithm for a data set with particular parameters.

Given a d-dimensional data set $\mathcal{D}$ with subspace dimension $\mathcal{A}$, a dimensionality reduction algorithm, such as PCA and FA, reduces the number of dimension to d^*, such that $d^* < d$, and a reduced data set $\mathcal{D}^*$ is produced with subspace dimension $\mathcal{A}^*$. Please note that, it does not necessary that $\mathcal{A}^* \subset \mathcal{A}$ because the dimensionality reduction algorithm might generate a new dimension for $\mathcal{A}^*$. In other words, there might exists a subspace $S^* = \{a_1^*, a_2^*, ..., a_{d^*}^*\}$ from $\mathcal{A}^*$ where $a_i^* \notin A_j$ for any $A_j \in \mathcal{A}$.

Thus, the top-R promotive subspace from PromoRank with original data set might differ from the top-R promotive subspace with reduced data set. As a consequence, they cannot be compared directly. In order to handle this, a simple mapping scheme is proposed based on the relationship between the original dimensions and reduced dimensions. Suppose that two original subspace dimensions, A_i and A_j, are reduced to a new subspace dimension A_k^*. Consequently, for a top-R promotive subspace contains $a_k^* \in A_k^*$, it will be compared with a subspace that has $a_i \in A_i$ and/or $a_j \in A_i$; together with the common other subspace dimensions. For example, let's assume that the original dimensions are {City, Country, Year}; then, after a dimensional reduction algorithm is performed on the data set, the new dimensions are {Location, Year} where *Location* is reduced from *City* and *Country*. Thus, if PromoRank considers a subspace {location=Lanna} where *Lanna* is reduced from *Chiang Mai* and *Thailand*, then, the subspace {location=Lanna} will be compared with the subspace {City=Chaing Mai}, {Country=Thailand} and {City=Chiang Mai, Country=Thailand}.

Nevertheless, performing dimensionality reduction algorithms incurs extra computational cost.However, dimensionality reduction algorithms such as PCA and FA have lower computational complexity than PromoRank. For instance, PCA that use Cyclic Jacobi's method has complexity of $O(d^3 + d^2n)$ [4]. The polynomial complexity of PCA is much lower than the exponential complexity of PromoRank. The experimental results in Section 5.5 confirms that the extra computational cost from a dimensionality reduction algorithm is lower than the performance gain from reducing dimensions.

5 Experimental Evaluation

To investigate the impact of dimensionality reduction algorithm on data sets, an experimental evaluation with four data sets, namely, **Top US private collage**[3], **NBA**[4], **Amazon affiliate income**[5] and **Market analysis**[6], was carried out.

A Java version of PromoRank was developed and tested on a computer with an Intel Core2 Duo 3GHz processor and 4GB of memory. The pruning optimization of PromoRank was disabled in the evaluation to remove the impact on the result.

This evaluation investigates three well-known dimensionality reduction algorithms, namely, principal component analysis (PCA), factor analysis (FA) and linear discriminant analysis (LDA). They reduces the number of variables, i.e., dimensions, by measuring the correlation among them. Then, the variables that highly correlate with the others are removed or combined into new variables. The differences among them are mainly the method that each of them use for correlate data, e.g., LDA does not order correlated dimensions by their variance, as in PCA, but instead focuses on class separability. The evaluation was performed on two parts; first part compares the Top-R promotive subspace of dimensional-reduced data sets and original data sets with PromoRank. For all data set, only top-5 promotive subspaces of each target object are considered. The result of this part is presented in Section 5.1, 5.2, 5.3 and 5.4 for Top US private college, NBA, Market Analysis data set and Amazon affiliate income, respectively. This part compares the performance of each algorithms for reducing dimensions of the data set. The second part, presented in Section 5.5, investigates the performance improvement from the dimensionality reduction.

5.1 Top US Private College Data Set

This data set consists of 100 tuples with 8 subspace dimensions, namely *State, Enrollment, Admission Rate, Admission Ratio, Student/faculty Ratio, 4yrs Graduated Rate, 6yrs Graduated Rate* and *Quality Rank*. This data set contains quantitative data, e.g., *6yrs Graduation Rate*, which cannot be used in PromoRank directly because, from Section 2, a subspace dimension has to be a set. Thus, quantitative data have to be converted to categorical data first. In this paper, the number of categories is set to ten. Each quantitative data will be linearly assigned to a category based on its value.

The result of PCA shows that there are two new principle components, i.e., dimensions, that represents four original dimensions, namely, *Grad Rate* and *Ratio*. *Grad Rate* strongly correlates, i.e., has low variance, with *4yrs Grad Rate* and *6yrs Grad Rate*. *Ratio*, on the other hand, strongly correlates with *Admission Ratio* and *Admission Rate*. For FA, there is a strong collation between *4yrs Graduation Rate* and *6yrs Graduation Rate*, so the former one is removed. Finally, LDA reduces the number of dimensions from ten to six.

[3] http://mathforum.org/workshops/sum96/data.collections/datalibrary/data.set6.html

[4] http://www.basketballreference.com

[5] http://wps.prenhall.com/esm_mcclave_statsbe_9/18/4850/1241836.cw/index.html

[6] http://www.stata.com

Table 3. Subspace ranking of Top US private college data set

Target	Original Data Set		FA		PCA		LDA	
Object	Subspaces	Ranks	Subspaces	Ranks	Subspaces	Ranks	Subspaces	Ranks
	{*}	1	{*}	1	{*}	1	{*}	1
CalTech	{State=CA}	1	{State=CA}	1	{State=CA	1	{State=CA}	1
	{4yrs Grad R.=70%}	1	{4yrs Grad R.=70%}	-	{Grad R.=85%}	1	{Grad R.=85%}	1
	{6yrs Grad R.=90%}	1	{6yrs Grad R.=90%}	1	{Grad R.=85%}	1	{Grad R.=85%}	1
	{*}	2	{*}	2	{*}	2	{*}	2
Rice	{Enrollment=2}	1	{Enrollment=2}	1	{Enrollment=2}	1	{Enrollment=2}	1
Uni.	{Enrollment=2, 6yrs Grad R.=90%}	1	{Enrollment=2, 6yrs Grad R.=90%}	1	{Enrollment=2, Grad R.=85%}	1	{Enrollment=2, Grad R.=85%}	1
	{6yrs Grad R.=90%}	2	{6yrs Grad R.=90%}	2	{Grad R.=85%}	2	{Grad R.=85%}	1★
	{*}	3	{*}	3	{*}	3	{*}	3
William	{State=MA}	1	{State=MA}	1	{State=MA}	1	{State=MA}	1
College	{Student/Fac. Rt.=80%}	1	{Student/Fac. Rt.=80%}	1	{Student/Fac. Rt.=80%}	1	{Ratio=40%}	1
	{4yrs Grad R.=90%, 6yrs Grad R.=100%}	1	{6yrs Grad R.=100%}	1	{Grad R.=95%}	1	{Grad R.=95%}	2 ★

Table 3 compares ranks (marked as *Ranks*) of Top-5 promotive subspaces (marked as *Subspaces*) from the original US private college data set and reduced data sets of three target objects. With Williams College as the target object, when the Top-5 promotive subspace of original data are {*4yrs Graduation Rate=90%, 6yrs Graduation Rate=100%* }. In the PCA reduced data, the comparable subspace is {*Graduation Rate=95%*} . First of all, these two subspaces are compared because *Graduation Rate* is the principle component of *4yrs Graduation Rate* and *6yrs Graduation Rate*. Then, to map these two subspaces, the average of the original data set categories, i.e. 95, is assigned to the reduced data set category. As a consequence, it is possible that , when compared with the other object in *O*, the rank of the target object in the reduced subspace can be different from the original subspace. The differences are marked by a star symbol. However, this mismatch is infrequently happened. Therefore, the result shows that the ranking of Top-5 promotive subspace is mostly maintained even after the dimensionality reduction is performed on the data. The results show that LDA, which can reduce the number of dimension (from ten to six) the most, maintains an acceptable ranking result compared with the original one. Thus, LDA is the most preferred for this data set.

5.2 NBA Data Set

This data set consists of 4,051 tuples with 12 subspace dimensions, namely *First Name, Last Name, Year, Career Stage, Position, Team, Games, Minutes, Assists, Block, Turnover* and *Coach*. The result from PCA and FA dictates that 6 dimensions, *Game, Minutes, Assists, Block, TurnOver* and *Coach* can be removed. Thus, after dimensional reduction, the reduced data set contains only six subspace dimensions, namely, *First Name, Last Name, Year, Career Stage, Position* and *Team*. On the other hand, LDA cannot be applied to this data set because the classification criterion cannot be met. In other words, LDA cannot distinguish between independent and dependent variables.

Table 4 shows results from NBA data set. After six dimensions are removed, the Top-5 promotive subspaces are hardly change in this data set. In particular, the result of

Table 4. Subspace ranking of NBA data set

Target	Original Data Set		FA		PCA		LDA	
Object	Subspaces	Ranks	Subspaces	Ranks	Subspaces	Ranks	Subspaces	Ranks
	{*}	1	{*}	1	{*}	1		
Kareem	{Pos.=Center}	1	{Pos.=Center}	1	{Pos.=Center}	1		
Abdul-	{League=N}	1	{League=N}	1	{League=N}	1	N/A	
Jabbar	{Team=LA Lakers, Year=1978}	1	{Team=LA Lakers, Year=1978}	1	{Team=LA Lakers, Year=1978}	1		
	{*}	2	{*}	2	{*}	2		
Michael	{Pos.=Forward}	1	{Pos.=Forward}	1	{Pos.=Forward}	1		
Jordan	{League=N}	2	{League=N}	2	{League=N}	2	N/A	
	{Team=Utah Jazz}	1	{Team=Utah Jazz}	1	{Team=Utah Jazz}	1		
	{*}	251	{*}	251	{*}	251		
LeBorn James	{Car. Stg.=Young, Pos.=Guard}	4	{Car. Stg.=Young, Pos.=Guard}	4	{Car. Stg.=Young, Pos.=Guard}	8★	N/A	
	{Car. Stg.=Young}	14	{Car. Stg.=Young}	14	{Car. Stg.=Young}	14		
	{League=N}	233	{League=N}	233	{League=N}	258★		

FA does not change at all. This result show that, there are some data set that cannot be improved by LDA. So, FA and PCA are the only available choices for such data set.

5.3 Stock Market Data Set

This data set consists of 5,891 tuples with 23 subspace dimensions, namely *Company Name, Industry Name, Ticket Symbol, SIC Code, Exchange Code, Size Class, Stock Price, Price/Piece, Trading Volume, Market Price, Market Cap, Total Debt, Cash, FYE Date, Current PE, Trailing PE, Firm Value, Enterprise Value, PEG Ratio, PS Ratio, Outstanding, Revenues* and *Payout Ratio*. Similar to the previous data set, this data set is converted to categorical data with ten categories for each subspace dimension.

The result from PCA dictates that a subspace dimension, *Price/Piece* can be removed, and there are two new principle components, namely, *Price* and *Forward PE*. *Price* strongly correlates with *Stock Price* and *Market Price*. *Forward PE*, on the other hand, strongly correlates with *Current PE* and *Trailing PE*. FA, on the other hand, can reduce only one dimension. *Stock Price* and *Market Price* are highly correlated, so the later is removed. Similar to the previous data set, LDA cannot improve this data set.

Table 5 shows results from Stock market data set. From the table, there is only few differences between original and reduced data set, marked by a star in the table. Even though, PCA incurs more changes than FA but they are small and acceptable. On the contrarily, PCA can reduces three dimensions, compared with one of FA, so it should perform more efficient. As a conclusion, in some data set, FA can reduce only a few dimension, so, PCA performs better for such data set.

5.4 Amazon Affiliate Income Data Set

To verify the previous observation, the three algorithms was applied to another data set, namely, Amazon affiliate income data set. This data set has 1,384 tuples and 12 dimensions, namely, *Product Line, ASIN, Seller, Date Shipped, Prices, Real Prices,*

Table 5. Subspace ranking of Stock market data set

Target Object	Original Data Set Subspaces	Ranks	FA Subspaces	Ranks	PCA Subspaces	Ranks	LDA Subspaces	Ranks
Bank of America	{*}	1	{*}	1	{*}	1		
	{Stock Price=$6, Mkt. Price=$6}	1	{Stock Price=$6}	1	{Price=$6}	1		
	{Size Class=10, FYE=31/12/2010}	1	{Size Class=10, FYE=31/12/2010}	1	{Size Class10, FYE=31/12/2010}	1	N/A	
	{Current PE=20, Trailling PE=12},	1	{Current PE=20, Trailling PE=12},	1	{Current PE=20, Trailling PE=12},	1		
Greenshift Corp	{*}	2	{*}	2	{*}	2		
	{Stock Price=$1, Mkt. Price=$1}	1	{Stock Price=$1}	1	{Price=$1}	1		
	{SIC Code=4953}	1	{SIC Code=4953}	1	{SIC Code=4953}	1	N/A	
	{Size Class=8, FSE=31/12/2010}	2	{Size Class=8, FSE=31/12/2010}	2	{Size Class=8, FSE=31/12/2010}	2		
AppTech Corp	{*}	3	{*}	3	{*}	3		
	{Stock Price=$0, Mkt. Price=$0}	5	{Stock Price=$0}	1★	{Price=$0}	1★		
	{Size Class=4}	1	{Size Class=4}	1	{Size Class=4}	1	N/A	
	{Ind. Nm.=Softw., Stock Price=$0}	1	{Ind. Nm.=Softw., Stock Price=$0}	1	{Ind. Nm.=Softw., Stock Price=$0}	2★		

Table 6. Subspace ranking of Amazon affiliate income data set

Target Object	Original Data Set Subspaces	Ranks	FA Subspaces	Ranks	PCA Subspaces	Ranks	LDA Subspaces	Ranks
Mr. Spreads. Excel 2007 Library	{*}	1	{*}	1	{*}	1		
	{Seller=Amazon}	1	{Seller=Amazon}	1	{Seller=Amazon}	1		
	{Date Shipped=Sep09}	7	{Date Shipped=Sep09}	7	{Date Shipped=Sep09}	7	N/A	
	{Date Shipped=Nov09, Items Shipped=1}	4	{Date Shipped=Nov09 Items Shipped=1}	4	{Date Shipped=Nov09 Items Shipped=1}	4		
Excel 2007 Power Prog. With VBA	{*}	2	{*}	2	{*}	2		
	{Ref. Fee Rt.=6},	3	{Ref. Fee Rt.=6},	3	{Fee=7}	2★		
	{Prod. Line=Books, Ref. Fee Rt.=8, Revenue=100}	4	{Prod. Line=Books, Ref. Fee Rt.=8, Revenue=100}	4	{Prod. Line=Books, Ref. Fee Rt.=8, Revenue=100}	1★	N/A	
	{Seller=3rd Party}	5	{Seller=3rd Party}	5	{Seller=3rd Party}	5		
Herman Miller Mirra Chair; Fully Loaded; Color; Graphite	{*}	3	{*}	3	{*}	3		
	{Seller=3rd Party}	1	{Seller=3rd Party}	1	{Seller=3rd Party}	1		
	{Seller=3rd Party, Date Shipped=Feb09, Items Shipped=1, Ref. Fee Rt.=8}	2	{Seller=3rd Party, Date Shipped=Feb09, Items Shipped=1, Ref. Fee Rt.=8}	2	{Seller=3rd Party, Date Shipped=Feb09, Items Shipped=1, Fee=10}	1★	N/A	
	{Item Shipped=1, Ref. Fee Rt.=8, Revenue=120}	2	{Item Shipped=1, Ref. Fee Rt.=8, Revenue=120}	2	{Item Shipped=1, Fee=10, Revenue=120}	1★		

Referral Fee Rate, Item Shipped, Revenue, Referral Fees and *URL*, Again, LDA cannot be used to improve this data set because the classification criterion cannot be met. FA can be used with this data set but it can remove only one dimension, i.e., Real Price.

Table 7. Performance comparison

Date Set	Execution Time			
	Original Data Set	with FA	with PCA	with LDA
Top Private College	2 seconds	1 second	1 second	1 second
NBA	20.5 minutes	16 minutes	15.2 minutes	-
Stock Market	124 minutes	108 minutes	99 minutes	-
Amazon Affiliate Income	36.5 minutes	29.3 minutes	21.1 minutes	-

PCA indicates two new component, *Prices* and *Fee*. *Prices* is strongly correlated with *Prices* and *Real Prices* in the original data set while *Fee* is strongly correlated with *Referral Fee Rate* and *Referral Fees*.

Even though, PCA incurs more changes in Top-5 promotive subspace ranking that FA, but the changes are small and acceptable. Therefore, PCA is preferred for this data set because it can reduce more dimensions thatn FA.

5.5 Performance Improvement

This section shows the performance of PromoRank. Table 7 shows the comparison between the execution time of PromoRank on the original data set and the execution time of PCA, FA and LDA to produce reduced data sets plus the execution time of PromoRank on the reduced data sets. The result shows that the dimensionality reduction algorithm can improve performance of PromoRank, for about 25% on a large data set.

6 Conclusion

Dimensionality reduction algorithms are introduced to reduce the dimensions of data set in order to improve the performance of PromoRank algorithm. The results confirm that the dimensionality reduction algorithm can reduce the execution time of PromoRank up to 25% while mostly maintains the ranking result. In particular, when a data set can met the classification criterion of LDA, then LDA is the best choices in terms of the number of reduced dimensions. However, if LDA is not eligible, FA should be evaluated next to see the number of dimensions it can reduce. If it can reduce many, then it is the next best choices. Finally, if FA can reduce only one or two dimensions, PCA should be the best choice because, in general, PCA can reduce more dimensions than FA.

References

1. Ahn, H.J., Kim, J.W.: Feature reduction for product recommendation in internet shopping malls. International Journal of Electronic Business 4(5), 432–444 (2006)
2. Ailon, N., Chazelle, B.: Faster dimension deduction. Commun. ACM 53(2), 97–104 (2010)
3. Fodor, I.: A survey of dimension reduction techniques. Tech. rep., Center for Applied Scientific Computing, Lawrence Livermore National Research Laboratory (2002)
4. Forsythe, G.E., Henrici, P.: The cyclic jacobi method for computing the principal values of a complex matrix. In: Transactions of the American Mathematical Society, pp. 1–23 (1960)

5. Kamishima, T., Akaho, S.: Dimension reduction for supervised ordering. In: Proceedings of the International Conference on Data Mining, pp. 330–339. IEEE Press, Hongkong (2006)
6. Kavilkrue, M., Boonma, P.: Improving efficiency of PromoRank algorithm using dimensionality reduction. In: Nguyen, N.T., Attachoo, B., Trawiński, B., Somboonviwat, K. (eds.) ACIIDS 2014, Part I. LNCS, vol. 8397, pp. 261–270. Springer, Heidelberg (2014)
7. Kotler, P., Keller, K.: Marketing Management. Prentice Hall (2008)
8. Symeonidis, P., Nanopoulos, A., Manolopoulos, Y.: Tag recommendations based on tensor dimensionality reduction. In: Proceedings of the ACM Conference on Recommender Systems, pp. 43–50. ACM, Lausanne (2008)
9. Wang, J., Zhang, Y.: Opportunity model for e-commerce recommendation: Right product; right time. In: Proceedings of the 36th International ACM SIGIR Conference on Research and Development in Information Retrieval, pp. 303–312. ACM, New York (2013)
10. Wu, T., Xin, D., Mei, Q., Han, J.: Promotion analysis in multi-dimensional space. In: Proceedings of the International Conference on Very Large Databases, pp. 109–120. VLDB Endowment, Lyon (2009)

Neuro – Genetic Approach for Predicting Maintainability Using Chidamber and Kemerer Software Metrics Suite

Lov Kumar and Santanu Ku. Rath

Dept. CS&E
NIT Rourkela, India
lovkumar505@gmail.com,
skrath@nitrkl.ac.in

Abstract. Accurate estimation of attributes such as effort, quality and risk is of major concern in software life cycle. Majority of the approaches available in literature for estimation are based on regression analysis and neural network techniques. In this study, Chidamber and Kemerer software metrics suite has been considered to provide requisite input data to train the artificial intelligence models. Two artificial intelligence (AI) techniques have been used for predicting maintainability viz., neural network and neuro-genetic algorithm (a hybrid approach of neural network and genetic algorithm). These techniques are applied for predicting maintainability on a case study i.e., Quality Evaluation System (QUES) and User Interface System (UIMS). The performance was evaluated based on the different performance parameters available in literature such as: Mean Absolute Relative Error (MARE), Mean absolute error (MAE), Root Mean Square Error (RMSE), and Standard Error of the Mean (SEM) etc. It is observed that the hybrid approach utilizing Neuro-GA achieved better result for predicting maintainability when compared with that of neural network.

Keywords: Artificial neural network, CK metrics suite, Maintainability, Genetic algorithm.

1 Introduction

Software quality is considered as one of the most important parameter of software development process. Software quality attributes that have been identified by ISO/IEC 9126 [7] are efficiency, functionality, maintainability, portability, reliability and usability. In recent years, maintainability plays a high priority role for achieving considerable success in software system and it is considered as essential quality parameter. The ISO/IEC 9126 [7] standard defines maintainability as the capability of the software product to be modified, including adaptation or improvements, corrections of the software to changes in environment and in requirements and functional specifications. In this paper, Maintainability is considered as the number of source of lines changed per class. A line change can be an 'addition' or 'deletion' of lines of code in a class [9].

H. Unger et al. (eds.), *Recent Advances in Information and Communication Technology 2015*,
Advances in Intelligent Systems and Computing 361, DOI: 10.1007/978-3-319-19024-2_4

Some of the Object-Oriented software metrics available in literature are as follows: Abreu MOOD metrics suite [1], Briand *et al.* [8], Halstead [6], Li and Henry [9], McCabe [11], Lorenz and Kidd [10] and CK metrics [4] suite etc. In this paper, CK metrics suite has been considered to provide requisite input data to design the models for predicting maintainability using ANN with Gradient descent learning method [2] and hybrid approach of ANN and genetic algorithm i.e., Neuro-genetic (Neuro-GA) [3]. CK metrics suite is a six metrics set. The important aspect of CK metrics suite is that it covers most of the feature of the Object-Oriented software i.e, size, cohesion, coupling and inheritance. WMC show size or complexity of class, DIT and NOC show the class hierarchy, CBO and RFC show class coupling and LCOM show cohesion.

The remainder of the paper is organized as follows: Section 2 shows the related work in the field of software maintainability and Object-Oriented metrics. Section 3 briefs about the methodologies used to predicting maintainability. Section 4 emphasizes on mining of CK metrics values from data repository. Section 5 highlights on the results for effort estimation, achieved by applying ANN and Neuro-GA techniques. Section 6 gives a note (comparison) on the performance of the designed models based on the performance parameters. Section 8 concludes the paper with scope for future work.

2 Related Work

It is observed in literature that software metrics are used in design of prediction models which serve the purpose of computing the prediction rate in terms of accuracy such as fault, effort, re-work and maintainability. In this paper, emphasis is given on work done on the use of software metrics for maintainability prediction. Table 1 shows the summary of literature review done on Maintainability, where it describes the applicability of numerous software metrics used by various researchers and practitioners in designing their respective prediction models. From table 1, it can be interpreted that many of the authors have used statistical methods such as regression based analysis and their forms in predicting the maintainability. But keen observation reveals that very less work has been carried out on using neural network models for designing their respective prediction models. Neural network models act as efficient predictors of dependent and independent variables due to sophisticated modeling technique where in they posses the ability to model complex functions. In this paper, two artificial intelligence techniques are applied to design the model to estimate the maintainability of the software product using CK metrics suite.

3 Proposed Work for Predicting Maintainability

This section highlight on the use of artificial intelligence techniques (AI) such as Artificial neural network (ANN) with Gradient descent learning method [2], hybrid approach of Artificial neural network (ANN) and genetic algorithm i.e., Neuro-genetic (Neuro-GA) [3] for predicting maintainability.

Table 1. Summary of Empirical Literature on Maintainability

Author	Prediction technique
Li and Henry (1993) [9]	Regression based models
Paul Oman (1994) *et al.* [13]	Regression based models
Don Coleman (1994) *et al.* [5]	Aggregate complexity measure, Factor analysis, Hierarchical multidimensional assessment model, polynomial regression models , principal components analysis.
Scott L. Schneberger (1997) [14]	Regression based models
Van Koten *et al.* (2006) [15]	Bayesian Network, Backward Elimination and Stepwise Selection, Regression Tree.
Yuming Zhou and Hareton Leung (2007) [16]	Artificial neural network, Multivariate linear regression, Multivariate adaptive regression splines, Regression tree and Support vector regression

3.1 Artificial Neural Network (ANN) Model

ANN is used for solving problems such as classification and estimation [12]. In this study, ANN is being used for predicting maintainability using CK metrics.

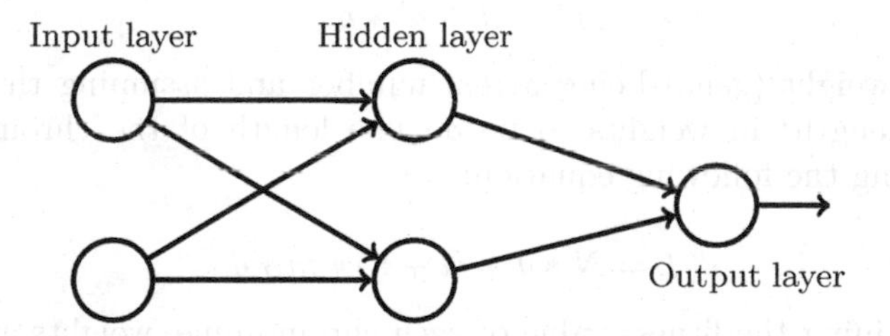

Fig. 1. Artificial neural network

Figure 1 shows the architecture of ANN, which contains three layers i.e., input layer, hidden layer and output layer. Here, for input layer, linear activation function is used and for hidden layer and output layer, sigmoidal function or squashed-S function is used.

A neural network can be represented as:

$$Y' = f(W, X) \tag{1}$$

where X is the input vector, Y' is the output vector, and W is the weight vector.

Gradient Descent Learning Method. Gradient descent learning method is one of the methods for updating the weights during learning phase [2]. It uses first-order derivative of total error to find the minima in error space. Normally

Gradient vector G is defined as the 1st order derivative of error function E_k, and the error function is represented as:

$$G = \frac{\mathrm{d}}{\mathrm{d}W}(E_k) = \frac{\mathrm{d}}{\mathrm{d}W}\left(\frac{1}{2}(y'_k - y_k)^2\right) \tag{2}$$

After obtaining the value of gradient vector G in each iteration, weighted vector W is updated as:

$$W_{k+1} = W_k - \alpha G_k \tag{3}$$

where W_{k+1} is the updated weight, W_k is the current weights, G_k is gradient vector, α is the learning constant, y and y' are the actual and expected output respectively.

3.2 Neuro-Genetic (Neuro-GA) Approach

Neuro-genetic (Neuro-GA) is a hybrid approach of ANN and genetic algorithm. In Neuro-GA, genetic algorithm is used for updating the weight during learning phase. A neural network with a configuration of 'l-m-n' is considered for estimation. The number of weights N required for this network can be computed using the following equation:

$$N = (l + n) * m \tag{4}$$

with each weight (gene) being a real number and assuming the number of digits (gene length) in weights to be d. The length of the chromosome L is computed using the following equation:

$$L = N * d = (l + n) * m * d \tag{5}$$

For determining the fitness value of each chromosome, weights are extracted from each chromosome using the following equation:

$$W_k = \begin{cases} -\frac{x_{kd+2}*10^{d-2}+x_{kd+3}*10^{d-3}+....+x_{(k+1)d}}{10^{d-2}} & \text{if } 0 <= x_{kd+1} < 5 \\ +\frac{x_{kd+2}*10^{d-2}+x_{kd+3}*10^{d-3}+....+x_{(k+1)d}}{10^{d-2}} & \text{if } 5 <= x_{kd+1} <= 9 \end{cases} \tag{6}$$

The fitness value of each chromosome is computed using following equation:

$$\boxed{F_i = \frac{1}{E_i} = \frac{1}{\sqrt{\frac{\sum_{j=1}^{j=N} E_j}{N}}} = \frac{1}{\sqrt{\frac{\sum_{j=1}^{j=N} T_{ji}-O_{ji}}{N}}}} \tag{7}$$

where N is the total number of training data set. T_{ji} and O_{ji} are the estimated and actual output of input instance j for chromosome i.

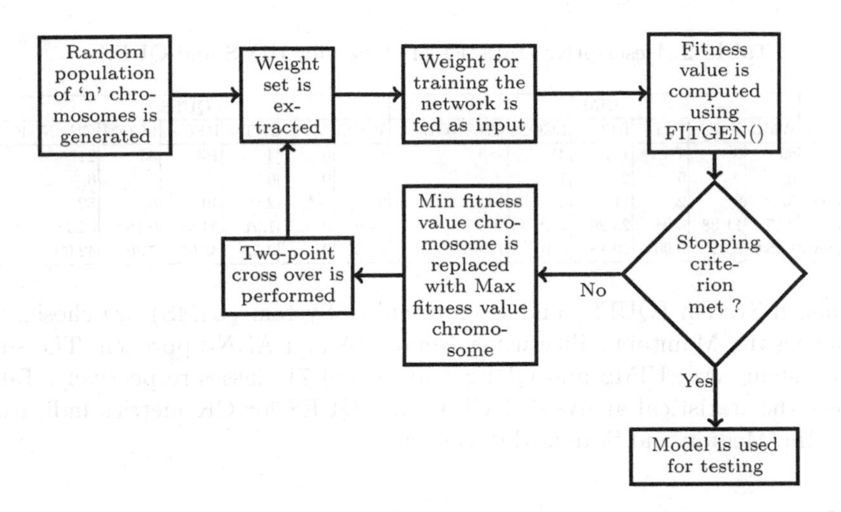

Fig. 2. Flow chart representing Neuro-GA execution

Figure 2 shows the block diagram for Neuro-GA approach. This block diagram represents the steps followed to design the model using Neuro-GA approach.

4 Metrics Set and Empirical Data Extraction

Metrics suites are defined for different goals such as effort estimation, fault prediction, re-usability and Maintainability. In this paper, CK metrics suite has been considered for Maintainability estimation i.e., the number of lines changed per class is considered as a criterion is determining the Maintainability. A line change can be addition or deletion of code in a class. The CK metrics values were extracted using Chidamber and Kemerer Java Metrics (CKJM) tool. This tool extracts Object-Oriented metrics by processing the byte code of compiled Java classes. Hence in this approach, Maintainability for a class is considered as a dependent variable and each of the CK metrics is an independent variable. In this analysis, we disregarded the CBO metric of the CK metrics suite for computing maintainability as it measures non-inheritance related coupling" [9]. Maintainability is assumed to be a function of WMC, DIT, NOC, RFC and LCOM. It is represented as:

$$Maintainability = Change = f(WMC, DIT, NOC, RFC, LCOM) \tag{8}$$

4.1 Case Study

In this paper, to analyze the effectiveness of the proposed approach, two commercial software products were used as case studies. Softwares such as Quality

Table 2. Descriptive statistics of classes for UIMS and QUES

	UIMS						QUES					
	WMC	NOC	DIT	RFC	LCOM	CHANGE	WMC	NOC	DIT	RFC	LCOM	CHANGE
Max.	69	8	4	101	31	289	83	0	4	156	33	217
Min.	0	0	0	2	1	2	1	0	0	17	3	6
Median	5	0	2	17	6	18	9	NA	2	40	5	52
Mean	2.15	11.38	0.94	23.20	7.48	46.82	14.95	0	1.91	54.38	9.18	62.18
Std Dev.	15.89	2.01	0.90	20.18	6.10	71.89	17.05	0	0.52	32.67	7.30	42.09

Evaluation System (QUES) and User Interface System (UIMS) are chosen for computing the Maintainability using Neuro-GA and ANN approach. The softwares systems viz., UIMS and QUES had 39 and 71 classes respectively. Table 2 shows the statistical analysis of UIMS and QUES for CK metrics indicating Max, Min, Median and Standard deviation.

5 Results

In this section, the relationship between value of metrics and the maintainability of class (changes in class). Six CK metrics are considered as input nodes and the output is the computed maintainability. The number of hidden nodes vary from six to thirty three. In this paper, 10-fold and 5-fold cross-validation concept has been considered in QUES and UIMS for comparing the models i.e., each fold contain nearly seven number of data samples. True error and estimate of true error determine the suitable model to be chosen for predicting maintainability. The most suitable number of hidden node in each fold is chosen on the basis of minimum deviation between true error and the estimate of true error. The number of hidden node in final model chosen for predicting maintainability is based on the median values of the hidden nodes in their respective folds.

The following sub-sections give a brief implementation details of the applied neural network techniques.

5.1 Artificial Neural Network (ANN) Model

In this paper, three layers of ANN are considered, in which five nodes act as input nodes, the number of hidden nodes vary from six to thirty three and one node acts as an output node. The network is trained using Gradient descent learning method unless and until the neurons achieve the threshold value of 'MSE' or reach maximum iteration limit (of 2000 epochs). Table 3 shows the various performance parameters which were used to evaluate the best suitable model to be designed for maintainability estimation. From Table 3 it can be interpreted that the high value of 'r' is indicate the pearson's correlation between the actual maintainability and estimated maintainability. Figure 3 shows the variance of MSE vs Number of iterations.

Table 3. Performance matrix

	r	Epochs	MRE	MARE	SEM
UIMS	0.8624	578	0.1820	0. 6931	0.0391
QUES	0.8674	2000	0.1580	0.4384	0.0184

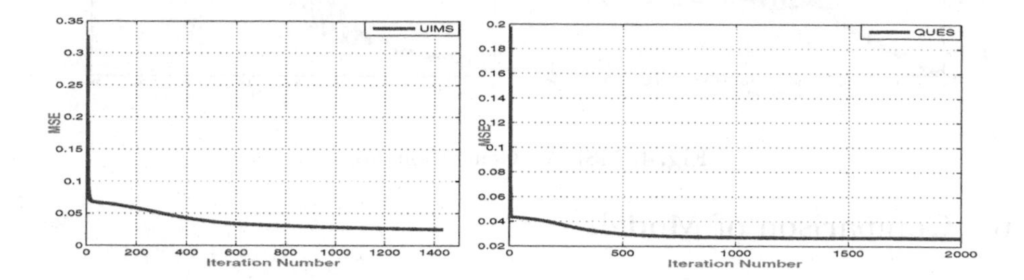

Fig. 3. MSE Vs No. of Iterations (epoch)

5.2 Neuro-Genetic (Neuro-GA) Approach

In this paper, 5-n-1 configuration of neural network is considered (5 numbers of input neurons, n numbers of hidden neurons, 1 output neuron). The total number of weights used in 5-n-1 configuration model are determined using equation 4 i.e., $(5+1)*n = 6*n$ (where 'n' represents the number of hidden nodes varying from six to thirty three), each weight is considered as a gene of length 5, so the length of one chromosome is calculated using equation 5 i.e., L=(5 + 1) * n * 5 = 30 * n.

In this study a population of size 50 is considered i.e., initially 50 chromosomes are randomly generated. The input-hidden layer and hidden-output layer weights of the network are computed using equation 6. Two-point cross-over operation is performed on the generated population. The execution of the algorithm converges when 95% of the chromosomes achieve same fitness values or reach maximum iteration limit (of 200 epochs).

Table 4 shows the various performance parameters which were used to evaluate the best suitable model to be designed for maintainability estimation. From Table 4 it can be interpreted that the high value of 'r' is indicate the pearson's correlation between the actual effort and estimated maintainability. Figure 4 shows the variance of number of chromosomes having same fitness value and generation number of UIMS and QUES.

Table 4. Performance matrix

	r	Epochs	MRE	MARE	SEM
UIMS	0.8108	76	0.1553	0.5331	0.0258
QUES	0.8227	74	0.1480	0.4180	0.0155

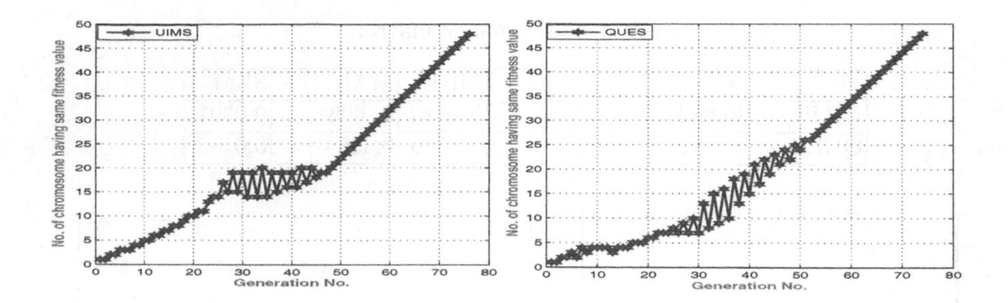

Fig. 4. MSE Vs Generation No.

6 Comparison of Models

Figure 5 shows the Pearson residual boxplots for neural network and Neuro-GA, allowing a visual comparison. The line in the middle of each box represents the median of the Pearson residual. Of two analysis, Neuro-GA in both UIMS and QUES has the narrowest box and the smallest whiskers, as well as the few number of outliers. Based on these boxplots, it is evident that Neuro-GA gave best estimation accuracy as compared to neural network with gradient descent algorithm.

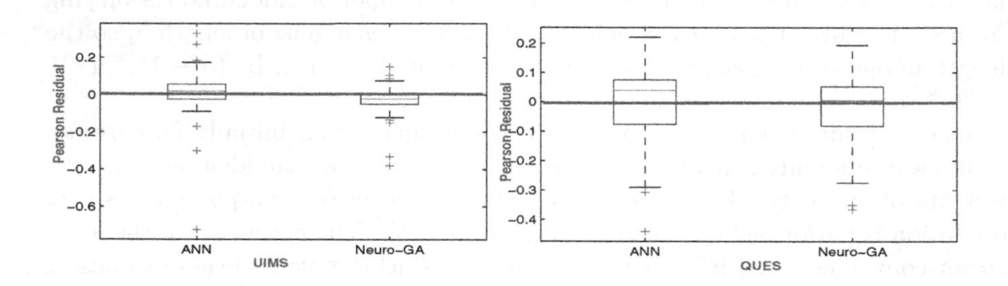

Fig. 5. Residual boxplot for UIMS and QUES

Apart from the comparative analysis done to find the suitable model which can predict the best software maintainability, this paper also makes the comparison of the proposed work with the work done by Yuming Zhou *et al.* [16] and Van Koten *et al.* [15]. Yuming Zhou *et al.* [16] and Van Koten *et al.* [15] have used same dataset i.e., UIMS and QUES for predicting maintainability based on different regression and neural network models . They have considered 'MMRE' as a performance parameter to compare the models designed for predicting maintainability of Object-Oriented software systems. Table 5 shows the MMRE value of the proposed work and the work done by Yuming Zhou *et al.* [16] and Van Koten *et al.* [15]. From Table 5, it can be observed that, in case of QUES software MMRE value is almost same but in case of UIMS software, the proposed approach obtained better prediction rate for maintainability.

Table 5. Performance based on MMRE for UIMS and QUES

MMRE			
Author	Technique	UIMS	QUES
Van Koten *et al.*[15]	Bayesian Network	0.972	0.452
	Regression Tree	1.538	0.493
	Backward Elimination	2.586	0.403
	Stepwise Selection	2.473	0.392
Zhou *et al.* [16]	Multivariate linear regression	2.70	0.42
	Artificial neural network	1.95	0.59
	Regression tree	4.95	0.58
	SVR	1.68	0.43
	MARS	1.86	0.32
	ANN	0.6931	0.4384
	Neuro-GA	0.5332	0.4180

7 Threats to Validity

Several issues affect the results of the experiment are:

- Two Object-Oriented systems, i.e., UIMS and QUES used in this study are design in ADA language. The models design in this study are likely to be valid for other Object-Oriented programing language, i.e., Java or C++. further research can extend to design a model for other development paradigms.
- In this study, only eleven set of software metrics are used to design a models. Some of the metrics which are widely used for Object-Oriented software are further considered for predicting maintainability.
- we only consider AI techniques for designing the prediction models to predict maintainability. Further, we can extend the work to reduce the feature using feature reduction techniques, i.e., PCA, RST, statistical test etc..

8 Conclusion

In this paper, an attempt has been made to use CK metrics suite in order to estimate software maintainability using gradient descent and hybrid approach of neural network and genetic algorithm. These approaches have the ability to predict the output based on historical data. The CK metrics are considered as input data to train the network and predicting software maintainability. The results reveal that the hybrid approach of Neuro-GA prediction model obtained low values of MAE, MARE, and RMSE when compared with those of gradient descent prediction model.

Further, work can be replicated by using hybrid approach of neural network and fuzzy logic. Also feature reduction techniques such as rough set and principal component analysis can be applied to minimize the computational complexity of the input data set.

References

1. Abreu, F.B.E., Carapuca, R.: Object-oriented software engineering: Measuring and controlling the development process. In: Proceedings of the 4th International Conference on Software Quality, vol. 186 (1994)
2. Battiti, R.: First and second-order methods for learning between steepest descent and newton's method. Neural Computation 4(2), 141–166 (1992)
3. Burgess, C., Lefley, M.: Can genetic programming improve software effort estimation. Information and Software Technology 43, 863–873 (2001)
4. Chidamber, S.R., Kemerer, C.F.: A metrics suite for object-oriented design. IEEE Transactions on Software Engineering 20(6), 476–493 (1994)
5. Coleman, D., Ash, D., Lowther, B., Oman, P.: Using metrics to evaluate software system maintainability. IEEE Computer 27(8), 44–49 (1994)
6. Halstead, M.: Elements of Software Science. Elsevier Science, New York (1977)
7. Jung, H.W., Kim, S.G., Chung, C.S.: Measuring software product quality: A survey of iso/iec 9126. IEEE Software 21(5), 88–92 (2004)
8. Briand, L.C., Wust, J., Daly, J.W., Porter, D.V.: Exploring the relationships between design measures and software quality in object-oriented systems. The Journal of Systems and Software 51(3), 245–273 (2000)
9. Li, W., Henry, S.: Maintenance metrics for the object-oriented paradigm. In: Proceedings of First International Software Metrics Symposium, pp. 88–92 (1993)
10. Lorenz, M., Kidd, J.: Object-Oriented Software Metrics. Prentice-Hall, Englewood (1994)
11. McCabe, T.J.: A complexity measure. IEEE Transactions on Software Engineering 2(4), 308–320 (1976)
12. McCulloch, W., Pitts, W.: A logical calculus of ideas immanent in nervous activity. Bulletin of Mathematical Biophysics 5(4), 115–133 (1943)
13. Oman, P., Hagemeister, J.: Construction and testing of polynomials predicting software maintainability. Journal of Systems and Software 24(3), 251–266 (1994)
14. Schneberger, S.L.: Distributed computing environments: effects on software maintenance difficulty. Journal of Systems and Software 37(2), 101–116 (1997)
15. Van Koten, C., Gray, A.: An application of bayesian network for predicting object-oriented software maintainability. Information and Software Technology 48(1), 59–67 (2006)
16. Zhou, Y., Leung, H.: Predicting object-oriented software maintainability using multivariate adaptive regression splines. Journal of Systems and Software 80(8), 1349–1361 (2007)

Finding the Critical Path with Loop Structure for a Basis Path Testing Using Genetic Algorithm

Jakkrit Kaewyotha and Wararat Songpan (Rungworawut)

Department of Computer Science, Faculty of Science, Khon Kaen University, Khon Kaen, Thailand
jakkrit_k@kkumail.com, wararat@kku.ac.th

Abstract. Path testing is strongest code coverage in white box testing techniques. The objective of path testing is design path generating of program under test. However, testing all paths does not mean that will find all defects in a program especially loop structure program problem. In this paper significantly achieved used genetic algorithm for automatic finding the critical path. The GA can analyze control flow graphs of program to finding defect path called critical path with loop structure. In addition, the approach designs automation of generating critical paths, which the algorithm can be adopted to find an optimal solution in five programs under test. The experimental results shown that can generate and recommend a set of critical defect path for five programs in different loop structure of program. Our proposed approach is effective to help developer to find critical paths which means the paths should be improved in a program previously.

Keywords: White Box, Path Testing, Genetic Algorithm.

1 Introduction

The software testing processing has spending time and high cost, according with the objective of testing the software in order to given the performance and accurate with the requirement. In the recently, the several software testing companies where interested for high accuracy and improved the testing quality to reduce errors in software also known as defects. The defect is weak point of the software's source code or the software function faults which brought the incorrect for the expectance output. The several importance factors for software testing both spend high cost and times. Many defects arise because programmers have forgotten to include some processing in their code, so there are no paths to execute. Some defects are also related to the order in which code segments are executed. Wherever, software testers have effort to decrease the times and cost for software testing to improve the performance and reduce in spending time. Therefore, the increasing of the automated software testing researches was growing which separated for two types: black box and white box techniques. Black box technique only is pointed the software input and output. In other hand, white box technique represented the testing for software structure which focuses on the software's source code. However, the problem is how to adopt this technique to

H. Unger et al. (eds.), *Recent Advances in Information and Communication Technology 2015*,
Advances in Intelligent Systems and Computing 361, DOI: 10.1007/978-3-319-19024-2_5

complex source code in order to real program can have both sequence structure and loop structure to implement source code completely. When source code is executed, it shows only error parameter within a line. Therefore, the basis path testing is used in this problem to analyze defects that found on whole paths. We propose an approach to generate the priority path of software which applied the genetic algorithm with loop structure that is the major contribute in this paper. In addition, we increased the parameter adjustment for find out the defect to solve loop structure program that is challenge. For example, population size, crossover, mutation and number of generation by using genetic algorithm. The paper is arranged as follows: Section 2 presents the related work. Section 3 gives definition of basis path testing and how to construct the control flow graph. Section 4 presents experiments an approach that uses genetic algorithm applied to path testing with case study. The experimental results and discussion are described in the section 5. Finally, the conclusion present in section 6.

2 Related Work

There are many research in software testing with genetic algorithm. Hermadi et al. [1] presented using genetic algorithm generate test cases for white box testing of software testing which test cases is built by path testing which this paper analyze to control parameters affect to performance of genetic algorithm for path testing. The rest of the paper shows population size very impact for path coverage, in order is allele range while number of generation and mutation rate is low impact in term of number of paths found. The paper showed only the experimental results with parameters adjustment. Mansour and Salame [2] presented generate test cases for execute specified paths in program by using two algorithms are simulated annealing algorithm and genetic algorithm. These algorithms are based on the formulation optimization of the path testing problem which includes integer and real value. This paper compare three algorithms are simulated annealing algorithm, genetic algorithm and Korel's algorithm with eight subject programs. The rest of paper shows simulated annealing algorithm and genetic algorithm are efficient than Korel's Algorithm in term number of executed path, simulated annealing algorithm tends to executed better than genetic algorithm in term number of executed path and genetic algorithm faster than simulated annealing algorithm. They focused on comparison of three algorithms with generated test cases for programs. However, the algorithm does not specify critical paths to suggest how to fix it. Srivastava and Kim [3] presented method for optimizing software testing efficiency by identifying the most critical path clusters in a program by developed variable length of genetic algorithm in order to optimize and select software path, moreover, they tried to use weight to control path generation. The problem is genetic algorithm generated many paths and redundant when occurred the same fitness values and also not likely cause of error. Nirpal and Kale [4] presented apply genetic algorithm to generate test cases to test selected path. The genetic algorithm is used to selected path a target and executes sequence of operator iteratively for test cases. This research experiment triangle classification program and used genetic algorithm to automatic generate test cases for path testing and compare with method generate test case produced by random. The experimental result shows genetic algorithm according to test data more effectively and quality of test cases by genetic algorithm is higher than test cases by produced by random. Ghiduk [5] presented genetic algorithm for generating test path and presented new technique for automatically generating a set of basis test paths. In this research, the basic structure of the software analyzes and design in form of control-flow graph (CFG) then converts CFG in form of dd-graph (DDG) to begin process of genetic algorithm a new technique was presented. This research aims to generated basis paths of each program which research was presented 10 programs. The experimental result shows their technique that can find path around 80% of actual basis paths of all subject programs which can be used testing paths in various parts of the technique in path testing and shows genetic algorithm of this paper effective in test path generation. Ngo and Tan [6] presented based approach to infeasible path detection for dynamic test data generation which is a method finding for infeasible path which can be apply with other dynamic path-oriented test data generation technique to increase efficiency the many techniques to finding infeasible path. This paper used control flow graph is representative structure software. The example software of paper is compute_grade and presented test data generation using infeasible path detection algorithm. This research was experiment to compare Bueno approach using 4 java system in experiment are PMD, JMathLib, GFP, SOOT. The experimental result shows the proposed method is very effective in finding infeasible paths with average precision of 96.02% และ recall of 100% in all the cases. The most papers solved with programs do not focus on loop structure. In difference, this paper showed how to solved the program with collected loop structure that is our challenge work.

3 Basis Path Testing

3.1 Path Testing

The path testing meaning the white box testing for assisted the tester who design the test case which measure by the cyclometic complexity of program graph. In addition, the cyclometic complexity measure could define the set of data for performing flow graph called directed graph and consists the node which including statement and edges represent the flow of control. For example, node i and j where the directed graph link between from i to j after i node is executed and j is computed immediately. Path testing is an approach to testing where you ensure that every path through a program has been executed at least once. The program graph is written from programming language which is transform to a directed graph called control flow graph (CFG). CFG represented the structure of program similar with the flowchart without the condition. The node of each control flow represented directed control flow graph including either assignment nodes or decision nodes. The statement defined by the line of code which the execution related with the path from started node to sequence node, which linkage between nodes are executed. For example a reserve number program performed by C language. The line of code i will be translated to node i of CFG as show in Fig 1. Between node i and j called edge and represented by e_n where n is consequence number execution. The nodes from 1 to 5 are sequence nodes and the edges also are executed from e_1 to e_5, next from node 6 to 8 repeat the loop with edge e_6, e_7, e_8 and exit loop in node 6 with e_9 and direct to node 9 and 10 with e_{10} respectively.

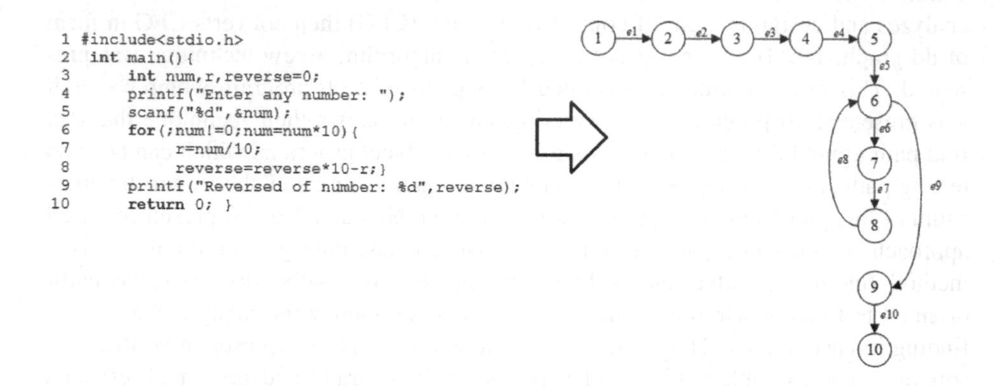

1 #include<stdio.h>
2 int main(){
3 int num,r,reverse=0;
4 printf("Enter any number: ");
5 scanf("%d",&num);
6 for(;num!=0;num=num*10){
7 r=num/10;
8 reverse=reverse*10-r;}
9 printf("Reversed of number: %d",reverse);
10 return 0; }
1
2
3
4
5
6
7
8
9
10
e1
e2
e3
e4
e5
e6
e7
e8
e9
e10

Fig. 1. Source code to CFG of Reserve number program

4 Experiments

In the process of finding the critical path could be compiled a genetic algorithm on a program under testing. In Fig 2 shows the applied genetic algorithm to path testing starts with the process of initial population generation which designs a set of many solutions which these population called chromosomes computed by fitness function. A critical path means the highest fitness function of those chromosomes. After completed fitness function computing then performs the selection process which the elitism method was selected. The elitism method is keeping the best chromosome according to fitness function. Then the Crossover state, random two chromosomes transform to parent and sharing a part of genes. The product of sharing stated called offspring and the mutation process was computed after completed the crossover process. Finally, swap some genes within an offspring. The algorithm can be repeated until the required number of generations is reached by population size. For example, using genetic algorithm find out critical path on reserve number program.

4.1 Initial Population

The number of population selected from all chromosomes which is random by the number of edge. The number of edge is randomized into each genes of a chromosome. The number of all edge depends on the program that translates to CFG. For example, reserve number program has 10 edges that are assigned randomly an integer only between 1 and 10 to each genes of a chromosome as show in Figure 3.

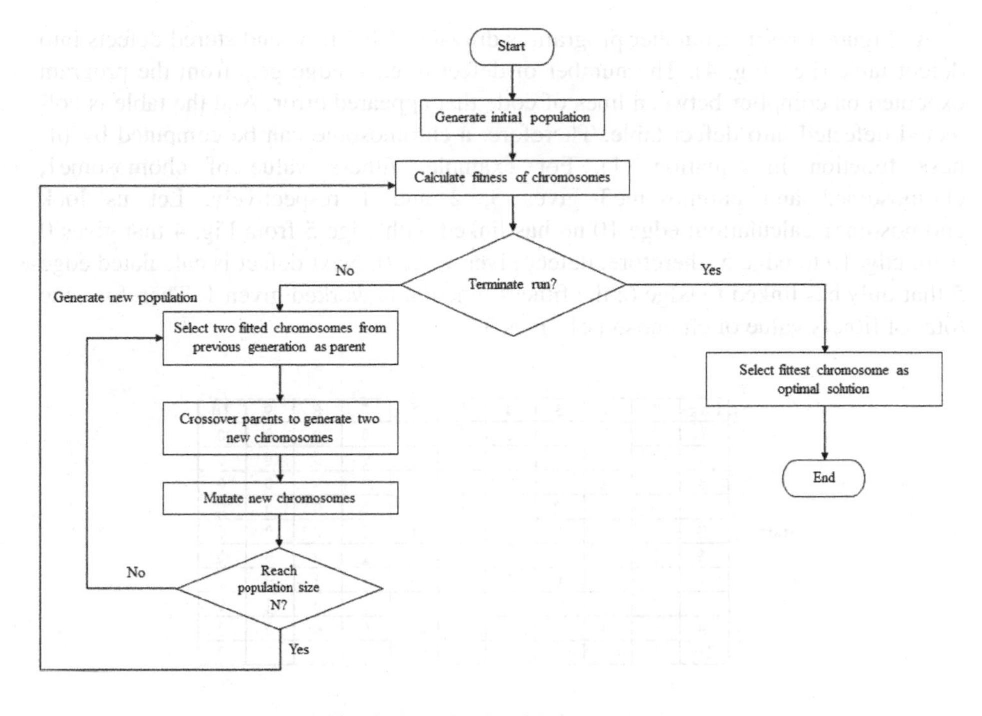

Fig. 2. Genetic algorithms process

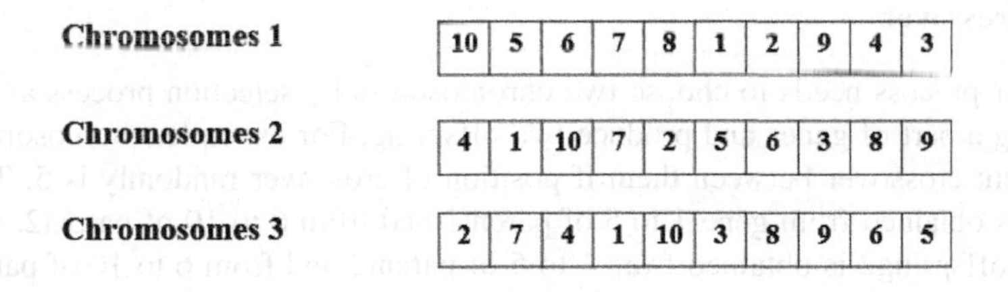

Fig. 3. Initial population generation

4.2 Fitness Function

The Fitness function is measure the fitness of chromosome of each chromosome where possible solution design. Each chromosome in population gives a fitness values. If the chromosome gives highest fitness value which means critical path. The fitness functions are represented follow as Eq.1

$$\text{Fitness Function} = \sum_{i=1}^{\#\text{Edges}} \text{Defect}(i)\,, \tag{1}$$

where edge i has linked with edge i+1

As Figure 1 reserve number program is drawn to CFG flow and stored defects into defect table (i.e., Fig. 4). The number of defect in each edge gets from the program executed on complier between lines of code that appeared error. And the table is collected defected into defect table. Therefore, a chromosome can be computed by fitness function in equation (1). For example, fitness value of chomosome1, chomosome2 and chomosome3 gives 3, 2 and 1 respectively. Let us look chomosome1 calculation; edge 10 no has linked with edge 5 from Fig. 4 that gives 0 from edge10 to edge 5, therefore, defect gives value 0. Next defect is calculated edge 5 that only has linked to edge 6; the fitness function is worked given 1. Therefore, the total of fitness value of chomosome1 gives 3.

	Edge	1	2	3	4	5	6	7	8	9	10
	1	0	0	0	0	0	0	0	0	0	0
	2	0	0	0	0	0	0	0	0	0	0
	3	0	0	0	0	0	0	0	0	0	0
	4	0	0	0	0	0	0	0	0	0	0
Start	5	0	0	0	0	0	1	0	0	0	0
	6	0	0	0	0	0	0	1	0	0	0
	7	0	0	0	0	0	0	0	1	0	0
	8	0	0	0	0	0	0	0	0	1	0
	9	0	0	0	0	0	0	0	0	0	0
	10	0	0	0	0	0	0	0	0	0	0

End

Fig. 4. Defect table of reserve number program

4.3 Crossover

Crossover process needs to choose two chromosomes by selection process as parents to sharing a part of genes and produce two offspring. For example, chomosome2 and 3 as parent crossover between them if position of crossover randomly is 5. The offspring1 is obtained from gene 1 to 5 of parent1 and from 6 to 10 of parent2. In addition, the offspring2 is obtained from 1 to 5 of parent2 and from 6 to 10 of parent1 in Fig. 5. However, for the algorithm will define a crossover rate. If the crossover is not greater than the rate, the parents will crossover process.

Parent 1 (Chromosomes 2)	4	1	10	7	2	5	6	3	8	9
Parent 2 (Chromosomes 3)	2	7	4	1	10	3	8	9	6	5
Offspring 1	4	1	10	7	2	3	8	9	6	5
Offspring 2	2	7	4	1	10	5	6	3	8	9

Fig. 5. Crossover process

4.4 Mutation

The mutation process will computed after the crossover process finished by swapping some genes within an offspring (e.g. position 2 and 8) as Fig. 6.

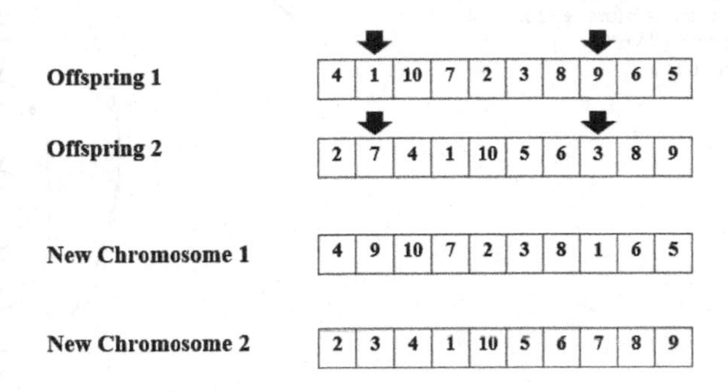

Fig. 6. Mutation process

5 Experimental Results

There are five programs for testing including as Reverse a number [7], Check prime number [8], Even number pyramid [9], Insert an element in an array [10] and Insertion sort [1,11,12], Which each program differences structure of program as show in Fig. 7-10. The objective of this experiment is to clarify the algorithm that discovers the defect path as critical path even more complex program of 5 programs.

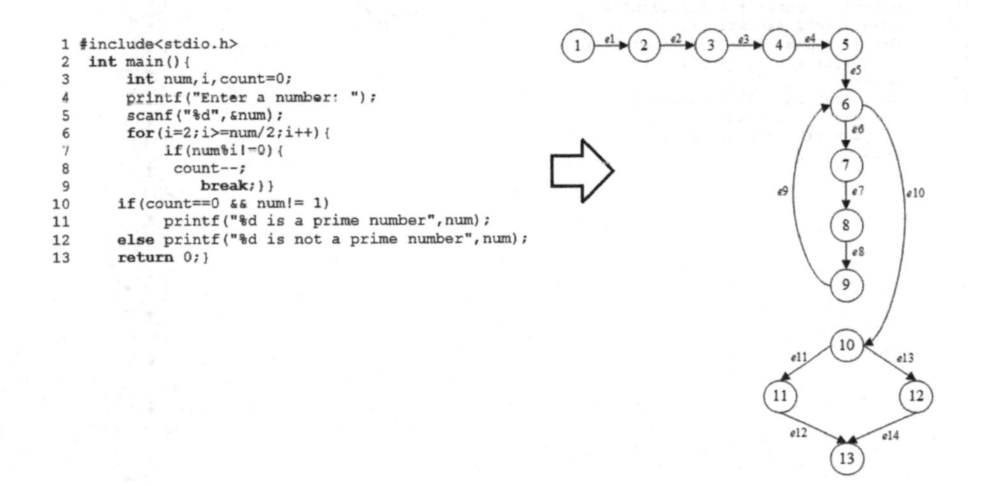

1 #include<stdio.h>
2 int main(){
3 int num,i,count=0;
4 printf("Enter a number: ");
5 scanf("%d",&num);
6 for(i=2;i>=num/2;i++){
7 if(num%i!=0){
8 count--;
9 break;}}
10 if(count==0 && num!= 1)
11 printf("%d is a prime number",num);
12 else printf("%d is not a prime number",num);
13 return 0;}
1 2 3 4 5 6 7 8 9 10 11 12 13
e1 e2 e3 e4 e5 e6 e7 e8 e9 e10 e11 e12 e13 e14

Fig. 7. CFG of Check prime number

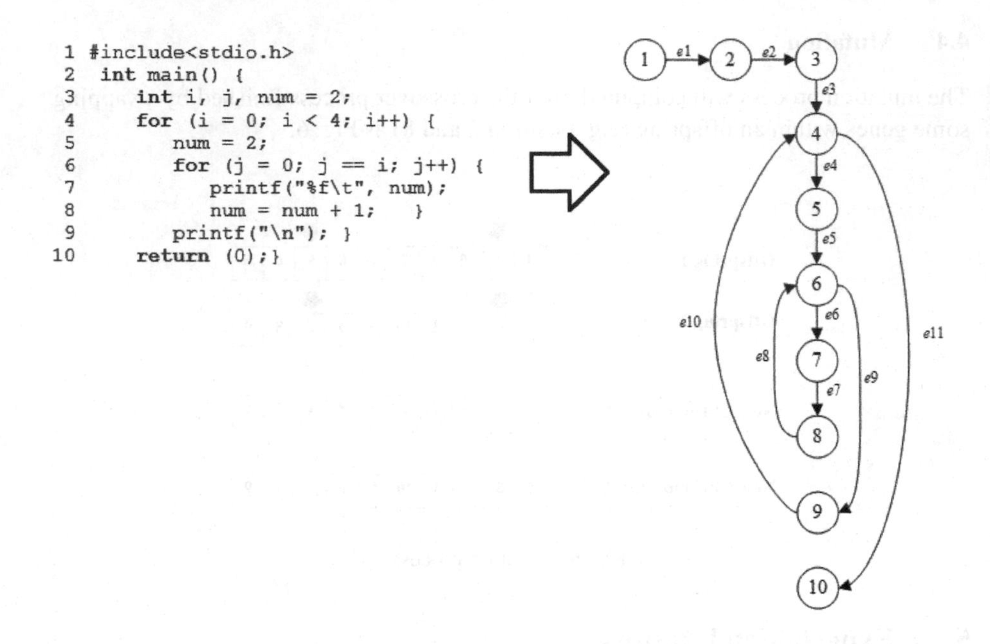

Fig. 8. CFG of Even number pyramid

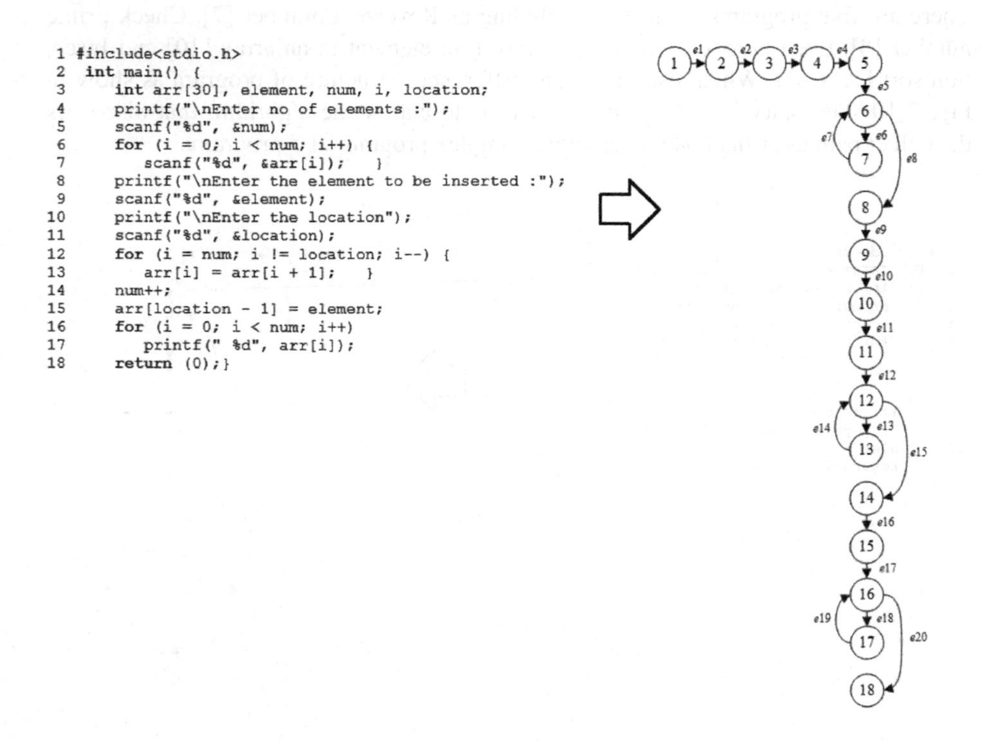

1 #include<stdio.h>
2 int main() {
3 int i, j, num = 2;
4 for (i = 0; i < 4; i++) {
6 for (j = 0; j == i; j++) {
7 printf("%f\t", num);
8 num = num + 1; }
9 printf("\n"); }
3 int arr[30], element, num, i, location;
4 printf("\nEnter no of elements :");
5 scanf("%d", &num);
6 for (i = 0; i < num; i++) {
7 scanf("%d", &arr[i]); }
8 printf("\nEnter the element to be inserted :");
9 scanf("%d", &element);
10 printf("\nEnter the location");
11 scanf("%d", &location);
12 for (i = num; i != location; i--) {
13 arr[i] = arr[i + 1]; }
14 num++;
15 arr[location - 1] = element;
16 for (i = 0; i < num; i++)
17 printf(" %d", arr[i]);
18 return (0);}
1 2 3 4 5 6 7 8 9 10 11 12 13 14 15 16 17 18
e1 e2 e3 e4 e5 e6 e7 e8 e9 e10 e11 e12 e13 e14 e15 e16 e17 e18 e19 e20

Fig. 9. CFG of Insert an element in an array

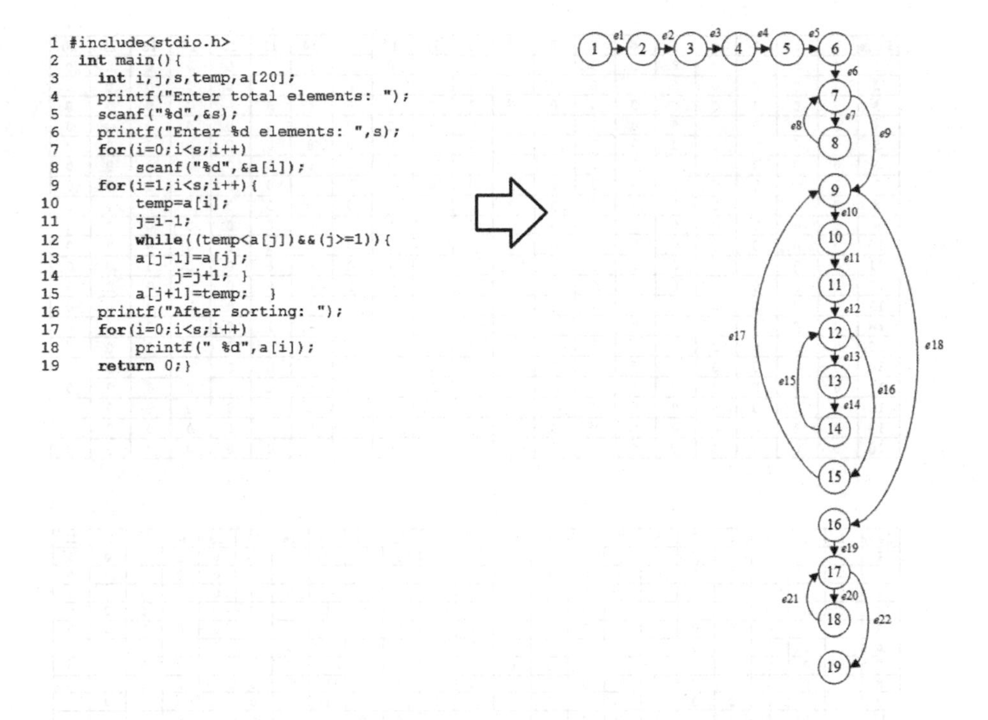

1 #include<stdio.h>
2 int main(){
3 int i,j,s,temp,a[20];
4 printf("Enter total elements: ");
5 scanf("%d",&s);
6 printf("Enter %d elements: ",s);
7 for(i=0;i<s;i++)
8 scanf("%d",&a[i]);
9 for(i=1;i<s;i++){
10 temp=a[i];
11 j=i-1;
12 while((temp<a[j])&&(j>=1)){
13 a[j-1]=a[j];
14 j=j+1; }
15 a[j+1]=temp; }
16 printf("After sorting: ");
17 for(i=0;i<s;i++)
18 printf(" %d",a[i]);
19 return 0;}

Fig. 10. CFG of Insertion sort

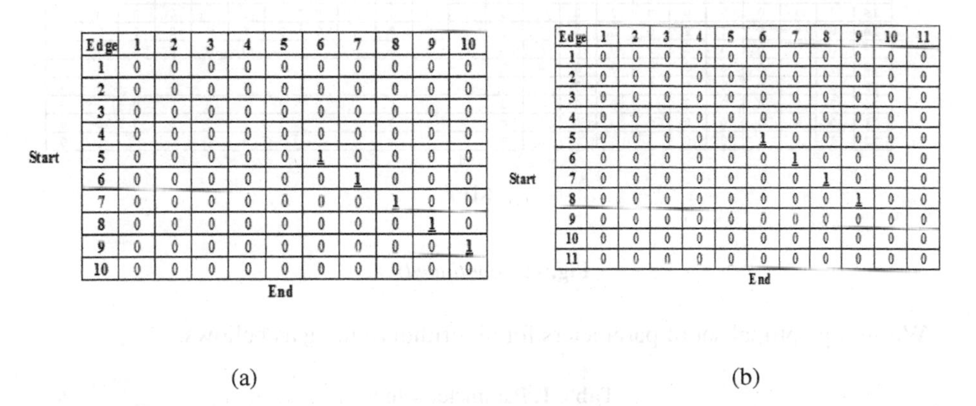

(a) Start / End

Edge	1	2	3	4	5	6	7	8	9	10
1	0	0	0	0	0	0	0	0	0	0
2	0	0	0	0	0	0	0	0	0	0
3	0	0	0	0	0	0	0	0	0	0
4	0	0	0	0	0	0	0	0	0	0
5	0	0	0	0	0	1	0	0	0	0
6	0	0	0	0	0	0	1	0	0	0
7	0	0	0	0	0	0	0	1	0	0
8	0	0	0	0	0	0	0	0	1	0
9	0	0	0	0	0	0	0	0	0	1
10	0	0	0	0	0	0	0	0	0	0

(b) Start / End

Edge	1	2	3	4	5	6	7	8	9	10	11
1	0	0	0	0	0	0	0	0	0	0	0
2	0	0	0	0	0	0	0	0	0	0	0
3	0	0	0	0	0	0	0	0	0	0	0
4	0	0	0	0	0	0	0	0	0	0	0
5	0	0	0	0	0	1	0	0	0	0	0
6	0	0	0	0	0	0	1	0	0	0	0
7	0	0	0	0	0	0	0	1	0	0	0
8	0	0	0	0	0	0	0	0	1	0	0
9	0	0	0	0	0	0	0	0	0	0	0
10	0	0	0	0	0	0	0	0	0	0	0
11	0	0	0	0	0	0	0	0	0	0	0

Fig. 11. Defect table of (a) Check prime number (b) Even number pyramid (c) Insert an element in an array (d) Insertion sort

Start (rows) / End (columns)

Edge	1	2	3	4	5	6	7	8	9	10	11	12	13	14	15	16	17	18	19	20
1	0	0	0	0	0	0	0	0	0	0	0	0	0	0	0	0	0	0	0	0
2	0	0	0	0	0	0	0	0	0	0	0	0	0	0	0	0	0	0	0	0
3	0	0	0	0	0	0	0	0	0	0	0	0	0	0	0	0	0	0	0	0
4	0	0	0	0	0	0	0	0	0	0	0	0	0	0	0	0	0	0	0	0
5	0	0	0	0	0	0	0	0	0	0	0	0	0	0	0	0	0	0	0	0
6	0	0	0	0	0	0	0	0	0	0	0	0	0	0	0	0	0	0	0	0
7	0	0	0	0	0	0	0	0	0	0	0	0	0	0	0	0	0	0	0	0
8	0	0	0	0	0	0	0	0	0	0	0	0	0	0	0	0	0	0	0	0
9	0	0	0	0	0	0	0	0	0	0	0	0	0	0	0	0	0	0	0	0
10	0	0	0	0	0	0	0	0	0	0	0	0	0	0	0	0	0	0	0	0
11	0	0	0	0	0	0	0	0	0	0	0	0	0	0	0	0	0	0	0	0
12	0	0	0	0	0	0	0	0	0	0	0	0	1	0	0	0	0	0	0	0
13	0	0	0	0	0	0	0	0	0	0	0	0	0	1	0	0	0	0	0	0
14	0	0	0	0	0	0	0	0	0	0	0	0	0	0	1	0	0	0	0	0
15	0	0	0	0	0	0	0	0	0	0	0	0	0	0	0	0	0	0	0	0
16	0	0	0	0	0	0	0	0	0	0	0	0	0	0	0	0	0	0	0	0
17	0	0	0	0	0	0	0	0	0	0	0	0	0	0	0	0	0	0	0	0
18	0	0	0	0	0	0	0	0	0	0	0	0	0	0	0	0	0	0	0	0
19	0	0	0	0	0	0	0	0	0	0	0	0	0	0	0	0	0	0	0	0
20	0	0	0	0	0	0	0	0	0	0	0	0	0	0	0	0	0	0	0	0

(c)

Start (rows) / End (columns)

Edge	1	2	3	4	5	6	7	8	9	10	11	12	13	14	15	16	17	18	19	20	21	22
1	0	0	0	0	0	0	0	0	0	0	0	0	0	0	0	0	0	0	0	0	0	0
2	0	0	0	0	0	0	0	0	0	0	0	0	0	0	0	0	0	0	0	0	0	0
3	0	0	0	0	0	0	0	0	0	0	0	0	0	0	0	0	0	0	0	0	0	0
4	0	0	0	0	0	0	0	0	0	0	0	0	0	0	0	0	0	0	0	0	0	0
5	0	0	0	0	0	0	0	0	0	0	0	0	0	0	0	0	0	0	0	0	0	0
6	0	0	0	0	0	0	0	0	0	0	0	0	0	0	0	0	0	0	0	0	0	0
7	0	0	0	0	0	0	0	0	0	0	0	0	0	0	0	0	0	0	0	0	0	0
8	0	0	0	0	0	0	0	0	0	0	0	0	0	0	0	0	0	0	0	0	0	0
9	0	0	0	0	0	0	0	0	0	0	0	0	0	0	0	0	0	0	0	0	0	0
10	0	0	0	0	0	0	0	0	0	0	0	0	0	0	0	0	0	0	0	0	0	0
11	0	0	0	0	0	0	0	0	0	0	0	0	0	0	0	0	0	0	0	0	0	0
12	0	0	0	0	0	0	0	0	0	0	0	0	1	0	0	0	0	0	0	0	0	0
13	0	0	0	0	0	0	0	0	0	0	0	0	0	1	0	0	0	0	0	0	0	0
14	0	0	0	0	0	0	0	0	0	0	0	0	0	0	1	0	0	0	0	0	0	0
15	0	0	0	0	0	0	0	0	0	0	0	0	0	0	0	1	0	0	0	0	0	0
16	0	0	0	0	0	0	0	0	0	0	0	0	0	0	0	0	0	0	0	0	0	0
17	0	0	0	0	0	0	0	0	0	0	0	0	0	0	0	0	0	0	0	0	0	0
18	0	0	0	0	0	0	0	0	0	0	0	0	0	0	0	0	0	0	0	0	0	0
19	0	0	0	0	0	0	0	0	0	0	0	0	0	0	0	0	0	0	0	0	0	0
20	0	0	0	0	0	0	0	0	0	0	0	0	0	0	0	0	0	0	0	0	0	0
21	0	0	0	0	0	0	0	0	0	0	0	0	0	0	0	0	0	0	0	0	0	0
22	0	0	0	0	0	0	0	0	0	0	0	0	0	0	0	0	0	0	0	0	0	0

(d)

Fig. 11. *(continued)*

We set up optimal set of parameters for algorithm running as bellows,

Table 1. Parameter setup

Programs	Population size	Crossover rate	Mutation rate	Number of generation
Reverse a number	2000	0.8	0.08	100
Check prime number	20000	0.8	0.08	100
Even number pyramid	2000	0.8	0.08	100
Insert an element in an array	2000	0.8	0.08	100
Insertion sort	20000	0.8	0.08	100

Table 2. Critical Paths of 5 programs in different structure

Programs	Path generation	Critical path	Fitness value
Reverse a number	*e*2-*e*4-*e*1- *e*3-*e*10- *e*5- *e*6- *e*7- *e*8-*e*9	*e*5- *e*6- *e*7- *e*8- *e*9	4
Check prime number	*e*4-*e*11- *e*14- *e*13- *e*12- *e*1- *e*5- *e*6- *e*7- *e*8- *e*9- *e*10- *e*3- *e*2	*e*5- *e*6- *e*7- *e*8- *e*9- *e*10	5
Even number pyramid	*e*3- *e*11- *e*2- *e*10- *e*1- *e*5- *e*6- *e*7- *e*8- *e*9- *e*4	*e*5- *e*6- *e*7- *e*8- *e*9	4
Insert an element in an array	*e*11- *e*3- *e*8- *e*17- *e*2- *e*1- *e*16- *e*12- *e*13- *e*14- *e*15- *e*20- *e*10- *e*18- *e*4- *e*5- *e*9- *e*7- *e*6- *e*19	*e*12- *e*13- *e*14- *e*15	3
Insertion sort	*e*6- *e*9- *e*10- *e*20- *e*2- *e*12- *e*13- *e*14- *e*15- *e*16- *e*4- *e*11- *e*19- *e*3- *e*8- *e*1- *e*17- *e*7- *e*22- *e*21- *e*18- *e*5	*e*12- *e*13- *e*14- *e*15- *e*16	4

In Table 2 presented the algorithm runs to solve finding critical path in five programs under test. For example, reverse a number program defect path is generated *e*2-*e*4-*e*1- *e*3- *e*10- *e*5-*e*6-*e*7-*e*8-*e*9, which expressed path defect *e*5- *e*6- *e*7- *e*8- *e*9 was generated represents the number of edge of program to analyze the critical path from reverse a number program. For example, number inputs 12345 for the reverse number program testing and expected result 54321 which difference with 29614540. Therefore, should consider solving this critical path in Fig. 12. The source code in line 6 is revised from num=num*10 to num=num/10, in line 7 is revised from r=num/10 to r=num%10 and in line 8 is revised from reverse*10-r to reverse*10+r

```
 1 #include<stdio.h>
 2 int main(){
 3     int num,r,reverse=0;
 4     printf("Entor any number: ");
 5     scanf("%d",&num);
 6     for(;num!=0;num=num/10){
 7         r=num%10;
 8         reverse=reverse*10+r;} }
 9     printf("Reversed of number: %d",reverse);
10     return 0; }
```

Fig. 12. Example of revision reverse a number program by GA recommends

In addition, the programs of Check prime number, Even number pyramid, Insert an element in an array and Insertion sort program are generated to the critical path. The path *e*4-*e*11- *e*14- *e*13- *e*12- *e*1- *e*5- *e*6- *e*7- *e*8- *e*9- *e*10- *e*3- *e*2 of check prime number program gives the edge of defect in program that is *e*5- *e*6- *e*7- *e*8- *e*9- *e*10. The analysis of defect output when input 5 expected result is 5 is prime number but defect output 5 is not a prime number. However, developer will revise line 6-8 following path recommendation. The even number pyramid program is generated critical path as *e*3- *e*11- *e*2- *e*10- *e*1- *e*5- *e*6- *e*7- *e*8- *e*9- *e*4 defect analysis is revised as line 6-8. The Insert an element in an array program is generated critical path as *e*11- *e*3- *e*8- *e*17- *e*2- *e*1- *e*16-*e*12- *e*13- *e*14- *e*15- *e*20- *e*10- *e*18- *e*4- *e*5- *e*9- *e*7- *e*6- *e*19 defect analysis is revised as line 12 and 13. And The Insertion sort program is generated critical path as *e*6- *e*9- *e*10- *e*20- *e*2- *e*12- *e*13- *e*14- *e*15- *e*16- *e*4- *e*11- *e*19- *e*3- *e*8- *e*1- *e*17- *e*7- *e*22- *e*21- *e*18- *e*5 defect analysis is revised as line 12-14.

6 Conclusion

The result of finding the critical path for software testing using genetic algorithm in that can supports the developer to analyze defects. Specifically, in this paper focused on the software including the loop structure which the most software was often found defect in loops structure. The newly found that successful how to apply genetic algorithm repository complex source code which adjustment of some parameter in the process as population size, crossover, mutation and number of generation. The fitness function is very robustness to loop structure due to the mainly finding critical path is to solve a continuous difficult finding critical path. Moreover, our approach helps to solve the path design problem in a complex program in different structure.

References

1. Hermadi, I., Lokan, C., Sarker, R.: Genetic Algorithm Based Path Testing: Challenges and Key Parameters. In: Proceeding of the 2010 Second World Congress on Software Engineering (WCSE), vol. 2, 241–244 (2010)
2. Mansour, N., Salame, M.: Data Generation for Path Testing. Software Quality Journal. 12(2), 121-136 (2004)
3. Srivastava, P.R., Kim, T.H.: Application of Genetic Algorithm in Software Testing. International Journal of Software Engineering and Its Applications, 3(4), 87-96 (2009)
4. Nirpal, P.B., Kale, K.V.: Comparison of Software Test Data for Automatic Path Coverage Using Genetic Algorithm. International Journal of Computer Science & Engineering Technology, 1(1), 12-16 (2010)
5. Ghiduk, A.S.: Automatic generation of basis test paths using variable length genetic algorithm. Information Processing Letters, 114(6), 304–16(2014)
6. Ngo, M.N., Tan, H.B.K.: Heuristics-based infeasible path detection for dynamic test data generation. Information and Software Technology, 50(7), 641–55 (2008)
   Kumar, R., `http://www.cquestions.com/2008/01/c-program-to-reverse-any-number.html`
7. Kumar, R., `http://www.cquestions.com/2012/02/check-given-number-is-prime-number-or.html`
8. Taral P., `http://www.c4learn.com/c-programs/program-even-number-pyramid-in-c.html`
9. Taral P., `http://www.c4learn.com/c-programs/program-insert-element-in-array.html`
10. Kumar, R., `http://www.cquestions.com/2009/09/insertion-sort-algorithm.html`
11. Ahmed, M.A., Hermadi, I.: GA-based multiple paths test data generator. Computers & Operations Research, 35(10), 3107–3124 (2008)

Improved Triangle Box-Counting Method for Fractal Dimension Estimation

Yothin Kaewaramsri and Kuntpong Woraratpanya

Faculty of Information Technology,
King Mongkut's Institute of Technology Ladkrabang, Bangkok, Thailand
benz.it@windowslive.com, kuntpong@it.kmitl.ac.th

Abstract. A fractal dimension (FD) is an effective feature, which characterizes roughness and self-similarity of complex objects. However, the FD in nature scene requires the effective method for estimation. The existing methods focus on the improvement of selecting the suitable height of box-counts. This cannot overcome the overcounting problem, which is a key factor to have an impact on the accuracy of the FD estimation. This paper proposes a more accurate FD estimation, an improved triangle box-counting method, to increase the precision of box-counts associated with box sizes. The triangle-box-partition technique provides the double precision for box-counts, thus it can solve the overcounting issue and enhance the accuracy of the FD estimation. The proposed method is evaluated its performance in terms of fitting error. The experimental results show that the proposed method outperforms the existing methods, including differential box-counting (DBC), improved DBC (IDBC), and box-counting with adaptable box height (ADBC) methods.

Keywords: Fractal Dimension, Fractal Dimension Estimation, Box Counting, Triangle Box-Counting, Improved Triangle Box-Counting.

1 Introduction

A fractal dimension (FD) is an effective measure for complex objects found in nature, such as coastlines, mountains, and clouds. The FD has been broadly applied in many applications such as pattern recognition, texture analysis, segmentation, and medical signal analysis [1]. Although the achievement in applying FD to broad research areas was reported, the accurate FD estimation is still a grand challenge.

Box counting (BC) is one of the most successful methods to estimate FD. Yu, L. et al. [2] and Bruno, O. et al. [3] applied the BC method to estimate FD for automatic coarse classification of iris images, and for plan identification, respectively. However, this method cannot apply to gray-scale images.

Many papers have proposed techniques to improve the FD estimation in various applications. Sarkar, N. et al. [4] introduced to the differential box-counting (DBC) method for gray-scale images. This method partitions an image into grids with size $s \times s$. On each grid, there is a column of boxes representing intensity levels. The heart of this method is box-counting, which determines from the difference of the maximum

H. Unger et al. (eds.), *Recent Advances in Information and Communication Technology 2015*,
Advances in Intelligent Systems and Computing 361, DOI: 10.1007/978-3-319-19024-2_6

and minimum intensity levels on an $s \times s$ grid. The DBC method commonly increases estimation errors, since it covers the maximum and minimum intensity levels in a box when the difference between intensity levels is less than the scale s. This phenomenon is called undercounting.

The improved DBC (IDBC) method was introduced by Li, J. [5]. This method was improved by modifying the conventional methods in three factors, (i) box-height selection, (ii) box-number calculation, and (iii) image intensity surface partition. These modifications can increase the accuracy of the FD estimation by reducing the error of box-counts.

A box-counting method with adaptable box-height [6] was proposed for applying to arbitrary size images. This method allows the height of a box at the top of each grid to be locally adaptable to the maximum and minimum gray-scales. The ratio technique is used to partition an image into grid sizes, so that common rectangle images are also supported. A drawback of this method is that in the large box size the difference of the maximum and minimum intensity levels is high. This leads to the inaccurate FD estimation.

Most of these approaches focus on the improvement of selecting the suitable height of box-counts to overcome the undercounting problem, but none of these approaches attempt to overcome the overcounting problem. Therefore, this paper proposes a more accurate FD estimation by using an improved triangle box-counting method to increase the precision of box-counts associated with box sizes. The triangle-box-partition technique is able to double the precision of box-counts, thus it can enhance the accuracy of the FD estimation.

The rest of this paper is organized as follows. Section 2 reviews and identifies problems of related works. The improved method is proposed in section 3. In section 4, the experimental results are reported and discussed. Finally, the conclusions are presented in section 5.

2 Related Works

Although many approaches have been proposed to improve the accuracy of FD estimation, some limitations have been unsolved. The following subsections describe existing methods and identify their drawbacks.

2.1 Box-Counting Method

A box-counting method defined by Russel et al. in 1980 [7] is a classical technique to estimate the FD for binary images with a box size of length s. A binary image is partitioned into grids with size $s \times s$ and is counted objects $N(s)$ on each grid size that varies from larger to smaller scales. The FD is approximate as

$$\mathrm{FD} \approx \frac{\log(N(s))}{\log(1/s)} \tag{1}$$

where $N(s)$ is the number of boxes in the scale s needed to completely cover the image. The steps to estimate the FD of binary images can be briefly described as follows. First, the image is repeatedly partitioned into different grid sizes that vary from larger to smaller scales. On each scale of grid size, s, the objects are counted and recorded the number of box-counts $N(s)$. Subsequently, a graph of $\log(N(s))$ versus $\log(1/s)$ is plotted. Finally, a least-square regression line through the data points on the graph is calculated. The slop of a regression line represents the estimated FD. Nonetheless, this method does not achieve the precision of box-counts as proved by [7].

2.2 Triangle Box-Counting Method

The triangle box-counting (TBC) method [7] was proposed to improve traditional box-counting (BC). This method simply divides square boxes provided by a grid into two equally triangle boxes to increase the precision of box-counts associated with box sizes and to fit the requirements of the minimum box covering. The TBC greatly improves the estimation accuracy. However, it only works well with binary images.

2.3 Differential Box-Counting Method

Sarkar et al. [4] proposed a differential box-counting (DBC) to estimate the FD of gray-scale images. In this method, a gray-scale image is represented with a three-dimensional (x,y,z) coordinate system, such that the (x,y) coordinate denotes an image plane and z denotes a gray scale. With this system, a square image $M \times M$ can be partitioned into $s \times s$ grids. Each grid contains a column of boxes whose size is $s \times s \times h$, where h is the height of a single box and is equal to $s \times G/M$, such that G is the total number of gray levels. Let maximum and minimum intensity values in $(i, j)^{th}$ grid locate at the box number l^{th} and k^{th}, respectively, the box-count $n_d(i, j)$ can be calculated by using (2) and the total number of box-counts N_d is determined by using (3).

$$n_d(i, j) = l - k + 1 \tag{2}$$

$$N_d = \sum_{i,j} n_d(i, j) \tag{3}$$

The FD of an image can be estimated by the linear least square fit of $\log(N_s)$ against $\log(1/s)$.

However, the DBC method still has a drawback. The box height h is proportional to the grid size s; the larger grid size, the higher box-height. This has a direct impact on the small difference of the maximum and minimum intensities; it leads to a low precision of counting n_d.

2.4 Improved Differential Box-Counting Method

Li, J. et al. [5] proposed an IDBC method by modifying the conventional method in three factors, (i) box-height selection, (ii) box-number calculation, and (iii) image intensity surface partition. These modifications can be briefly summarized as follows.

(i) Box height selection: Let a standard deviation of an image be σ and a constant be a. A new scale of box height is determined by using (4). The box height of the IDBC method is much smaller at different scale h.

$$h' = \frac{h}{1 + 2a\sigma} \tag{4}$$

(ii) Box-number calculation: Let maximum and minimum intensity levels of the $(i,j)^{th}$ block be l and k, respectively, the number of boxes is calculated as in (5).

$$n'_d = \begin{cases} ceil\left(\frac{l-k}{h'}\right) & ; \quad l \neq k \\ 1 & ; \quad l = k \end{cases} \tag{5}$$

(iii) Image intensity surface partition: The following two partition schemes are required to ensure that the image intensity is covered.

- Partition an $M \times M$ image into grids with $s \times s$ pixels, and allow a grid boundary overlapping with an adjacent row and column.
- Give a new scale $r = s-1$ of a block with $s \times s$ pixels, and represent the maximum distance between two pixels in a block.

This method helps improve accuracy of FD estimation by reducing the error caused by an undercounting problem.

2.5 Box-Counting Method with Adaptable Box Height

A box-counting method with adaptable box height [6] was introduced to estimate FD of images whose size is not a square. It uses the ratio-based box-counting method. For instance, by given an image I with $M \times N$ pixels, a ratio r, such that $r \geq 2$, is used to calculate a grid size $m \times n$, where $m = M/r$ and $n = N/r$. The new grid size $m \times n$ can be categorized into four cases as summarized in Table 1.

Table 1. Four possible patterns of grid partitions based on a ratio r

Case	$m \times n$	$m \times (N - n \times r)$	$(M - m \times r) \times n$	$(M - m \times r) \times (N - n \times r)$
$M = m \times r, N = n \times r$	$r \times r$	-	-	-
$M = m \times r, N > n \times r$	$r \times r$	$r \times 1$	-	-
$M > m \times r, N = n \times r$	$r \times r$	-	$1 \times r$	-
$M > m \times r, N > n \times r$	$r \times r$	$r \times 1$	$1 \times r$	1

After partitioning an image into grids, the box size is equal to $m \times n \times h''$, where h'' represents a box height. To count the number of boxes in each column, the box height is calculated by $h'' = G/r$. The number of boxes in the $(i,j)^{th}$ grid has to be real numbers instead of integers. The box-counts are calculated by using equation (6).

$$n_d'' = \left(\frac{I_{Max}(i,j) - I_{Min}(i,j)}{h''} + 1 \right) \times \frac{S(i,j)}{m \times n} \tag{6}$$

where $S(i,j)$ is the area of the $(i,j)^{th}$ grid. In above equation, it does not perform ceiling function to allow n_d'' to be a real number. The total number of boxes of an entire image with ratio r is $N_d''=\Sigma_{i,j}n_d''(i,j)$. In addition, the upper limit H of the box size is defined as in (7).

$$H = \min\{\sqrt[3]{mn}, m, n\} \tag{7}$$

where m and n denote the grid sizes. The range of the box size is $2 \leq r \leq H$.

As mentioned in [4-6], most of them mainly attempt to overcome the box height calculation to eliminate the undercounting problem, but none of these approaches attempt to improve the precision of box-counts to eliminate the overcounting problem. Therefore, this paper proposes the improved triangle box-counting method to increase the precision of box-counts, so that the overcounting problem is solved and accuracy of FD estimation is improved.

3 Proposed Method

In this section, the square box partition is analyzed in order to point out the key factor affecting to the accuracy of FD estimation. Then the improved triangle box-counting algorithm is described in the last subsection.

3.1 Square Box Partition Analysis

Most of FD estimation methods use a square box partition technique, which is simple and practical. However, this technique does not provide the precision for box-counts as proved by [7]. Furthermore, for the large grid in size, the box-count in (2) fails to overcounting situation, when the maximum and minimum intensity levels fall into the same grid. This is a key factor that directly affects to the estimation accuracy. Fig. 1 is an example to explain this phenomenon. It shows a square box whose size is 4×4 pixels. The maximum and minimum intensity values, 189 and 28, are located at the upper and lower of that grid. Suppose that the box height is 4, the box-count n_d calculated by using (2) is 42, where l and k are equal to 48, ceiling(189/4)), and 7, ceiling(28/4), respectively. The greater difference of maximum and minimum intensity levels on each grid leads to the higher error of FD estimation. This problem can be solved by using a triangle partition technique as described in the following subsection.

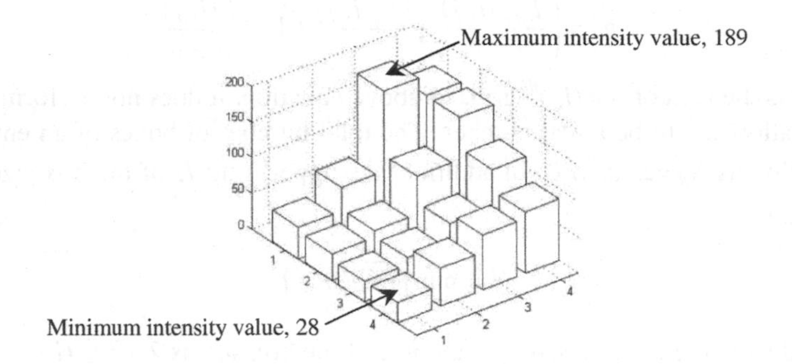

Fig. 1. An example of a grid with 4×4 pixels containing the great difference of maximum and minimum intensity values

3.2 Improved Triangle Box-Counting Algorithm

In this subsection, an improved triangle box-counting method is proposed in order to increase the accuracy of fractal dimension estimation. The algorithm is as follows.

Step 1. Partition an image with $M\times M$ pixels into square grids with box size s varying from 2 to $M/2$.

Step 2. Calculate box height $h = s\times G/M$, where G is the total number of gray levels. Each box size can be computed by $s\times s\times h$.

Step 3. Apply triangle box partition technique to each box in Step 2. The results are shown in Fig. 2. There are two patterns, p_1 and p_2, such that p_1 is composed of upper and lower right diagonals while p_2 is composed of upper and lower left diagonals.

Step 4. Determine the box-count n_d of each pattern in Fig. 2 using (1), such that l and k can be computed by I_{Max}/h and I_{Min}/h, where I_{Max} and I_{Min} are the maximum and minimum intensities of each grid.

Step 5. Compute p_1 by averaging the box-counts of upper and lower right diagonals, whereas compute p_2 by averaging the box-counts of the upper and lower left diagonals.

Step 6. Obtain each box-count $n_d(i\ ,j)$ by using (8).

$$n_d(i\ ,j) = max\{p_1, p_2\} \tag{8}$$

Step 7. Use (3) to calculate the total number of box-counts N_d.

As mentioned to the drawback of the square box partition in previous subsection, the great difference of maximum and minimum intensity levels on each grid can be reduced by dividing a square box into two equally triangle boxes as shown in Fig. 2. This example can be viewed as two patterns, upper and lower right diagonals, p_1, and upper and lower left diagonals, p_2. In the upper right diagonal, the maximum and minimum intensity values are 189 and 44, respectively. Suppose that the box height h

is 4, the box-count n_d of the upper right diagonal is equal to 38, such that l and k are equal to 48, ceiling(189/4) and 11, ceiling(44/4), respectively. In the same way, the maximum and minimum intensity values in the lower right diagonal are 85 and 28, respectively. Thus, the box-count n_d of the lower left diagonal is equal to 16. Then p_1 is computed by averaging 38 and 16. The result is 27.

On the other hand, the box-counts n_d of the lower right diagonal is equal to 42 and of the upper right diagonal is equal to 34. Then p_2 is calculated by averaging 42 and 34. The result is 38. Finally, in this case the box-count $n_d(i, j)$ of this image box equals to p_2 based on the criterion in (8). When compared with the square box partition as explained in previous subsection, the box-count $n_d(i, j)$ is 38, which is less than 42. This proves that the triangle box partition can overcome the overcounting problem.

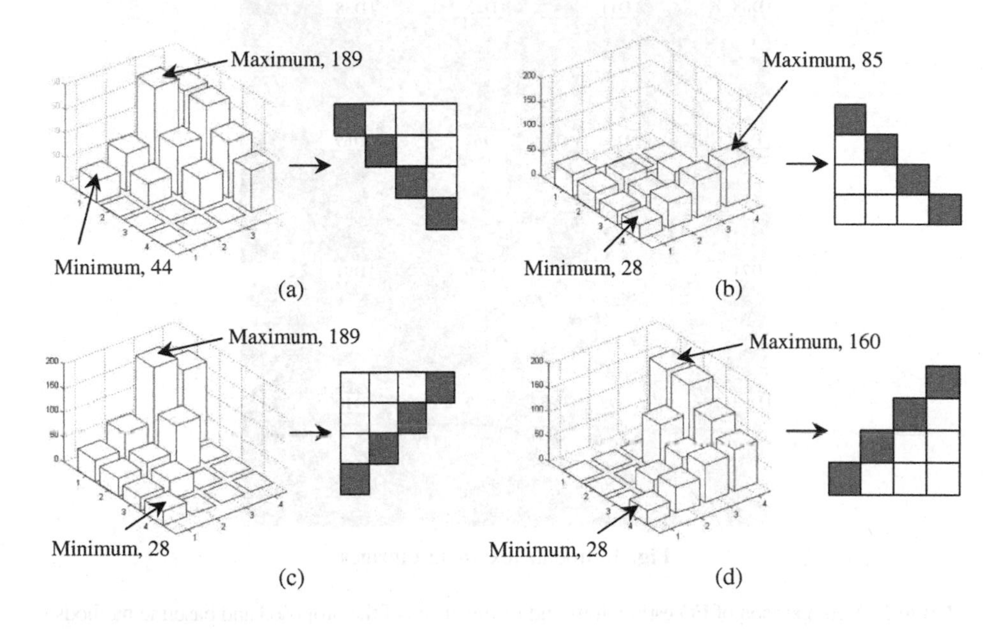

Fig. 2. Illustration of dividing an image box into two triangle boxes with four patterns: (a) upper right diagonal, (b) lower right diagonal, (c) upper left diagonal, and (d) lower left diagonal

4 Experimental Results

In order to evaluate the efficiency of the improved triangle box-counting method in terms of accuracy of FD estimation, the experiments are set up as follows. 16 Brodatz texture images with 512×512 pixels as shown in Fig. 7 are used to evaluate the performance of algorithms. The baseline algorithms are DBC [4], IDBC [5], and ADBC [6] methods implemented with simulation programming. The accuracy of FD estimation is evaluated by fitting error E as defined in (8) to measure the least square linear fit of $\log(N_s)$ versus $\log(1/s)$.

$$E = \frac{1}{n}\sqrt{\sum_{i=1}^{n}\frac{(mx_i + c - y_i)^2}{(1+m^2)}} \tag{8}$$

where m, x_i, y_i, and n denote a slop of graph, a value of $\log(1/s)$, a value of $\log(N_d)$, and a number of samples, respectively.

The triangle box partition is applied to baseline algorithms, including DBC, IDBC, and ADBC, and is implemented with simulation programming. Therefore, there are three improved triangle box-counting algorithms. That is, the DBC with triangle box partition (TDBC), IDBC with triangle box partition (TIDBC), and ADBC with triangle box partition (TADBC) algorithms. As illustrated in Table 2, the experimental results reveal that all proposed algorithms outperform the conventional DBC, IDBC, and ADBC algorithms. A comparison of accuracy illustrates that the TDBC, TIDBC, and TADBC methods can estimate fractal dimension with much smaller fitting error when compared to DBC, IDBC, and ADBC methods, respectively.

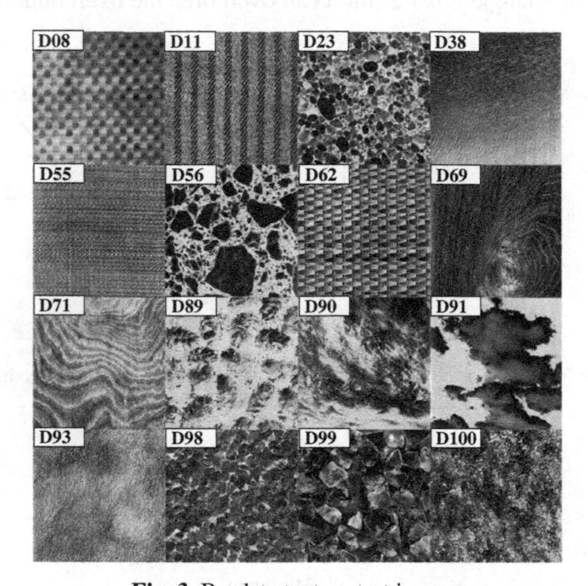

Fig. 3. Brodatz texture test images

Table 2. A comparison of FD estimations and fitting errors of the proposed and baseline methods

Test Images	DBC [4]		TDBC		IDBC [5]		TIDBC		ADBC [6]		TADBC	
	FD	Fit. Err.	FD	Fit. Err.	FD	Fit. Err.	FD	Fit. Err.	FD	Fit. Err.	FD	Fit. Err.
d08	2.43	0.00894	2.44	**0.00870**	2.55	0.00820	2.55	**0.00742**	2.52	0.00104	2.51	**0.00097**
d11	2.65	0.01111	2.64	**0.01047**	2.74	0.01104	2.73	**0.01037**	2.80	0.00075	2.78	**0.00074**
d23	2.50	0.01657	2.50	**0.01635**	2.60	0.01561	2.60	**0.01506**	2.68	0.00118	2.66	0.00119
d38	2.64	0.01131	2.63	**0.01056**	2.72	0.00998	2.72	**0.00929**	2.77	0.00086	2.75	**0.00083**
d55	2.64	0.01268	2.63	**0.01204**	2.73	0.01270	2.72	**0.01199**	2.83	0.00084	2.81	**0.00081**
d56	2.52	0.01559	2.51	**0.01491**	2.62	0.01503	2.61	**0.01424**	2.69	0.00121	2.68	**0.00118**
d62	2.46	0.01887	2.46	**0.01836**	2.55	0.01649	2.55	**0.01617**	2.63	0.00118	2.62	0.00121
d69	2.55	0.01099	2.55	**0.01076**	2.64	0.00990	2.64	**0.00915**	2.67	0.00073	2.65	**0.00072**
d71	2.60	0.01116	2.59	**0.01036**	2.65	0.00809	2.65	**0.00720**	2.66	0.00071	2.65	0.00071
d89	2.47	0.01707	2.47	**0.01667**	2.55	0.01443	2.55	**0.01357**	2.59	0.00123	2.57	**0.00117**
d90	2.43	0.01639	2.42	**0.01527**	2.52	0.01356	2.51	**0.01241**	2.55	0.00109	2.53	**0.00103**
d91	2.29	0.01504	2.28	**0.01403**	2.38	0.01053	2.38	**0.00939**	2.33	0.00127	2.32	**0.00120**
d93	2.67	0.00839	2.67	**0.00808**	2.76	0.00805	2.75	**0.00768**	2.79	0.00092	2.78	**0.00090**
d98	2.42	0.01516	2.41	**0.01403**	2.54	0.01447	2.53	**0.01332**	2.58	0.00128	2.56	**0.00121**
d99	2.45	0.01689	2.44	**0.01588**	2.52	0.01341	2.52	**0.01258**	2.54	0.00121	2.53	**0.00115**
d100	2.66	0.01139	2.66	**0.01121**	2.75	0.01152	2.75	**0.01117**	2.83	0.00081	2.82	**0.00077**

5 Conclusions

In this paper, an improved triangle box-counting approach has been proposed to increase the accuracy of fractal dimension estimation for gray-scale images. This method provides the double precision for box counts by simply dividing square boxes into two equally triangle boxes. Thus, it can solve the overcounting issue and enhance the accuracy of the FD estimation. The proposed method is evaluated its performance in terms of fitting error. Based on the experimental results, the proposed method outperforms the existing methods, including differential box-counting (DBC), improved DBC (IDBC), and box-counting with adaptable box height (ADBC) methods.

References

1. Lopes, R., Betrouni, N.: Fractal and multifractal analysis: A review. Med. Image Anal. 13(4), 634–649 (2009)
2. Yu, L., Zhang, D., Wang, K., Yang, W.: Coarse iris classification using box-counting to estimate fractal dimensions. Pattern Recogn. 38(11), 1791–1798 (2005)
3. Bruno, O., Plotze, R., Falvo, M., Castro, M.: Fractal dimension applied to plant identification. Inform. Sciences 178(12), 2722–2733 (2008)
4. Sarkar, N., Chaudhuri, B.: An Efficient Differential Box-Counting Approach to Compute Fractal Dimension of Image. IEEE Transactions on Systems, Man, and Cybernetics 24(1), 115–120 (1994)
5. Li, J., Du, Q., Sun, C.: An improved box-counting method for image fractal dimension estimation. Pattern Recogn. 42(11), 2460–2469 (2009)
6. Long, M., Peng, F.: A Box-Counting Method with Adaptable Box Height for Measuring the Fractal Feature of Images. Radioengineering 22(1), 208–213 (2013)
7. Woraratpanya, K., Kakanopas, D., Varakulsiripunth, R.: Triangle-box Counting Method for Fractal Dimension Estimation. ASEAN Engineering Journal Part D 1(1), 5–16 (2012)

Prioritized Multiobjective Optimization in WSN Redeployment Using Waterfall Selection

Rungrote Kuawattanaphan and Paskorn Champrasert

Department of Computer Engineering,
Chiang Mai University, Thailand
{rungrote,paskorn}@eng.cmu.ac.th

Abstract. This paper proposes and evaluates a novel evolutionary selection called *waterfall selection* in a multi-objective optimization evolutionary algorithm for a priority-based wireless sensor nodes redeployment problem. Since, there are a variety of sensor node types in a target area. Each sensor node may have different objectives (E.g., network lifetime, data transmission, and success rate). Practically, the objectives of sensor nodes are prioritized after the deployment process. This paper focuses on the redeployment process of wireless sensor nodes to achieve their prioritized objectives simultaneously. Simulation results show that the proposed novel *waterfall selection* in multi-objective optimization evolutionary algorithm seeks to the solutions that conform to the prioritized objectives in timely manner and outperforms a *NSGA-II* evolutionary algorithm for multi-objective optimization.

Keywords: WSN, Redeployment, Multi-objective optimization, Prioritized Objective.

1 Introduction

A Wireless sensor network (WSN) contains a group of small sensor nodes. These sensor nodes can communicate and transmit sensing data to each other. Wireless sensor nodes can be deployed on a disaster risk area to serve as the intermediate nodes. Mostly, the initial deployment of sensor nodes is processed in random fashion. For example, into a disaster management application, a group of small sensor nodes can be dropped from a helicopter in the desired area [3]. After the deployment process, each sensor node starts exchanging some information with its neighbors and send its information to a base station (BS). The data transmission route can be generated by the BS using a routing protocol in the wireless sensor network (WSN) [1,2]. Then, the sensing data from the sensor node automatically transmit to the BS via intermediate nodes on the data transmission route.

However, the random fashion deployment leads to several problems such as short network lifetime and low data transmission success rate. A data transmission route may happen to be a bottleneck route; shortly, the intermediate nodes will run out of battery energy. The network lifetime is short [1]. Also; some sensor nodes may not be used and cannot transmit data to the base station because

H. Unger et al. (eds.), *Recent Advances in Information and Communication Technology 2015*,
Advances in Intelligent Systems and Computing 361, DOI: 10.1007/978-3-319-19024-2_7

they are isolated in the unwanted area [10]. Moreover, in heterogeneous wireless sensor network, sensor nodes sense and transmit variety sensor data types. Each sensor node requires specific set of objectives. The set of objectives are different among sensor nodes. The objectives can be ordered by their priority due to the importance of their sensing data. Some of sensing data must be guaranteed to be arrived at the BS because it is a critical data. On the other hand, some of sensing data are delivered without haste. Thus, the prioritized multi-objective optimization problem will be considered. These problems can be solved by a redeployment process when some nodes are moved to the new locations in order to create more data transmission routes and eliminate the isolated networks. Finding the new optimal locations of redeployment process, this problem is proven to be an NP-complete problem [7]. To overcome this issue, an evolutionary algorithm(EA) with the redeployment process will be applied. An evolutionary algorithm is one of heuristic techniques that can be used to solve a NP-complete problem [11] and also can be used to seek a set of optimal solutions in the multi-objective problems [5].

This research proposes to apply a novel selection mechanism in an evolutionary algorithm to address the WSN priority-based redeployment problem. A novel selection mechanism , called *waterfall selection*, is proposed to apply in the wireless sensor node redeployment process. It heuristically seeks the Pareto optimal sensor node new locations. The *waterfall selection* is designed to improve the offspring creation process by considering objective priority of the sensor nodes. Its performance is evaluated through simulations. Simulation results show the comparison of the result from *waterfall selection* and the results from a well-known existing evolutionary algorithm for multi-objective optimization, NSGA-II [4].

2 The Problems in WSN Priority-Based Deployment

The objectives in this research are considered in three aspects: the network lifetime(NT), the data transmission success rate(SC) and the moving cost(MC). The network lifetime is a time that the sensor nodes can send their sensing data to the BS. The moving cost is a cost when some sensors move to new locations and the data transmission success rate is a ratio of send and receive sensing data. The sensor node stations (T_1, T_2, T_3) in the Fig 1 have the same for all three above objectives and each objective have different priority number. The objectives in this station are prioritized by the importance of their sensing data. For example, in flash flood monitoring WSN [8], the water level data is a critical data. This sensing water level data must be guaranteed to be arrived at the BS to investigate an occurrence of a flash flood. Thus, the highest priority of this water level sensor is the data transmission success rate; the network lifetime is considered as the second order priority. On the other hand, the weather monitoring sensor stations which sense temperature, humidity, and rain fall level can be deployed in the same WSN but in the different region. Normally, these weather monitoring stations are deployed in an area that are difficult to reach.

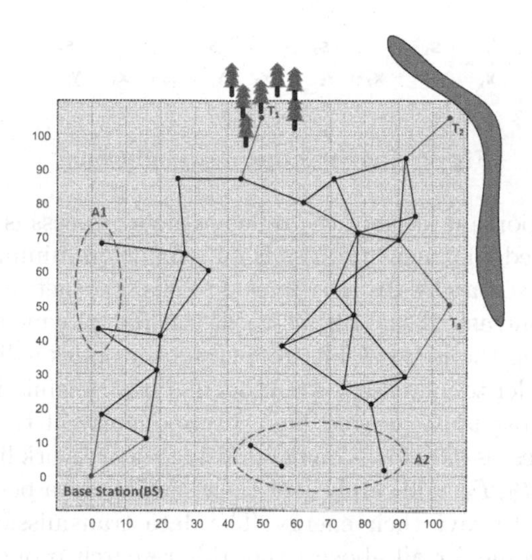

Fig. 1. An Example of WSN Priority-Based Deployment Problem

It is hard to replace the batteries. The weather sensing data is not critical. Thus, the highest priority of this weather monitoring station is the network lifetime; the data transmission success rate is considered as the second order priority. Thus, the priority number of objectives in sensor nodes are different. This research emphasize about priority-based in each objective.

In Fig 1, this is a normal WSN for environmental data sensing. Three sensor node stations(T_1, T_2, T_3) are considered to send the data to the BS. The other sensor nodes are the intermediate node. The sensor node stations (T_1) is a weather monitoring station, (T_2) and (T_3) are the flash flood monitoring station. The sensor node stations have the same objectives as network lifetime, the data transmission success rate and moving cost and each objective have different priority number as shown in the Table 1. After random deployment, the sensor nodes are crowded in the areas $A1$ and $A2$ and the sensor node stations used the same data transmission route in order to send their sensing data to the BS. An isolated problem and bottleneck route will be occurred. These problem can be solved by redeployment process. The redeployment process must improve not only the overall network performance but also the objective priority of each sensor node.

Table 1. Priority Number of Each Objective for Sensor Node Stations in Fig 1

	Sensor Node Station		
Priority Number	T_1	T_2	T_3
1 ($HighestPriority$)	NT	SC	MC
2	SC	MC	SC
3	MC	NT	NT

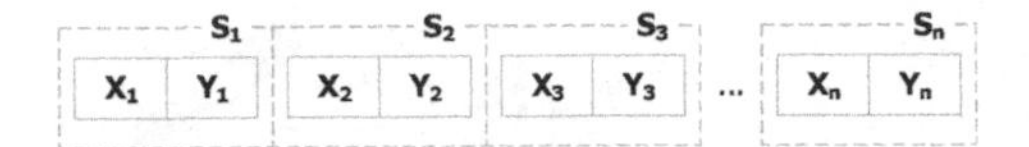

Fig. 2. The Structure of an Individual

However, an important issue of the redeployment process is the moving cost. If the goal of the redeployment process is only focus on minimizing the moving cost, all of the sensor nodes are not moved. Thus, the network lifetime is not lasting long. In contrast, if the objective of the redeployment process is only focus on maximizing the network lifetime, all sensor nodes will be moved to the new locations in order to increase the number of data transmission routes. In this case, the moving cost must be very high. Consequently, if the objective of the redeployment process is only focus on maximizing the network lifetime. All sensor node stations (T_1, T_2, T_3) will rarely sent their sensing data because this sensing node station want to save their energy. The data transmission success rate is very low. To overcome for all above issue, this research proposes an algorithm that considers the network lifetime, the data transmission success rate and the moving cost as the prioritized objectives simultaneously.

3 *Waterfall Selection* for WSN Priority-Based Redeployment

In order to improve the network lifetime, the data transmission success rate and minimize the moving cost as the prioritized objectives simultaneously, the multi-objective optimization approach is considered. For an example, assume that there are 10×10 grids represent as disaster area and there are 10 sensor nodes in the WSN. There are 100 positions which can be deployed for each sensor node. Therefore, the node placement combination is huge $(10 \times 10)^{10}$. So, the brute-force technique may not be suitable for node placement of a WSN. An evolutionary algorithm (EA) is one of heuristic techniques that can be used to solve a NP-complete problem [11] and also can be used to seek a set of optimal solutions in the multi-objective problems [5]. After EA is finished, the set of node properties (i.e., the set of solutions) are provided to the decision makers. The decision maker will select one of the solutions to develop the node redeployment process. This section describes the design of the study selection operators in a multi-objective optimization evolutionary algorithm.

3.1 Individuals

Each individual represents a set of nodes' positions in (x,y) coordinate. It consists of multiple segments, each of which represents a sensor node in the WSN. Therefore, the number of segments in each individual is equal to the total number of sensor nodes in the WSN. Fig 2 visualizes the structure of an individual. S_1 to S_n represent the first to n^{th} sensor nodes.

3.2 Optimization Objectives

This research considers prioritized multi-objective optimization in WSN redeployment problem. The objectives consist of three objectives as follow:

Network Lifetime (F_{nt}): The network lifetime(NT) in this research is defined as the time that each sensor node station starts to send their sensing data to the BS until the time that the sensing data cannot reach to the BS. This research seeks to maximize the network lifetime. The network lifetime can be calculated as Equation 1.

$$F_{nt} = Time_{last} - Time_{first} \tag{1}$$

Moving Cost (F_{mc}): The moving cost(MC) in this research is defined as the total cost when some of sensor nodes have to move to the new locations. This research seeks to minimize the moving cost. The total moving cost is a summation of each sensor node moving distances as described as Equation 2.

$$F_{mc} = \sum_{i=1}^{N} \sqrt{(x_{new} - x_{old})^2 + (y_{new} - y_{old})^2} \tag{2}$$

where N is the total number of sensor nodes, (x_{old}, y_{old}) is a current sensor node position and (x_{new}, y_{new}) is a new sensor node position.

Data Transmission Success Rate (F_{sc}): The data transmission success rate (SC) in this research is defined as the ratio of the number of received packets at the BS to the number of sent packets from each sensor node station. The data transmission success rate shows the WSN throughput performance. This throughput performance can be measured as Equation 3. This research seeks to maximize the data transmission success rate.

$$F_{sc} = \frac{\sum_i (P_{BS})_i}{\sum_i (P_{Total})_i} \tag{3}$$

where i is the number of sensor node stations, P_{BS} is the number of packets in each sensor node station received by the BS and P_{Total} is the total number of generate packet from each sensor node station in the target area.

3.3 Evolutionary Optimization Process

Waterfall Selection runs on the BS after random deployment. *Waterfall Selection* performs its evolutionary optimization process to adjust node properties. After *Waterfall Selection* is finished, the set of solutions are provided to the decision makers. The decision maker will select one of the solutions to develop the node redeployment process.

Algorithm 1 shows the algorithmic structure of evolutionary optimization in *Waterfall Selection.* The initial population (P^0) consists of μ individuals that

```
   output: the set of solutions (Q)
 1 parameter: g=number of each generation;
 2            μ=total number of individuals;
 3 g ← 0;
 4 P^0 ← Randomly generated μ individuals;
 5 Q^0 ← Null;
 6 while g != g_max do
 7   while |Q^g| != μ do
 8     p1 ← WaterfallTournament(P^g);
 9     p2 ← WaterfallTournament(P^g);
10     q1, q2 ← Crossover(p1, p2);
11     Q^g ← Q^g ∪ {q1, q2};
12   end
13   Q^g ← Mutation(Q^g);
14   P^{g+1} ← WaterfallSelection(P^g ∪ Q^g);
15   g ← g++;
16 end
```

Algorithm 1. Evolutionary Optimization in *Waterfall Selection*

contain randomly-generated node positions. In each generation (g), a pair of individuals, called parents (p_1 and p_2), are chosen from the current population P^g using a *waterfall tournament* operator (`WaterfallTournament()`). A *waterfall tournament* operator randomly takes two individuals from P^g, compares them based on their fitness values and order by the priority number, and chooses a superior one (i.e., the one whose fitness is higher) as a parent.

A pair of parents (p_1 and p_2) reproduce two offspring (q_1 and q_2) by using a crossover operator (`crossover()`). The offspring is mutated with a mutation operator (`mutation()`). The crossover and mutation operators change the node positions. The two offspring are created. These operators (`WaterfallTournament()`, `crossover()`, and `mutation()`) are repeated until the population of offspring ($|Q^g|$) reaches the population size (μ). Then, the *Waterfall Selection* is performed to select μ individuals for the next generation. The $|Q^g|$ offspring population is combined to the parent population P^g. Thus, the population size is 2μ (i.e., $P^g \cup Q^g$). Then, the (`WaterfallSelection()`) operator will be executed and selects the top μ individuals from $P^g \cup Q^g$ as the next generation population (P^{g+1}). This selection operator is performed by ordering of the fitness values and the priority number of each individual. The *Waterfall Selection* terminates its process when the number of the generations (g) reaches its predefined value (g_{max}).

3.4 Waterfall Selection Operation

The *waterfall selection* is designed to select a good parent or a good offspring with priority-based in multi-objectives optimization problem. The *waterfall selection* is divided into two procedures. First is called the *waterfall tounament* procedure. Algorithm 2 describes the pseudo code of the *waterfall tournament*

```
   input  : Population (Q)
   output: Parent individual (p)
 1 parameter: i =number of sensing nodes station;
 2            j^th=number of objectives priority;
 3            n =total number of sensing node station;
 4 r1, r2 ← random selection from Q;
 5 j ← 1;   // Begin at First Priority
 6 while p isNull do
 7  |  for i ← 1 to n do
 8  |  |  w1 ← PriorityValue(i, j, r1);
 9  |  |  w2 ← PriorityValue(i, j, r2);
10  |  |  i++;
11  |  end
12  |  p = DominationRanking(w1, w2);
13  |  j++;
14 end
```

Algorithm 2. *Waterfall Tounament* Procedure

procedure. The input for this procedure is a population(Q). Two individuals (r_1, r_2) are selected by random technique from a population(Q). In each priority(j), the objective values(fitness value)(w_1, w_2) is calculated from each individual(r_1, r_2) by operator `PriorityValue()`. Then, a domination ranking operator (`DominationRanking()`) selects the winner (the highest fitness value). A domination ranking technique is described in [5]. The winner is become a parent in Algorithm 1. Second is called the *waterfall selection* procedure. The pseudo code of the *waterfall selection* procedure is quite similar to the *waterfall tournament.* A little difference between the *waterfall tournament* and the *waterfall selection* procedure is adding crowding distance procedure(`CrowdingDistance()`) after line number 12 in Algorithm 2. The crowding distance procedure is used for select the best individual in the same rank. The crowding distance procedure is described in [5].

4 Simulation Evaluation

This section shows simulation configurations and a set of simulation results to evaluate how *waterfall selection* contributes to search for appropriate node position that optimize the three objectives and three priority numbers of priority-based problem.

4.1 Simulation Configurations

The *waterfall selection* simulations were carried out on the modified jMetal[6]. The WSN simulator has been implemented and use the Gradient-based routing protocol(GBR)[12] as a WSN routing protocol. The simulator is combined with battery energy consumption of wireless sensor nodes and the media loss [13] in this simulation is set to be 5%. The sensing node station generate sensing data one packet per second.

The *waterfall selection* enter a group of individuals to the WSN simulation. Each individual compounds with a set of sensor node positions (x,y) . Then, the WSN simulator performs the WSN operations and returns the network lifetime, the data transmission success rate and the moving cost value to the jMetal. The simulation terminates its evolutionary optimization process when the number of the generations reaches its maximum predefined value.

It is assumed that the simulated wireless sensor network is initial randomly deployed sensor nodes in disaster risk area. The simulated wireless sensor network consists of 29 nodes in maximum. Sensor nodes are placed to the disaster area size $100 \times 100\ m^2$. The physical properties of each wireless sensor node shows as Table 2, the priority table of the sensing node stations show as Table 1 and the simulation configurations in *EA* (*waterfall selection* and NSGA-II used the same config) show as Table 3

Table 2. Sensor Node Types

Type	#	Communication Range	Sensing Range
BS	1	50(m)	-
Sensor Node Station	3	25(m)	10(m)
Intermediate Node	25	25(m)	-

Table 3. The *EA* Simulation Configurations

Configuration	*EA*
# number of independent runs	16
μ	100
g_{max}	2,000
mutation rate	$1/n$
crossover rate	0.9
degree of SBX crossover	20
degree of polynomial mutation	20

4.2 Simulation Results

The simulation results are discussed in this section. The results from 16 independent runs of *waterfall selection* and *NSGA-II* are selected and compared in three metrics 1) The solution at last generations for all priority number of objectives 2) $\mathcal{C}$-metric which compared the obtained solutions of the algorithms, 3) The comparison of optimal solutions from each algorithm.

In jMetal[6], the default objectives is set to find minimum value. Thus, this research have to adjust the NT and the SC objectives. The NT and the SC objectives are re-formulated to $100 - NT$ and $1/SC$ respectively. Thus, if any algorithm which can find near the minimum value is better than the others.

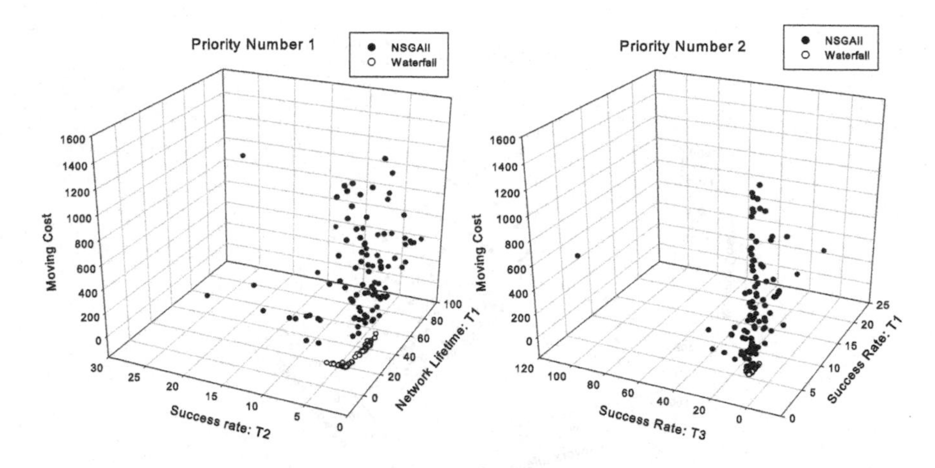

Fig. 3. *Waterfall Selection* versus NSGA-II in priority number 1 and 2

Fig 3 and Fig 4, the solutions from *waterfall selection* is near minimum value (the bottom right conner is a minimum value) than the solutions from *NSGA-II* for all priority number.

Table 4. $\mathcal{C}$-metric

Priority Number	$\mathcal{C}$(*Waterfall*,NSGA-II)	$\mathcal{C}$(NSGA-II, *Waterfall*)
1	0.98	0.00
2	0.47	0.00
3	0.78	0.00

$\mathcal{C}$-metric [14] represents how the individuals of an algorithm outperform the individuals of the other algorithm. Table 4 shows the $\mathcal{C}$(*Waterfall*,NSGA-II) and $\mathcal{C}$(NSGA-II, *Waterfall*) at generation 2,000. The result shows that at final generation $\mathcal{C}$(*Waterfall*,NSGA-II) is greater than $\mathcal{C}$(NSGA-II, *Waterfall*) for all priority number of objectives. This result means that the solutions from *Waterfall* dominate 98% of solutions from NSGA-II in priority number 1, 47% in priority number 2, 78% in priority number 3, and the solutions from NSGA-II cannot dominate any solutions of *Waterfall* in all priority number of objectives. Thus, the combination of genetic operations in *Waterfall* contribute to better solutions compared to the NSGA-II algorithm.

Table 4.2 shows average number of percentage which solutions from *waterfall selection* are superior than the solutions from *NSGA-II* and prioritized for all objectives. The priority number is the same as Table 1. The first order priority of sensor node station(T_1) is a NT. The solutions from *waterfall selection* can increase the NT to 52.8% from initial random deployment while the solutions from *NSGA-II* can increase the NT only 9.3%. The second order of sensor node station(T_1) is a SC. The *waterfall selection* can increase to 11.2% but the *NSGA-II* is decrease -6.5%. In sensor node station(T_1), the first order(NT) is increased

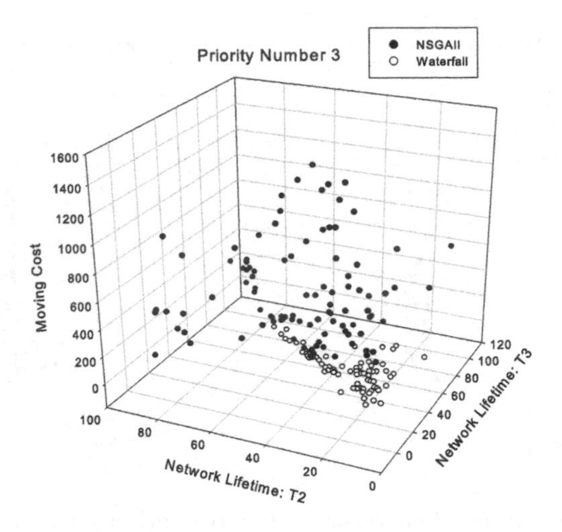

Fig. 4. *Waterfall Selection* versus NSGA-II in priority number 3

Table 5. Comparison of priorities

	T_1					T_2					T_3				
	Initial	*NSGA-II*		*waterfall*		Initial	*NSGA-II*		*waterfall*		Initial	*NSGA-II*		*waterfall*	
P_1	NT	NT	%	NT	%	SC	SC	%	SC	%	MC	MC	%	MC	%
	57	62.3	9.3	87.1	52.8	0.65	0.52	-20	0.75	15.3	0	503.8	-	18.3	-
P_2	SC	SC	%	SC	%	MC	MC	%	MC	%	SC	SC	%	SC	%
	0.62	0.58	-6.5	0.69	11.2	0	503.8	-	18.3	-	0.46	0.48	4.3	0.70	52.1
P_3	MC	MC	%	MC	%	NT	NT	%	NT	%	NT	NT	%	NT	%
	0	503.8	-	18.3	-	58	51.1	-11.9	65.9	13.6	58	57.5	-0.8	59.9	3.32

Note: T is a sensor nodes station, P is a priority number

up to 52.8% and the second order is increased up to 11.2%. This is ordered by priority number. It can be seen that the *waterfall selection* emphasize in order number of priority. Similarly with the sensor node stations(T_2, T_3), the *waterfall selection* can improve not only each of objectives versus initial random deployment but also achieve prioritized objective.

Fig 5 shows how the redeployment process improves the objectives of WSN. Fig 1 represents the initial nodes' positions from the random deployment process. Fig 5(left) represents one of the solutions after 2,000 generation evolution of *NSGA-II* and Fig 5(right) represents one of the solutions after 2,000 generation evolution of *waterfall selection*. Obviously, in Fig 5(right), there are more date transmission routes than that of the initial deployment and solution from *NSGA-II*. Since the data transmission load is distributed among the sensor nodes, the network lifetime of WSN is increased. Also, the data transmission rate is enlarged. However, only a few of the sensor nodes are moved to the new locations, which means the moving cost is small.

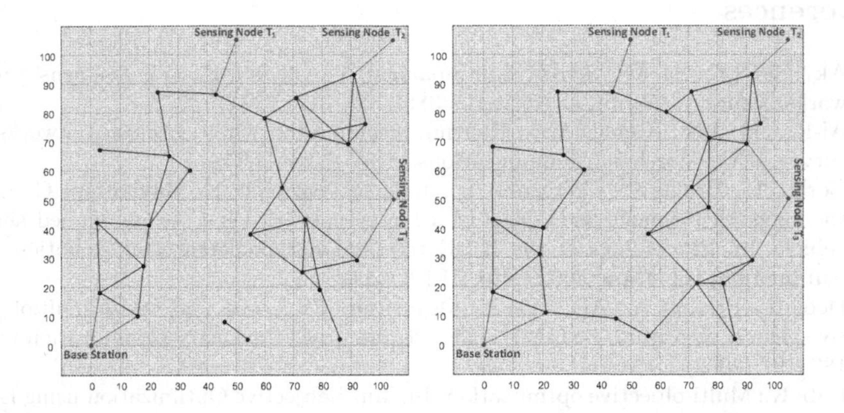

Fig. 5. An Example Solution from *NSGA-II*(Left) versus *Waterfall Selection*(Right)

5 Related Work

In [9], this paper investigates the proposed evolutionary algorithm to find appropriate sets of node locations for wireless sensor node redeployment. Simulation results show that the genetic operators in *FBEA* work properly and are able to find an appropriate set of node locations while handle the moving cost and data transmission success rate. The number of individuals that violate the constraints reduce faster than *NSGA-II* does. However, sensing node station in this research has only one station and does not take into account in prioritized multi-objective optimization problem.

6 Conclusion

This research investigates the proposed a novel *waterfall selection* operation in the evolutionary algorithm to find appropriate sets of node positions for prioritized multi-objective optimization problem. This selection mechanism selects a good offspring and evolve them via genetic operator in evolutionary algorithm. Simulation results show that the *waterfall selection* work properly and are able to find an appropriate set of node positions while corresponds with maximum the network lifetime, minimum the moving cost and maximum the data transmission success rate. The solutions from *waterfall selection* outperforms NSGA-II in all priority number of objectives while *waterfall selection* and NSGA-II take on the same execution time.

Acknowledgement. This work was supped by the graduate school, Chiang Mai university, Chiang Mai, Thailand. The authors also would like to thank the graduate school and department of computer engineering Chiang Mai university for their help and support in making this research possible.

References

1. Akyildiz, I.F., Su, W., Sankarasubramaniam, Y., Cayirci, E.: Wireless sensor networks: a survey. Computer Networks 38(4), 393–422 (2002)
2. Al-Karaki, J.N., Kamal, A.E.: Routing techniques in wireless sensor networks: a survey. IEEE Wireless Communications 11(6), 6–28 (2004)
3. Corke, P., Hrabar, S., Peterson, R., Rus, D., Saripalli, S., Sukhatme, G.: Autonomous deployment and repair of a sensor network using an unmanned aerial vehicle. In: Proceedings of the IEEE International Conference on Robotics and Automation, vol. 4, pp. 3602–3608. IEEE (2004)
4. Deb, K., Pratap, A., Agarwal, S., Meyarivan, T.: A fast and elitist multiobjective genetic algorithm: NSGA-II. IEEE Trans. on Evolutionary Computation 6(2), 182–197 (2002)
5. Deb, K.: Multi-objective optimization. In: Multi-objective Optimization using Evolutionary Algorithms, pp. 13–46 (2001)
6. Durillo, J., Nebro, A., Alba, E.: The jmetal framework for multi-objective optimization: Design and architecture. In: CEC 2010, Barcelona, Spain, pp. 4138–4325 (2010)
7. Efrat, A., Har-Peled, S., Mitchell, J.S.: Approximation algorithms for two optimal location problems in sensor networks. In: Proceedings of the 2nd International Conference on Broadband Networks, pp. 714–723. IEEE (2005)
8. Jankoo, S., Champrasert, P., Suntaranont, B.: Fuzzy logic control mechanism for flash flood monitoring station. In: 2014 IEEE Sensors Applications Symposium (SAS), pp. 349–354. IEEE (2014)
9. Kuawattanaphan, R., Kumrai, T., Champrasert, P.: Wireless sensor nodes redeployment using a multiobjective optimization evolutionary algorithm. In: Proceedings of the TENCON International Conference on IEEE Region 10 Conference, pp. 1–6. IEEE (2013)
10. Mahfoudh, S., Minet, P., Laouiti, A.: Overview of deployment and redeployment algorithms for mobile wireless sensor networks. Procedia Computer Science 10, 946–951 (2012)
11. Papadimitriou, C.H.: Computational complexity. John Wiley and Sons Ltd. (2003)
12. Schurgers, C., Srivastava, M.B.: Energy efficient routing in wireless sensor networks. In: Military Communications Conference, MILCOM 2001. Communications for Network-Centric Operations: Creating the Information Force, vol. 1, pp. 357–361. IEEE (2001)
13. Zhao, J., Govindan, R.: Understanding packet delivery performance in dense wireless sensor networks. In: Proceedings of the 1st International Conference on Embedded Networked Sensor Systems, pp. 1–13. ACM (2003)
14. Zitzler, E., Thiele, L.: Multiobjective Evolutionary Algorithms: A Comparative Case Study and the Strength Pareto Approach. IEEE Transactions on Evolutionary Computation 3(4), 257–271 (1999)

Statistical Feature Fusion for Sassanian Coin Classification

Seyyedeh-Sahar Parsa[1], Maryam Rastgarpour[1], and Mohammad Mahdi Dehshibi[2]

[1] Department of Computer Engineering, Islamic Azad University, Saveh Branch, Saveh, Iran
{sahar.parsa,m.rastgarpour}@iau-saveh.ac.ir
[2] Pattern Research Center (PRC), Iran
dehshibi@iranprc.org

Abstract. Ancient coins classification has attracted increasing attention for the benefits which it brings to numismatic community. However, high between-class similarity and, in the meantime, high within-class variability make the problem a particular challenge. This issue highlights the importance of extracting discriminative features for ancient coins classification. Therefore, in this paper, the capability of statistical feature fusion was examined. First, a representation of the coin image based on the phase of the 2-D Fourier transform of the image is using so that the adverse effect of illumination was eliminated. The phase of the Fourier transform preserves the locations of the edges of a given coin image. The problem of unwrapping is avoided by considering two functions of the phase spectrum rather than the phase directly. Then, BDPCA approach which can reduce the dimension of the phase spectrum in both column and row directions is used and an entry-wise matrix norm calculates the distance between two feature matrices so as to classify coins. Extensive experiments are conducted on a database of Sassanian coins in order to compare the performance of proposed method with the other feature extraction method which are used in other works. The results show the proposed method is promising.

Keywords: Ancient coins classification, BDPCA, Cultural Heritage, Entry-wise matrix norm, phase representation.

1 Introduction

Nowadays, applications of machine vision can be found in every aspect of life [12-16]. Ancient coins classification is one of the most important activities in the fields of cultural heritage and numismatics which can be benefited by machine vision, pattern recognition and other related fields. More precisely, an accurate automated coins classification system significantly improves the classification accuracy, speeds the process, and reduces the processing time. In addition, such a system can be used for classifying and documenting large collections of unclassified coins which are being kept in the museums. Furthermore, illegal trade of stolen coins can be detected and prevented as the great majority of them are being sold through the Internet and, therefore, manual tracking of the trade is almost impossible.

H. Unger et al. (eds.), *Recent Advances in Information and Communication Technology 2015*,
Advances in Intelligent Systems and Computing 361, DOI: 10.1007/978-3-319-19024-2_8

Classification of ancient coins is not a trivial task and encounters many difficulties. On the one hand, irregular shape of coins caused by manual minting, fractures, and erosions leads to considerable variation within class samples which, in some cases, can be increased by prototyping the same person at different ages with various clothes, hairstyles, crowns and decorations. On the other hand, between-class similarity is also high, especially in cases in which the rulers belong to the same dynasty and therefore their coins follow almost similar patterns. Fig. 1 shows examples of mentioned challenges in Sasanian coins. It is worth noting that the coins with the same person prototyped on their obverses, who is often a ruler or a king, usually go to the same class.

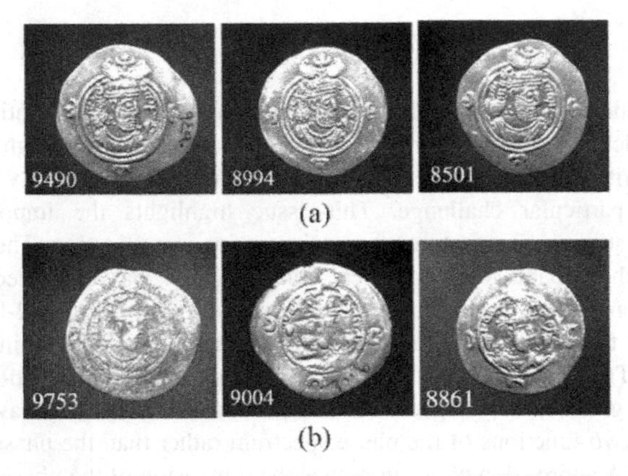

Fig. 1. (a) Similar coins belonging to three different classes (b) Three dissimilar coins from the same class

Developed ancient coins classification methods have mainly utilized local features in order to build a discriminative feature space. For example in [1], different local descriptors are used as feature extractors for ancient coins recognition and it was observed in the course of experiments that SIFT descriptor has the capability of producing a promising result. In another attempt, Arandjelovic [2] can achieve the rate of 57% in Roman's coins classification with a view to utilizing the visual words and locally biased directional histograms. Although local features are typically suggested for this application, global features are less taken into account and have been overshadowed by local features. Although the latter work conducted experiments on a database with 65 classes and examined different features for comparing the end results, it is similar to Kampel work in performing local features and the experiments are not reproducible. The reason is that the detail of feature's parameters and data distributions corresponded to between/within classes are left in doubt. Allahverdi et al. [3], [4] did the first attempt in ancient Persian coins classification in the Sasanian's era. They used statistical methods in their works; these methods were Eigen analysis and Discrete Cosine Transform, respectively, in conjunction with Support Vector Machine. They achieved the recognition rate of 21.7% and 86.21%, in turn. In addition to ancient coins, several works have been reported on classifying modern coins.

For instance, Huber et al. [5] utilize Eigen analysis in order to construct a discriminative model which has the capability of classifying coins from their diameter and thickness. Another identification system for matching EURO coins is presented by Khashman et al. [6] in which a neural network is trained with images relate to the both side of EURO in different rotated positions. In [7], angular and distance information of coin's edge image are encoded in histograms; then, a 3-nearest neighbor approach is utilized on two sides of the coin to construct a classification pilot so-called COIN-O-MATIC.

In this paper, we explored the effectiveness of the dimensional reduction approach in a phase of the 2-D Fourier transform representation of the coin image. Bi-Directional Principal Component Analysis (BDPCA) approach [8] is used because its power in reducing the dimension of the phase spectrum in both column and row directions. Finally, entry-wise matrix norm calculates the distance between two feature matrices so as to classify coins. In order to compare the features comprehensively, extensive experiments are conducted on a database of Sasanian coins. The experiments include performance of each group while considering obverse and reverse sides of coins, and capability of each one in making distinction between samples of each class and other classes.

The organization of this paper is as follows. Section 2 briefly describes the proposed method. Experimental results are illustrated in Section 3. Finally, in section 4, results are discussed and a conclusion is drawn.

2 Proposed Method

Feature extraction is a key step in any pattern recognition problem. In particular, performance of a method is directly dependent on how discriminative the extracted features are. An appropriate feature extraction method has to minimize the overlap or similarity between classes while maximizing the similarity within each class. This is a particular challenge especially in cases in which the nature of the problem itself leads to a large overlap between classes like ancient coins classification as foreshadowed. Therefore, exploring discriminative feature extraction methods for the purpose of ancient coins classification is of great importance.

2.1 Preprocessing and Region Extraction

The areas of the coins under study have to be extracted from cluttered background images. In fact, backgrounds of images under this study have been cluttered with some 4-digit numbers, as is shown in Fig.1, indicating the coins' record IDs. These numbers have to be removed from the backgrounds as they can introduce correlation and similarities between images and therefore will affect the classification rate. To do this, Sobel operator is firstly applied to create a binary gradient mask which represents lines of high contrast; i.e. edges; in the image. This mask is later enhanced by being dilated as well as filling holes and removing small undesired objects. Finally, the binary mask is applied to the image and the coin's region is extracted. This process is illustrated in Fig. 2.

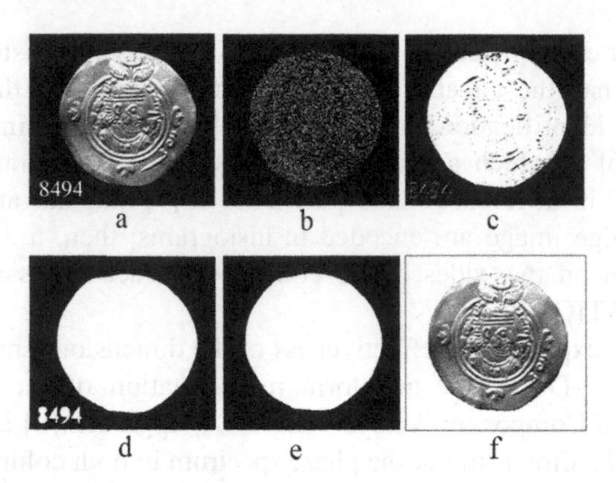

Fig. 2. Preprocessing steps. (a) Original image; (b) Binary gradient mask; (c) Dilated gradient mask; (d) Gradient mask after filling holes; (e) Gradient mask after removing undesired connected components; (f) Segmented image.

2.2 Phase of the Fourier Transform

The most common way of representing an image in the spatial domain is by a two-dimensional array of positive numbers, corresponding to the gray levels of the pixels. An image can also be represented in the frequency domain as the discrete Fourier transform (FT) of the two-dimensional array of pixels [9]. The Fourier representation involves complex numbers, i.e., the magnitude and phase parts. The relative importance of the magnitude and phase of the FT of a signal/image under different situations was studied in [10], [11]. It is difficult to visualize how the information in these two components are related, because the magnitude and phase are not directly comparable.

Let us represent an image by $x\,[n_1, n_2]$, $n_1 = 0, 1, \ldots, R - 1$, $n_2 = 0, 1, \ldots, C - 1$. Here R and C are the number of rows and columns of the given image, respectively. The discrete Fourier transform (DFT) [9] of $x\,[n_1, n_2]$ is given by:

$$\begin{aligned} X[k_1, k_2] &= \mathrm{DFT}\,\{x[n_1, n_2]\} = X_r\,[k_1, k_2] + X_i\,[k_1, k_2] \\ &= |X[k_1, k_2]| \times \exp[\theta[k_1, k_2]] \end{aligned} \tag{1}$$

where $|X[k_1, k_2]| = ((X_r\,[k_1, k_2])^2 + (X_i\,[k_1, k_2])^2)^{0.5}$, and $\theta[k_1, k_2] = \arctan\{X_i\,[k_1, k_2]/\,X_r\,[k_1, k_2]\}$ are the magnitude and the phase of the DFT, respectively. The real and imaginary parts of the DFT are denoted by X_r and X_i, respectively. The original image can be obtained from $X[k_1, k_2]$ by inverse DFT relation [9] which is abbreviated as IDFT. The information contained in the magnitude and phase of the DFT can be visualized using either magnitude-only synthesis of the image or the phase-only synthesis of the image, as are shown in Fig. 3. The phase-only image gives more crucial features as compared to the magnitude-only image. But computation of the phase spectrum using *arctan* leads to the problem of phase wrapping [9]. One way to resolve this problem is to use a function of the phase spectrum as follows:

$$exp\left[j\theta[k_1,k_2]\right] = cos\left[\theta[k_1,k_2]\right] + j \times sin\left[\theta[k_1,k_2]\right] = \frac{X[k_1,k_2]}{|X[k_1,k_2]|} \quad (2)$$

$$cos\left[\theta[k_1,k_2]\right] = \frac{X_r[k_1,k_2]}{|X[k_1,k_2]|}, sin\left[\theta[k_1,k_2]\right] = \frac{X_i[k_1,k_2]}{|X[k_1,k_2]|} \quad (3)$$

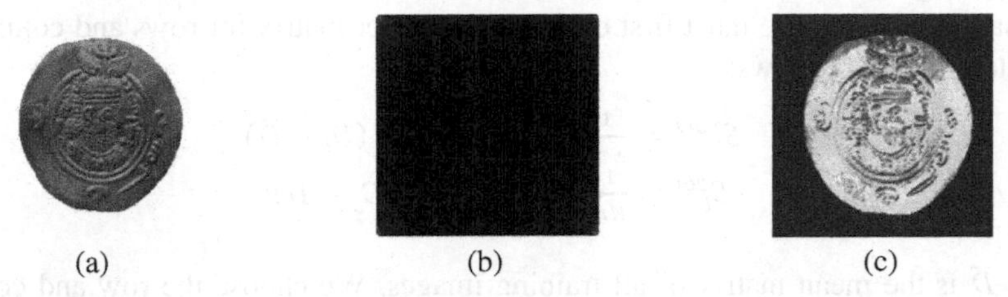

Fig. 3. (a) Gray-level coin image. (b) Magnitude-only synthesis of coin image. (c) Phase-only synthesis of coin image

2.3 BDPCA with Entry-Wise Matrix Norm

The cosine and sine functions of the phase spectrum accentuate the high frequency components. Hence they emphasize noise also. The effect of noise can be reduced using Eigen analysis. Let the training coin images for the coin i be denoted by set D_i. The cosine and sine functions of the phase spectrum are computed using Eq. 3. The DFT of a real image exhibits conjugate symmetry [9]. Hence only the non-redundant coefficients (the shaded region in Fig. 4) of the cosine and sine functions of the phase spectrum are used in the Eigen analysis.

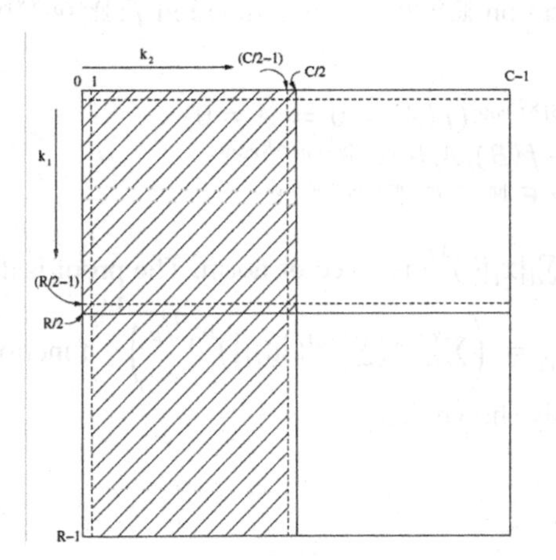

Fig. 4. DFT coefficients $X[k_1, k_2]$ in the shaded area determine the remaining coefficients

Here, we use an extend model of Eigen analysis method, so-called Bi-Directional PCA, for achieving the aim of noise reduction as well as the feature extraction. Then, an entry-wise matrix distance metric is used to calculate the distance of two feature matrices. BDPCA directly extracts feature matrix Y from image matrix D by,

$$Y = W_{col}^T D W_{row} \tag{4}$$

where W_{row} and W_{row} are the column and row projectors. In order to calculate these projectors, given a training set $\{D_1, ..., D_N\}$ where N is the number of samples and each sample is mxn, we must first define total scatter matrix for rows and columns of the data matrix as follows:

$$S_t^{row} = \frac{1}{Nm} \sum_{i=1}^{N} (D_i - \overline{D})^T (D_i - \overline{D}) \tag{5}$$

$$S_t^{col} = \frac{1}{Nn} \sum_{i=1}^{N} (D_i - \overline{D})(D_i - \overline{D})^T \tag{6}$$

where $\overline{D}$ is the mean matrix of all training images. We choose the row and columns eigenvectors corresponding to the first k_{row} and k_{col} largest eigenvalues of S_t^{row} and S_t^{col} to construct the row (W_{row}) and column (W_{col}) projectors, respectively. Finally we use Eq. 4 to extract feature matrix Y from image D. BDPCA just produces a feature matrix, with which doing a classification task needs to define a matrix distance. Therefore, we use a sort of entry-wise matrix norm so as to increase the recognition rate. This distance is defined as follows and its efficiency is subjected to several experiments:

$$d(A,B) = \left(\sum_{j=1}^{k_{row}} \left(\sum_{i=1}^{k_{col}} \left(a_{ij} - b_{ij}\right)^p \right)^{q/p} \right)^{1/q} \tag{7}$$

It is worth mentioning $A = (a_{ij})_{k_{col} \times k_{row}}$ and $B = (b_{ij})_{k_{col} \times k_{row}}$ are two feature matrixes. A matrix norm on $\mathbb{R}^{k_{col} \times k_{row}}$ is a function $f: \mathbb{R}^{k_{col} \times k_{row}} \to \mathbb{R}$ with the following properties [6]:

1. $f(A) \geq 0, A \in \mathbb{R}^{k_{col} \times k_{row}} (f(A) = 0 \Leftrightarrow A = 0)$
2. $f(A + B) \leq f(A) + f(B), A, B \in \mathbb{R}^{k_{col} \times k_{row}}$
3. $f(\alpha A) \leq |\alpha| f(A), \alpha \in \mathbb{R}, A \in \mathbb{R}^{k_{col} \times k_{row}}$

Theorem 1: $\|x\|_p = (\sum_i |x_i|^p)^{1/p}$ is a vector norm. The proof is discussed in [6].

Theorem 2: The $\|A\|_{p,q} = \left(\sum_{j=1}^{k_{row}} \left(\sum_{i=1}^{k_{col}} \left(|a_{ij}|\right)^p \right)^{q/p} \right)^{1/q}$ function is a matrix norm.

Proof: It can be easily shown that

– $\|A\|_{p,q} \geq 0$,
– $\|A\|_{p,q} = 0 \Leftrightarrow A = 0$,
– $\|\alpha A\|_{p,q} = |\alpha| . \|A\|_{p,q}$.

Now, we prove $\|A+B\|_{p,q} \leq \|A\|_{p,q} + \|B\|_{p,q}$.

$$\|A + B\|_{p,q} = \left(\sum_{j=1}^{k_{row}} \left(\sum_{i=1}^{k_{col}} (a_{ij} + b_{ij})^p \right)^{q/p} \right)^{1/q} \leq \left(\sum_{j=1}^{k_{row}} \left(\|a^{(j)}\|_p + \|b^{(j)}\|_p \right)^q \right)^{1/q}$$

where $a^{(j)}$ and $b^{(j)}$ denote the jth column vectors of A and B, respectively. From **Theorem 1**, we know $\left(\sum_{j=1}^{k_{row}}\left(\|a^{(j)}\|_p\right)^q\right)^{1/q}$ and $\left(\sum_{j=1}^{k_{row}}\left(\|b^{(j)}\|_p\right)^q\right)^{1/q}$ are vector norms for a and b, respectively. Therefore, the following property is reasonable:

$$\left(\sum_{j=1}^{k_{row}}\left(\|a^{(j)}\|_p+\|b^{(j)}\|_p\right)^q\right)^{1/q} \le \left(\sum_{j=1}^{k_{row}}\left(\|a^{(j)}\|_p\right)^q\right)^{1/q}+\left(\sum_{j=1}^{k_{row}}\left(\|b^{(j)}\|_p\right)^q\right)^{1/q}$$

$$=\left(\sum_{j=1}^{k_{row}}\left(\sum_{i=1}^{k_{col}}(a_{ij})^p\right)^{q/p}\right)^{1/q}+\left(\sum_{j=1}^{k_{row}}\left(\sum_{i=1}^{k_{col}}(b_{ij})^p\right)^{q/p}\right)^{1/q}=\|A\|_{p,q}+\|B\|_{p,q}$$

Consequently, $\|A\|_{p,q}$ is a matrix norm and can be used to find out an input image belongs to which class.

2.4 Eigenanalysis Using Fourier Phase

We only used the non-redundant coefficients (the shaded region in Fig. 4) of the cosine and sine functions of the phase spectrum in the Eigenanalysis. Let $\mathbf{\Psi}^c \in \mathbb{R}^{N\times m} = [\psi_1^c, \dots, \psi_m^c]$ and $\mathbf{\Psi}^s \in \mathbb{R}^{N\times m} = [\psi_1^s, \dots, \psi_m^s]$be the eigenvectors corresponding to the m, where $m = min\{k_{row}, k_{col}\}$, largest eigenvalues derived from W_{row} and W_{row} of the given training images, respectively. Here, $N = (row \times col/2) + 2$. The eigenvectors are used to represent the image approximately in cosine and sine domain with respect to Eq. 4. These new representations are used for matching in a coin recognition task. The effect of noise is reduced as only the first m ($m \le N$) coefficients are considered for matching. The proposed representations have another advantage in the context of Eigenanalysis, as the size of resulting scatter matrixes is approximately one fourth as compared to the scatter matrixes $[(row \times col) \times (row \times col)]$ obtained using gray level values of the coin images. Thus, the estimation of the eigenvectors may be more accurate for same number of training images.

3 Experiments

This section describes experiments with the proposed method. The extracted cosine and sine features are investigated in the Bi-Directional PCA algorithm so as to find the right classes. In addition, the classification rate of the system is compared with classification rates obtained from applying different feature extraction and classification methods. It is worth mentioning that in these experiments, both sides of coins were considered.

3.1 Database

The experiments are conducted on a database with 570 coins images in JPEG format with resolution 1014×1014 pixels. The images are equally distributed in three classes. In other words, there are 95 coins, i.e. 190 images, in each class. Each class corresponds to a

Sasanian king namely, Khosrow I, Khosrow II and Hormizd IV. Two third of the data are allocated to training set and the remaining part is considered as test set. Fig. 5 depicts sample coin images of each class in the database.

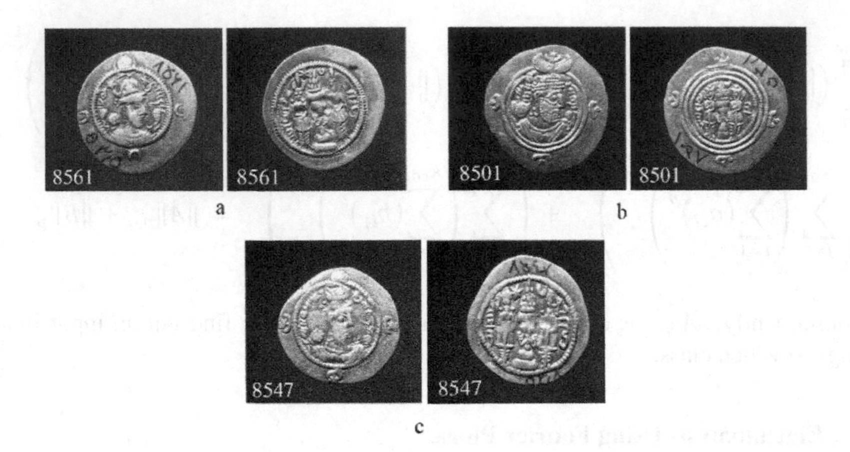

Fig. 5. Database samples. Obverse and reverses of (a) Khosrow I; (b) Khosrow II; (c) Hormozd IV.

3.2 Significance of DFT Coefficients and BDPCA Features

The spacing of the edges will be inversely proportional to the frequency in the phase of the Fourier transform. As a result, the low frequency DFT coefficients correspond to events/edges separated by large spacing, and the high frequency DFT coefficients for events/edges separated by small spacing. The effect of the different DFT coefficients can be seen in the phase-only synthesis of the image.

Matching true class coin images having some variation with respect to training images can be improved by removing some high frequency DFT coefficients. This is an advantage because noise and events with small spacing are given less importance. Experiments were conducted by considering only the first k DFT coefficients along both the axes of the $X^c[k_1, k_2]$ and X^s $[k_1, k_2]$ representations of the given training coin image. Only non-redundant coefficients are used for Eigenanalysis. The recognition performance is improved by removing some high frequency DFT coefficients.

In general the performance increases with m, but after some value of m the performance reaches a maximum value. The performance can be improved further by removing some high frequency DFT coefficients. In fact the performance improves from 79.31% to 87.35% using k 128 for choosing the number of DFT coefficients and decreasing the size of feature vector to $k_{row} = 4$, $k_{col} = 18$, and setting p_1=2 and p_1= 0.25.

Performance comparison with other methods is given in Table 1. The results show that the proposed method performs better than the existing methods.

Table 1. Average recognition rate in %

Methods	Set of reference coin images		
	Khosrow I	Khosrow I	Hormozd IV
Principal Component Analysis (PCA)	68.96	82.75	44.82
2D PCA	79.31	82.75	65.51
Bi-Directional PCA	79.31	93.10	86.20
Indipent Component Analysis (ICA)	62.06	82.75	62.06
Discrete Cosine Transform + SVM	79.31	93.10	65.51
Wavelet + SVM	65.51	93.10	79.31
Proposed	82.75	93.10	86.20

4 Conclusion

Ancient coins classification is one of the most important activities in the fields of cultural heritage and numismatics which can be benefited by machine vision, pattern recognition and other related fields. This paper presented an efficient ancient coins classification method. This method utilizes a representation of the coin image based on the phase of the 2-D Fourier Transform (FT) of the image so that the adverse effect of illumination was eliminated. The phase of the Fourier transform preserves the locations of the edges of a given coin image. The problem of unwrapping is avoided by considering two functions of the phase spectrum rather than the phase directly. Then, BDPCA approach which can reduce the dimension of the phase spectrum in both column and row directions is used and an entry-wise matrix norm calculates the distance between two feature matrices so as to classify coins.

Effect of different number of FT as well as BDPCA coefficients was examined in order to find the best choice. The highest classification rate of 87.35 % was obtained with 128 DFT coefficients and 4×18 Eigenvectors where were extracted from row and column scatter matrixes, respectively. In addition, we have compared performance of the system with 6 other cases. It was observed in the course of experiment that the proposed method outperformed others in spite of the smaller feature vector.

References

1. Kampel, M., Zaharieva, M.: Recognizing ancient coins based on local features. In: Proceeding of the 4th International Symposium on Visual Computing, pp. 11–22 (2008)
2. Arandjelovic, O.: Automatic attribution of ancient Roman imperial coins. In: Proceeding of the 2010 IEEE Conference on Computer Vision and Pattern Recognition (CVPR), pp. 1728–1734 (2010)
3. Allahverdi, R., Dehshibia, M.M., Bastanfard, A., Akbarzadeh, D.: EigenCoin: sassanid coins classification based on Bhattacharyya distance. Computer Science 1, 1151–1160 (2012)

4. Allahverdi, R., Bastanfard, A., Akbarzadeh, D.: Sasanian coins classification using discrete cosine transform. In: Proceeding of the 16th CSI International Symposium on Artificial Intelligence and Signal Processing (AISP), pp. 278–282 (2012)
5. Huber, R., Ramoser, H., Mayer, K., Penz, H., Rubik, M.: Classification of coins using an eigenspace approach. Pattern Recognition Letters 26, 61–75 (2005)
6. Khashman, A., Sekeroglu, B., Dimililer, K.: Intelligent coin identification system. In: Proceeding of the Computer Aided Control System Design, 2006 IEEE International Conference on Control Applications, 2006 IEEE International Symposium on Intelligent Control, pp. 1226–1230 (2006)
7. Van Der Maaten, L.J., Poon, P.: Coin-o-matic: A fast system for reliable coin classification. In: Proceeding of the Muscle CIS Coin Competition Workshop, Berlin, Germany, pp. 7-18 (2006)
8. Zuo, W., Wang, K., Zhang, D.: Bi-directional PCA with assembled matrix distance metric. In: Proceeding of the IEEE International Conference on Image Processing, pp. II-958–II-961 (2005)
9. Oppenheim, A.V., Schafer, R.W., Buck, J.R.: Discrete-time signal processing. Prentice-Hall, Englewood Cliffs (1989)
10. Yegnanarayana, B., Saikia, D., Krishnan, T.: Significance of group delay functions in signal reconstruction from spectral magnitude or phase. IEEE Transactions on Acoustics, Speech and Signal Processing 32(3), 610–623 (1984)
11. Oppenheim, A.V., Lim, J.S.: The importance of phase in signals. Proceedings of the IEEE 69(5), 529–541 (1981)
12. Bastanfard, A., Nik, M.A., Dehshibi, M.M.: Iranian face database with age, pose and expression. In: Proceeding of the International Conference on Machine Vision, pp. 50–55 (2007)
13. Dehshibi, M.M., Allahverdi, R.: Persian Vehicle License Plate Recognition Using Multiclass Adaboost. International Journal of Computer and Electrical Engineering 4, 355–358 (2012)
14. Dehshibi, M.M., Bastanfard, A.: A new algorithm for age recognition from facial images. Signal Processing 90(3), 2431–2444 (2010)
15. Dehshibi, M.M., Bastanfard, A., Kelishami, A.A.: LPT: Eye Features Localizer in an N-Dimensional Image Space. In: Proceeding of the 2010 International Conference on Image Processing, Computer Vision, & Pattern Recognition, pp. 347–352 (2010)
16. Dehshibi, M.M., Fazlali, M., Shanbehzadeh, J.: Linear principal transformation: toward locating features in N-dimensional image space. Multimedia Tools and Applications 72(3), 2249–2273 (2014)

Preliminary Experiments with Ensembles of Neurally Diverse Artificial Neural Networks for Pattern Recognition

Abdullahi Adamu[1], Tomas Maul[1], Andrzej Bargiela[2], and Christopher Roadknight[2]

[1] University of Nottingham - Malaysia Campus, Malaysia
[2] University of Nottingham - UK Campus, United Kingdom

Abstract. Although there have been a few approaches to achieve the goal of fault tolerance by diversifying redundancy of the individual networks that make up a neural network ensemble, some of which include ensembles of neural networks of different sizes, and ensembles of different models of neural networks such as Radial Basis Function Networks and Multilayer Perceptron, there is yet to be an empirical study on hybrid neural networks that makes use of a diverse set of transfer functions, which we would expect to be able to exhibit diverse network architectures, and thus possibly more diverse error patterns. In this paper, we present an approach that uses transfer function diversity to achieve significant results on ensembles. The results show that even with relatively small networks having 5 hidden nodes, and a relatively small ensemble size of just 10 members, the ensemble is able to get competitive results on the Iris data set. It also capable of obtaining competitive results with 20 ensemble members of relatively small networks on other popular data sets such as the Diabetes, Sonar, Hepatitis, and Australian Credit Card problems. In addition to that, it is shown that these results can be achieved with a simple sorting and selection of the Top N solutions of the population, in contrast to other methods of selecting ensemble members that can be computationally expensive, such as selection of the Pareto-front, or hill climbing methods of selection.

Keywords: Hybrid Neural Networks, Artificial Neural Networks, Transfer function Optimization, Pattern Recognition.

1 Introduction

Fault tolerance through redundancy is a concept that is used in many fields, for example in software engineering where one of the methods for achieving reliability is through redundant diverse implementations[19]. The aim is to minimize the failure of a system by combining several redundant, but diverse components designed for a single task in such a way that they are unlikely to all fail at

H. Unger et al. (eds.), *Recent Advances in Information and Communication Technology 2015*,
Advances in Intelligent Systems and Computing 361, DOI: 10.1007/978-3-319-19024-2_9

once. This concept has also gained interest in the field of Artificial Neural Networks as Ensembles of Artificial Neural networks, where Neural Networks have diverse designs resulting in an improved generalization ability when combined. The research findings from earlier studies [6,19,14,20,10,15,18] show that neural networks combined as ensembles exhibit an improved generalization ability provided there is diversity in the bias of its members. In other words, if there are differences in the limitations of the learners, then an ensemble of such learners should result in increased learning accuracy. In addition to these empirical findings, it is also proven [6] that the error of an ensemble is guaranteed to be the same or better than the average error of its components.

The importance of diversity in ensembles is quite intuitive. If the individual solutions in an ensemble were all identical, the performance of the ensemble would not differ from any of the members of the ensemble. However, if all the members of the ensemble were different from each other such that the decision boundaries that they project onto the input-space are varied yet accurate, then one can expect that the averaged decision boundary of the learners is likely to be significantly more accurate, or at the very least the same. The significance also depends on some other factors such as the method used to combine the outputs of the members, the number of members in the ensemble, and the accuracy of the members in the ensemble.

Generally, most of the promising methods for creating diversity in ensembles could be categorized into three according to their area of focus [6]: data set, model, and training algorithm. In the first case, some approaches found in the literature use re-sampling and pattern distortion methods to achieve some variations in the training data set. These variations then implicitly cause behavioral differences in members of the ensemble. Popular re-sampling methods include bagging and boosting. In bagging, random samples of the data set are used to train each member. Boosting is similar to bagging, however it considers the distribution of the subsets while sampling. Another method used by some studies that was highlighted by Brown et. al. [6] is to re-sample the features of each pattern in the training data set. On the other hand, distortion methods used include the addition of Gaussian noise, or non-linear transformations of the training patterns in the data set. One of the non-linear transformation approaches found [6,19] to be effective was to stimulate a randomly generated neural network with the training pattern and then use its output as the distorted pattern. Gaussian noise was also found to be helpful [6,19]. In the case of diversity creation methods focusing on models [6], some methods use a mixture of models. The most popular includes the use of homogeneous models with varied parameters [6], such as neural networks of different sizes within the ensemble, or neural networks with different types of architectures such as Multilayer Perceptrons and Radial Basis Function networks [6]. Other methods [6] use heterogeneous models such as the use of decision trees and neural network within an ensemble. In the final category are methods which focus on the training algorithm, some of which include the use of different training algorithms [19], and the introduction of an additional term in the objective function [15,18], such as in neural network ensembles trained by evolutionary algorithms [6].

There are some areas that have been deficient with regards to experiments. One such area is the topic of model diversification approaches for neural network ensembles. Intuitively, it makes sense that if we are aiming for error diversity within our neural network ensemble, an equally likely approach to the others that could yield significant diversity is an approach that is explicit. By *explicit*, we mean an approach that takes a direct method, such as the combination of diverse architectural models of neural networks. According to our knowledge, there has yet to be experiments with ensembles using hybrid artificial neural networks implementing a diverse transfer functions set; which we would expect to increase error diversity in ensembles. This was also highlighted in the thorough survey of ensembles by Brown [6]. The only partially related work done so far was by Partridge who used pure models of Multilayer Perceptron's (MLP) and Radial Basis Functions (RBF) in an ensemble to achieve diversity. However, even that work was suggested to be a preliminary study by [6].

In this paper, we experiment with hybrid neural network ensembles implementing a diverse transfer function set, also known as Neural Diversity Machine Ensembles (NeuDiME). This is unlike other approaches found in the literature which have used a mixture of pure models as reported by Brown et. al [6]. We explore the performance of this approach in different circumstances; specifically, we test it on popular pattern classification problems such as the Iris, Sonar, Hepatitis, Diabetes, and the Australian credit card problem, which have been used as benchmarks of choice in most literature. We also compare this approach with other approaches found in the literature, and also analyze two accurate yet different strategies evolved for some of the problems (i.e. diabetes and Iris)

The contributions of this paper are as follows: firstly, the paper presents the application of hybrid neural networks in ensembles and provides a study on some of the effects of transfer function diversity in neural network ensembles. Secondly, it shows that this neural network framework is able to develop different strategies for a problem that can be used in ensembles without explicit diversity maintenance that can be expensive, such as selection of the Pareto front for use as ensembles, or fitness sharing. It also shows that this approach can evolve a relatively smaller ensemble of compact networks that has a competitive performance. Finally, it shows how neural diversity can result in diverse classifiers by analyzing two computational strategies evolved for the diabetes problem.

It should be noted that the term "weight function" is used interchangeably with the term "input combination function"and activation function. Likewise, the term, "node function" is used interchangeably with the term "output function". Additionally, the term "node" is used interchangeably with the term "neuron" and "unit".

2 Neural Diversity Machines

In this section, the Neural Diversity Machines approach proposed by Maul [16] is explained, starting from transfer function diversity - the core of the framework, then followed by weight and architectural evolution.

2.1 Transfer Function Diversity

A transfer function of a neural network is composed of the input combination function and an output function; the input combination function computes the input values being transferred into a neuron by other neurons in the network connected to it using weighted connections, while the output function can be considered as a mathematical model of its biological counterpart, which determines the output value of the neuron.

Traditional and well established Artificial Neural Networks such as Radial Basis Functions (RBF) and Multilayer Perceptrons (MLP) use predetermined transfer functions for the nodes of each layer as part of their distinct architectural properties. Often, the nodes of each layer are homogeneous in their transfer function. A Radial Basis Function uses a combination of a distance-based input combination function (e.g. Euclidean distance) and a radial basis output function, which is typically a Gaussian, Multiquadratic, thin-plate-spline, or inverse Multiquadratic. In the case of a Multilayer Perceptron, its typically an inner-product input combination function accompanied with a sigmoid output function. Though it's proven that both RBF networks (RBFN) and Multilayer Perceptrons are able to approximate any function, provided that the complexity of the network's model is matched with the complexity of the problem [4,8], it's not proven that the model will be optimal or practical. By *optimal*, we are specifically emphasizing the use of a minimal number of hidden units to learn a problem.

There is no one-size-fits-all in the choice of transfer functions [9,13,16]. A certain problem might be more suited for an MLP unit, whereas another may require RBF units. Thus, there is a need for adapting the choice of transfer function.

Several studies have approached this problem using hybrid models [16,9,13] which implement different basis functions, either as a single hybrid layer, or as several pure layers and have been found to perform better than their canonical models (i.e. RBF or MLP).

This work differs because it is particularly motivated from the biological standpoint of the benefits of neural diversity found in biological neural networks, which includes increased representational capacity that contributes significantly to their efficiency [5]. This was replicated by having different classes of functions in the transfer functions pool, such as radial-basis units, projection-basis units and higher-order units. However, due to the flexibility allowed in the combination of activation and output functions during optimization, other unconventional transfer functions were also evolved.

Table 1. List of some input combination and output functions used by Neural Diversity Machines and their visualization color codes

Input Combination Functions	Color Code
Inner-Product ($ap_i = \sum_{i=0}^{n} w_i i_i + w_{bias}$)	Red Solid Edge
Higher-Order Product ($ap_i = \prod_{i=0}^{n} cw_i * i_i$)	Yellow Solid Edge
Higher-Order Subtractive ($ap_i = \sum_{i=1}^{n} \lvert x_0 - x_i \rvert$)	Yellow Dashed Edge
Euclidean Distance ($ap_i = \sqrt{\sum_{i=0}^{n} (w_i - i_i)^2}$)	Magenta Dashed Edge
Standard Deviation ($ap_i = stdDev(w_i i_i, w_{i+1} i_{i+1} ... w_n i_n)$)	Blue Solid Edge
Min ($ap_i = min(w_i i_i, w_{i+1} i_{i+1} ... w_n i_n)$)	Gray Dashed Edge
Max ($ap_i = max(w_i i_i, w_{i+1} i_{i+1} ... w_n i_n)$)	Black Dashed Edge

Table 2. List of Output functions for Neural Diversity Machines and their visualization color codes

Output functions	Color Codes
Linear ($O_i = \alpha * ap_i$)	Yellow Node Outline
Hyperbolic tangent ($O_i = \frac{1-e^{-\alpha * ap_i}}{1+e^{-\alpha * ap_i}}$)	Cyan Node Outline
Sigmoid ($O_i = \frac{c}{1+e^{-\alpha * ap_i}}$)	Red Node Outline
Gaussian ($O_i = e^{\frac{-(ap)^2}{width}}$)	Blue Node Outline
Gaussian II ($O_i = e^{\frac{-(ap)^2}{width}} if\ O_i > \theta\ then\ O_i = 1.0$)	Dark-Blue Node Outline

3 Neural Diversity Machine Ensembles (NeuDiME)

Neural Diversity Machine Ensembles is an ensemble made up of Neural Networks that conforms to the framework of Neural Diversity Machines [16]. It uses a set of diverse input combination functions (see input combination set in table 1) and output functions (see output function set in table 2) which enables the optimization algorithm to find near optimal transfer functions for each node from the transfer function set, thus increasing the likelihood of having accurate ensemble members. In addition to that, the diverse transfer function set should result in relatively more diverse computational strategies [6] for a problem as compared to an approach using pure models. Subsequently, this is expected to result in the desired balance between bias and variance. After all, it is intuitive that an ensemble made up of different yet accurate models will likely yield more *useful* diversity. By *useful*, we are referring to the diversity that results in significant improvements in the generalization ability of the ensemble. This differs from other approaches which have attempted the use of neural networks of different sizes within an ensemble or a mixture of pure models [6]. The stages for the optimization algorithm are as follows:

1. Solutions (of various hidden node sizes) are created.
2. Solutions are evaluated.
3. Differential Evolution is applied.
4. Other Evolutionary Operators are applied (i.e. Mutation and Cross-over).
5. Finally, ensemble members are selected from the population.

6. Ensemble is evaluated.
7. Weak solutions are eliminated.
8. Stopping condition is checked.
9. If the stopping condition is met, the ensemble is evaluated on the test set and the error is used as its generalization performance.

3.1 Ensemble Member Selection

In this study, we use the Top N solutions for convenience. It lacks the relatively higher computational cost of other selection methods such as the use of the Pareto-front [2], or selection of members using a hill climbing approach while maintaining diversity with fitness sharing [7,12]. In addition to that, it's relatively simple and has been used in later works by Opitz & Shavlik [17].

The method of selection selects the Top N solutions from the population after sorting by fitness, where N is the desired size of the ensemble. The sorting prior to selection helps in improving the chances of picking the N solutions with the most generalization ability.

However, this selection method assumes there is already diversity in the population such that the top N solutions are diverse after sorting by accuracy. In the case of NeuDiME, there is already diversity introduced by using a diverse set of transfer function combinations to pick from for all the neural networks. In addition to that, there is also the added diversity of neural networks of different sizes.

4 Experiments

In this section of the report, we present the benchmarks, performance measures, and finally the experimental setup.

4.1 Experimental Setup

The benchmarks used consisted of some problems commonly used in the ensemble literature: Iris, Sonar, Australian credit card, Hepatitis and Diabetes retrieved from the machine learning repository [3]. In order to conform to the common measure of generalization ability found in the literature [8,14,10], 10 fold cross-validation was used.

The optimization parameters for the Global Stochastic Optimization (GSO) algorithm used to optimize the neural network ensembles is given as in Table 5.

5 Results and Discussion

In this section we present the results of the neural network ensembles, NeuDiME, on the popular data sets and a comparison with various other methods that include: Memetic Pareto Artificial Neural Networks(MPANN), Cooperative Neural Network Ensemble(CNNE), Ensemble with Negative Correlation Learning (EENCL), Diverse and Accurate Ensembles(DIVACE).

Table 3. List of some of the benchmarks retrieved from the UCI Machine Learning Repository[3]

Benchmarks	Features	Classes	Examples
Iris	4	3	150
Sonar	60	2	208
Card (Australian)	51	2	690
Diabetes	8	2	768
Hepatitis	19	2	155

Table 4. Experimental setup for the data sets showing the maximum number of hidden units allowed per ensemble member, the number of members in the ensemble and the number of folds used for K-fold cross-validation

Benchmarks	Max Hidden units	Members(Ensemble)	Folds (K-fold CV)
Iris	5	10	10
Sonar	5	10	10
Diabetes	5	20	10
Hepatitis	5	20	10
Card (Australian)	5	20	12

5.1 Performance on Some Popular Benchmarks

The results produced by the Neural Diversity Machine Ensemble were competitive. It achieved the best performance in 3 out of 5 of the benchmarks when compared with the 5 other methods. However, it did not perform as well in the Australian Credit Card data set. This could be for a variety of reasons, which include the relatively higher dimensionality of the Australian Credit Card Data set (i.e input space being $\{x\}^{51}$, and sample size being 690). Typically, problems of higher dimensionality in terms of their input space require relatively complex solutions. As the dimensions of the problem increases, solutions have to account for these new dimensions by making the decision of which tuple of dimensions have more information gain over others, and form decision boundaries through these dimensions that separate the classes accurately. As one can imagine, as the dimensionality increases, this problems gets consistently harder. In the case of NeuDiME, its increased access to a wider variety of solutions can be mixed news. The advantage is that it increases its chances of finding computational strategies with creative hypotheses that describes the problem with simplicity and usually better generalization [8][13]. The disadvantage is that this can present more local minima. In the case of these results for the Australian Credit card and Sonar, the dimensionality of the benchmarks played some role in the slight performance difference; while the Card benchmark had 51 inputs, the Sonar had 60. In addition to having relatively more local minima, one can also expect that convergence might also be relatively slower as a result of the increased computational capacity.

Table 5. The optimization parameters used for the neuroevolution of the ensemble members

Optimization Parameter(s)	Value(s)
Max Iterations	100
Population size	30
Percent to eliminate	0.3
Min cost (elimination)	0.66
Min age (elimination)	3
Cross Over	True
Probability of Cross Over	0.2
Differential Evolution (DE) Iterations	3
DE alpha	0.2
Gene range	[-0.9, 0.9]
Probability of Mutation	0.2
Gaussian Mutation (Mean,Std)	(0.0, 0.2)

Table 6. Comparison of test error with other learning methods - Results of MPANN, EENCL, CNNE, DIVACE and Optiz and Shavlik as reported in their findings. In the case of Opitz et. al the best results of the five techniques experimented with were used. The Testing Error of EENCL was used as reported in [11]. As for MPANN, the best of the three results as reported in [2], while results of CNNE and DIVACE were retrieved as reported in [12,7], respectively.

	Australian Credit Card	Iris	Sonar	Diabetes	Hepatitis
MPANN	0.135	-	-	0.23	-
EENCL	0.138	-	-	0.221	-
CNNE	**0.092**	-	-	0.198	-
DIVACE	0.138	-	-	0.226	-
Opitz & Maclin [17]	0.137	0.039	**0.129**	0.233	0.178
NeuDiME	0.221	**0.027**	0.181	**0.174**	**0.115**

5.2 Interesting Computational Strategies Evolved for Prediction

In this subsection of the results and discussion, we present statistics of transfer function use for some of the data sets used in the experiments. These include the probabilities of combining the possible input combination functions with the possible output functions in the fittest members of the ensemble, and the associated error with each combination. In addition to that, we also reveal some of the diverse strategies used by the members of the ensemble towards solving the diabetes problem by looking at their choice of transfer function, connectivity and weights.

Diabetes. The diabetes problem showed an emphasis on strategies that relied on using standard deviation and output function such as the identity function, and hyperbolic tangent. The least error was associated with the combination of Euclidean distance and Gaussian II, which is essentially a variant of a radial

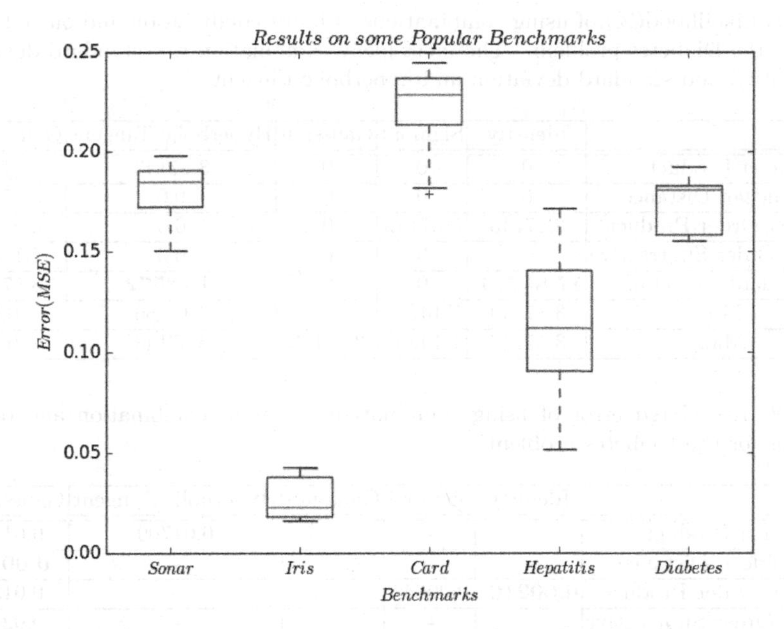

Fig. 1. Results on the popular data sets of NeuDiME - Test error averaged over 10-fold cross-validation,except in the case of Australian credit card problem which was set to 12-fold cross-validation for the sake of comparability with published results

basis function unit. Higher-order product combined with identity function also showed a relatively lower associated error. However, considering these were some of the least likely to be evolved for the problem, they should be expected to have lower errors. The tables below present the statistics.

In the following discussion, we will present a study of two diverse strategies evolved for NeuDiME on the diabetes problem. This reveals the ability of NeuDiME to exhibit diverse neural computation strategies that are accurate. One of the most accurate strategies evolved for the diabetes problem was a fully connected network consisting of four hidden units, implementing the following transfer functions found in Figure 2.

It seems that the range of the feature values (min and max) is somehow important in this strategy. This probably explains why min and max functions were transferring information to the projection unit (i.e. the perceptron output unit). In an effort to understand why this is essential for this strategy, we compared the min and max of the raw data set: most of the time, the max value corresponds to the 2nd feature of the data set, i.e. *glucose concentration reading.* While the min value usually corresponded to either the 4th or the 5th feature, i.e. *skin fold thickness* and *serum insulin reading*, respectively. Based on the connection weights of these features to the hidden layers using min and max as a relay; the 4th feature (skin fold thickness) was given more weight as compared to the rest. The 2nd feature (glucose concentration) and 5th feature (serum insulin reading) were both given a medium weight, perhaps to normalize its values with the rest

Table 7. Likelihood(%) of using combinations of input combination and output functions for the Diabetes problem - The most likely combination was standard deviation and identity, and standard deviation and hyperbolic tangent

	Identity	Sigmoid	Gaussian	Hyperbolic Tangent	Gaussian II
Inner Product	0	0	0	3.57143	3.57143
Euclidean Distance	0	0	0	0.0	3.57143
Higher Order Product	3.57143	3.57143	0	0.0	3.57143
Higher Order Subtractive	0	0	0	0.0	7.14286
Standard Deviation	**17.85714**	0	0	**14.28572**	3.57143
Min	3.57143	7.14286	0	7.14286	0.0
Max	3.57143	7.14286	3.57143	3.57143	0.0

Table 8. Associated error of using combinations of input combination and output functions for the Diabetes problem

	Identity	Sigmoid	Gaussian	Hyperbolic Tangent	Gaussian II
Inner Product	-	-	-	0.01709	0.01161
Euclidean Distance	-	-	-	-	**0.00125**
Higher Order Product	**0.00210**	0.00644	-	-	0.01319
Higher Order Subtractive	-	-	-	-	0.02434
Standard Deviation	0.03476	-	-	0.05334	0.00405
Min	0.01357	0.00539	-	0.01066	-
Max	0.00500	0.02989	0.02224	0.00543	-

of the features, as they usually have the highest values. The most important feature relayed by these relay units based on their weights to them was *age*.

Interestingly, *age* and *skin fold thickness* are actually regarded as highly correlated to diabetes. The American Diabetes Association for example regards age as one of the leading contributing factors that increases the risk of a person having type-2 diabetes[1].

In general, this strategy seems to be taking advantage of the variety of input combination function and output functions to extract important features using unusual combinations of transfer functions such as the minimum feature value, weighted variance of between features, proximity of the feature vector to the centre of the RBF unit, and maximum feature value, and finally used these features for training a simple perceptron in the output layer. In other words, its using the hidden layer for feature selection, then taking advantage of the reduced dimensionality to train a simple hyperbolic-tangent perceptron.

Another strategy evolved was a fully connected network with two hidden units where one adopted a min input combination function and a sigmoid output function, while the other adopted a standard deviation output function with a hyperbolic tangent output function. The output unit differed from the other strategy, consisting of a max input combination which has a winner-take-all effect on the hidden nodes. Its hyperbolic tangent output function has another norming effect.

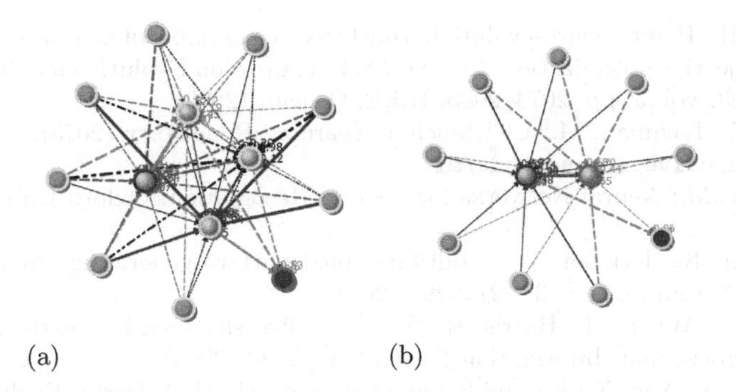

Fig. 2. Visualizations of two models that evolved different strategies for the Diabetes problem

Once again, this strategy uses the min input combination function, however in this case, it is coupled with a sigmoid output function, which has a normalizing effect on the output value - restricting it between 0.0 and 1.0. In addition to that, in this case it seems to act as a threshold for the other hidden node using the weighted variance. This is because of the choice of using max as the input combination function by the output unit.

In general, the normalizing effect of the output functions of both hidden nodes allows the hidden node using the min input combination to essentially resemble the role of a bias node that is dependent on the features.

6 Conclusion

In conclusion, this paper has presented the application of hybrid neural networks in the field of ensembles and has shown that neural diversity in an artificial neural network framework is able to exhibit different computational strategies for a problem that can be used in ensembles without the need of other explicit, and usually computationally expensive diversity maintenance. This is shown by the two different strategies for the diabetes problem. It has also been shown that this approach can evolve relatively small ensembles of compact networks that have a competitive performance. The limitations include the increased local minima, as a result of the increased access to the search space. In addition, relatively slower convergence is also a concern. However, considering the increased possibility for gain of generalization and efficiency, it is arguable that these properties are significant motivations for further research.

References

1. Lower Your Risks: Age, Race, Gender & Family History (2013), `http://www.diabetes.org/are-you-at-risk/lower-your-risk/nonmodifiables.html`

2. Abbass, H.: Pareto neuro-evolution: constructing ensemble of neural networks using multi-objective optimization. In: The 2003 Congress on Evolutionary Computation, CEC 2003, vol. 3, pp. 2074–2080. IEEE, Cancun (2003)
3. Bache, K., Lichman, M.: UCI Machine Learning Repository (2013), http://archive.ics.uci.edu/ml
4. Bishop, C.M.: Neural Networks for Pattern Recognition. Oxford University Press (1995)
5. Briggman, K., Kristan, W.: Multifunctional pattern-generating circuits. Annual Review of Neuroscience 31, 2710294 (2008)
6. Brown, G., Wyatt, J., Harris, R., Yao, X.: Diversity creation methods: a survey and categorisation. Information Fusion 6(1), 5–20 (2005)
7. Chandra, A., Yao, X.: Ensemble Learning Using Multi-Objective Evolutionary Algorithms. Journal of Mathematical Modelling and Algorithms 5(4), 417–445 (2006)
8. Gutierrez, P., Hervas, C., Carbonero, M., Fernandez, J.: Combined projection and kernel basis functions for classification in evolutionary neural networks. Neurocomputing 72(13-15), 2731–2742 (2009)
9. Gutiérrez, P.A., Hervás-Martínez, C.: Hybrid Artificial Neural Networks: Models, Algorithms and Data. In: Cabestany, J., Rojas, I., Joya, G. (eds.) IWANN 2011, Part II. LNCS, vol. 6692, pp. 177–184. Springer, Heidelberg (2011)
10. Hansen, L., Salamon, P.: Neural network ensembles. IEEE Transactions on Pattern Analysis and Machine Intelligence 12(10), 993–1001 (1990)
11. Higuchi, T.: Evolutionary ensembles with negative correlation learning. IEEE Transactions on Evolutionary Computation 4(4), 380–387 (2000)
12. Islam, M.M., Yao, X., Murase, K.: A constructive algorithm for training cooperative neural network ensembles. IEEE Transactions on Neural Networks 14(4), 820–834 (2003)
13. Jankowski, N., Duch, W.: Optimal transfer function neural networks. In: 9th European Symposium on Artificial Neural Networks, ESANN 2001, Bruges, vol. (I), pp. 101–106 (2001)
14. Krogh, A., Vedelsby, J.: Neural network ensembles, cross validation, and active learning. In: Advances in Neural Information Processing Systems, pp. 231–238 (1995)
15. Liu, Y., Yao, X.: Ensemble learning via negative correlation. Neural Networks: The Official Journal of the International Neural Network Society 12(10), 1399–1404 (1999)
16. Maul, T.: Early experiments with neural diversity machines. Neurocomputing 113, 36–48 (2013)
17. Opitz, D., Shavlik, J.: Generating Accurate and Diverse Members of a Neural-Network Ensemble. In: Advances in Neural Information Processing Systems, vol. 8, pp. 535–541 (1996)
18. Perrone, M., Cooper, L.: When networks disagree: Ensemble methods for hybrid neural networks. In: Neural Networks for Speech and Image Processing, pp. 126–142 (1993)
19. Sharkey, A., Sharkey, N.: Combining diverse neural nets. The Knowledge Engineering Review 12(3), 231–247 (1997)
20. Wu, Z., Chen, Y.: Genetic algorithm based selective neural network ensemble. In: Proceedings of the Seventeenth International Joint Conference on Artificial Intelligence, IJCAI 2001, vol. 1, pp. 797–802 (2001)

A Comparative Study on Sensor Displacement Effect on Realistic Sensor Displacement Benchmark Dataset

Lumpapun Punchoojit and Nuttanont Hongwarittorrn

Department of Computer Science
Faculty of Science and Technology, Thammasat University, Thailand
l.punchoojit@gmail.com, nth@cs.tu.ac.th

Abstract. Activity Recognition (AR) research is growing and plays a major role in various fields. The approach of using wearable sensors for AR is well-accepted, as it compensates the need to install cameras in image processing approach which can lead to privacy violation. Using wearable sensors can suffer from one disadvantage – sensor displacement. There have been a number of research which studies sensor displacement problem. However, the conclusion cannot be made as which classifier is better than another in recognizing displacement data, as the prior experiments were performed under different conditions and focused on different parts of the body. This work aims to evaluate recognition performance of different algorithms – SVM, C4.5, and Naïve Bayes – on ideal-placement and displacement data on whole body activities, by adopting REALDISP dataset to make such evaluation. The accuracy of all algorithms on ideal placement data was above 90%, where SVM yielded the highest accuracy. Displacement data were tested against classification models constructed from ideal-placement data. The results shows that there was a dramatic drop in recognition performance. The accuracy of all algorithms on displacement data was between 50-60%, and C4.5 could handle displacement data the best.

Keywords: Activity Recognition, Sensor Displacement, Naïve Bayes, SVM, C4.5.

1 Introduction

Activity Recognition (AR) is an active research area in recent years. It plays a major role in diverse applications. In pervasive healthcare, AR supports preventive or chronic healthcare, cognitive assistance, and elderly monitoring [1, 2]. In security, AR supports the identification of terrorist actions and threats, or monitoring individual's activities in security sensitive areas, such as hospitals, and banks [3]. On mobile devices, AR supports user's activities monitoring, and enabling screen rotation [4].

According to Ugulino et. al, there are two approaches commonly used for activity recognition: image processing and the use of wearable sensors [1]. Image processing approach does not require users to put an equipment on their bodies; however, this approach encounters some limitations, including the requirement to install cameras, and image quality which can be affected by environmental conditions, such as lighting [1]. The installation of cameras may controversially violate user's privacy

H. Unger et al. (eds.), *Recent Advances in Information and Communication Technology 2015*,
Advances in Intelligent Systems and Computing 361, DOI: 10.1007/978-3-319-19024-2_10

[1]. The second approach for AR is the use of wearable sensors. This approach compensates the limitations of image processing approach [1]. Even though this approach requires users to wear the equipment through a period of time, general public is more likely to accept it [1], as the equipment can be easily turned-off or removed [2]. Wearable sensors also offer real-time activity information for AR [2].

Despite having many strengths, using wearable sensors for activity recognition can suffer from one major disadvantage – sensor displacement. The implementation of wearable sensors requires the equipment to be attached at predefined positions, and the classification model is built by assuming constant sensor positions [5]. However, this can hardly be maintained in real-life condition. As a result, the model may fail to classify the activities from the sensor data. Misclassification may lead to unwanted consequence; for instance, it could be very dangerous in elderly pervasive healthcare system if the classifier failed to recognize elderly falling. Displacement of sensors can be caused by either sensor loose fitting or displacing by the users themselves [5].

There have been a number of research which studies the sensor displacement; some of them proposes heuristics to improve classifier's robustness. The research may, however, focus on different parts of the body or a different set of activities. For instance, Chavarriaga et. al [6] study the effects of sensor displacement and propose an unsupervised adaptive classifier to tackle the problem. The experiment in this study focused on three activity scenarios: HCI gestures, fitness, and daily living scenarios. The sensors were installed on an participant's arm in HCI gesture scenario, on a leg in fitness scenario, and on both arms and the back in daily living scenario [6]. Kunze and Lukowicz investigate the effect of displacement on onbody activity recognition systems [7]. They presented a set of heuristics which increase the robustness of the recognition with the respect of sensor displacement [7]. In the experiment, they focused on locomotion activities (e.g. walking, running, walking up hill, biking, etc) and gym exercise activities (e.g. shoulder press, arm extension, arm curl, etc). The sensors were installed participant's legs in locomotive activity experiment, and the sensors were installed on the arms in gym exercise experiment [7].

Regarding the previous studies, the conclusion cannot be made as which classifier is better than others for recognizing sensor displacement data, as the experiments were performed under different conditions and were focusing on different parts of the body or different sets of activities. In addition, none of the studies has explicitly described performance degradation of the classifier on ideal-placement and displacement data. Therefore, this current work aims to examine performance of three popular recognition algorithms – SVM, C4.5, and Naïve Bayes – on ideal-placement and displacement data on whole body activities, by adopting REALDISP dataset to make such evaluation. The main objective is to evaluate which of the algorithms would outperform others on the displacement problem.

2 Related Works

2.1 REALDISP Dataset

REALDISP (REAListic sensor DISPlacement) dataset lends itself as a benchmark dataset for activity recognition, whether in ideal, real-life, or extreme displacement conditions.

The dataset was collected to investigate the effects of sensor displacement in activity recognition [8], which can be either caused by loose fitting of sensors or displacement by users themselves. This dataset was created by Banos et. al [5, 9]. The dataset was built upon three scenarios: ideal-placement, self-placement, and induced-displacement [8]. The dataset covers a wide range of physical activities and locations of wearable sensors [9].

Ideal-Placement data was generated when the sensors were positioned to predefined locations by the instructor (i.e. research team). The data from ideal-placement can be considered as the training set for the recognition model [10]. Self-Placement data was induced when the users place the sensors on their body parts specified by the research team. Data from self-placement may slightly differ from the ideal-placement one; however, the difference is considered to be too trivial [10]. Induced-Displacement occurred when the sensors were misplaced by rotations or translations with respect to the ideal setups. In REALDISP dataset, the induced-displacement data was generated by intentionally displacement of sensors by the research team [10].

Activity data were collected from 17 subjects. Thirty-three physical activities were included in the dataset, as listed in table 1.

Table 1. Activity Set

#	Activity	#	Activity	#	Activity
1	Walking	12	Waist rotation	23	Shoulders high amplitude rotation
2	Jogging	13	Waist bends	24	Shoulders low amplitude rotation
3	Running	14	Reach heels backwards	25	Arms inner rotation
4	Jump Up	15	Lateral bend	26	Knees to the breast
5	Jump Front & Back	16	Lateral bend arm up	27	Heels to the back side
6	Jump Sideways	17	Repetitive forward stretching	28	Knees bending
7	Jump legs/arms opened/closed	18	Upper trunk and lower body opposite twist	29	Knees bend forward
8	Jump rope	19	Arms lateral elevation	30	Rotation on the knees
9	Trunk twist (arms outstretched)	20	Arms frontal elevation	31	Rowing
10	Trunk twist (elbows bended)	21	Frontal hand claps	32	Elliptic bike
11	Waist bends forward	22	Arms frontal crossing	33	Cycling

The measurement of the whole body was measured in activities 1 – 3, 5 – 8, and 31 – 33 [10]. Activities focused on trunk were measured in activity 9 – 18, upper extremities in 19 – 25, and lower extremities in activity 26 – 29 [10]. All activities were measured by 9 sensors; each of which measured four sensor modalities: acceleration, rate of turn (gyroscope), magnetic field, and orientation [9]. Each sensor provided tri-directional measurements; except for orientation that was measured in quaternion format [10]. The sensors were installed on nine different parts of the subject's body:

1) left calf, 2) left thigh, 3) right calf, 4) right thigh, 5) back, 6) left lower arm, 7) left upper arm, 8) right lower arm, and 9) right upper arm [9]. Altogether, this makes up 117 attributes.

2.2 Recognition Algorithms

This section gives an overview of algorithms used for the comparison in this paper. The algorithms are selected based on the survey of activity recognition algorithms conducted by Ugulino et. al [1]. They are three algorithms which are commonly used for recognition tasks: SVM, C4.5, and Naïve Bayes.

SVM

SVM (Support Vector Machines) is a supervised learning algorithm used for binary classification of both linear and non-linear data [11]. When data are linearly separable, SVM would search for maximum marginal hyperplane, or a decision boundary that best separate the tuples of one class from another. Hyperplane with larger margin is expected to be more accurate in classification [11]. When data are not linearly separable, SVM uses a nonlinear mapping to transform the original data into a higher dimension feature space. Then, SVM searches for a linear separating hyperplane in the new feature space [11].

C4.5

C4.5 is an algorithm for decision tree induction, presented in the year 1993 [12]. C4.5 uses top-down approach to construct a classification model. It starts with all training tuples at the root node of the tree, then an attribute would be selected to partition the tuples [13]. The process would be repeated as the tree is being built [11].

An attribute selection measure is required to select the attribute that best split the tuples [11]. Attribute selection measure is a heuristic for selecting the splitting criterion that best separate the tuples of class-labeled training tuples into individual classes [11]. Specifically, the attribute that yields 'pure' partitions would be selected. A pure partition means that all tuples in that particular partition belong to the same class [11]. In other words, the selected attribute minimizes an information entropy applied to tuple partition [13]. For C4.5, it uses gain ratio as splitting criteria; the splitting would stop when the number of instances to be split is below the threshold.

Naïve Bayes

Naïve Bayes classification, or simple Bayesian Classifier, is in the family of Bayes Classification methods [11]. Classifiers in this family are statistical classifiers; meaning that they can predict membership probabilistic [11]. Naïve Bayes, like other Bayesian Classifiers, is based on Bayes' Theorem [11].

Naïve Bayes presumes conditionally independence of the classes. It determines the probability that an instance would belong to a particular class - posterior probability (i.e. $P(C_i|X)$). Posterior probability is calculated from another three prior probability values: $P(C_i)$, $P(X)$, and $P(X|C_i)$. $P(C_i)$ is the probability that an instance belongs to class C_i, regardless of X. $P(X)$ is the probability that an instance has attribute values

X, regardless of C_i. $P(X|C_i)$ is the probability that an instance has attribute values X, given an instance belongs to class C_i [11]. $P(C_i|X)$ is calculated by the following equation:

$$P(Ci|X)=\frac{P(X|Ci)P(Ci)}{P(X)} \quad (1)$$

3 Comparative Experiments

3.1 Methodology

Although REALDISP contains activity data from 17 subjects, only some subjects had the data on both ideal-placement scenario and displacement scenario. In this paper, data from subject number 2 were selected for the analysis, as the data on both scenarios of this subject were available.

In this paper, only whole body activities were selected for the analysis. There were altogether 10 activities, including: 1) Walking, 2) Jogging, 3) Running, 4) Jump front and back (jump 1), 5) Jump Sideways (jump 2), 6) Jump legs/arms open/closed (jump 3), 7) Jump rope (jump 4), 8) Rowing, 9) Elliptic bike, and 10) Cycling.

Ideal-placement data were used to construct classification models by using classification algorithms described in section 2.2. Ten-fold cross validation was employed in every model construction. Sensor displacement dataset was used to test against the constructed models, to examine the effect of sensor displacement on activity recognition.

3.2 Results

This section describes the recognition performance of SVM, C4.5, and Naïve Bayes algorithms on ideal-placement data and displacement data. Confusion matrices for recognition performance under each condition for each algorithm are also provided.

Ideal Placement

The recognition performance of each algorithm on ideal-placement data was highly accurate. There were some differences; however, they were very small.

Table 2. Ideal-Placement Recognition Accuracy

SVM	C4.5	Naïve Bayes
99.95%	99.61%	91.73%

The accuracy of SVM is the highest among the three algorithms (99.95%). Accuracy of C4.5 is slightly lower than that of SVM (99.61%). The accuracy of Naïve Bayes is the lowest, compared to the other two algorithms (91.73%). Confusion matrix of each algorithm is elaborated as follows:

Table 3. Confusion Matrix (SVM)

Walking	Jogging	Running	Jump 1	Jump 2	Jump 3	Jump 4	Rowing	E. Bike	Cycling	
15976	0	0	0	0	0	0	0	0	0	Walking
0	**15124**	7	0	0	0	0	0	0	0	Jogging
0	13	**11656**	0	0	0	0	0	0	0	Running
0	0	0	**3603**	6	0	0	0	0	0	Jump 1
0	0	0	12	**3644**	0	0	0	0	0	Jump 2
0	0	0	0	0	**3785**	1	0	0	0	Jump 3
0	0	0	0	0	1	**2022**	0	0	0	Jump 4
0	0	0	0	0	0	0	**8203**	0	0	Rowing
0	0	0	0	0	0	0	0	**10040**	0	E. Bike
0	0	0	0	0	0	0	0	0	**11111**	Cycling

Table 3 describes the confusion matrix of classification results of SVM on ideal-placement data. Recognition accuracy on walking, rowing, elliptic bike, and cycling was 100% accurate. There was some instances in running and jogging activities that were misclassified as one another, and some particular jumping activities that were classified as other types of jumping.

Table 4. Confusion Matrix (C4.5)

Walking	Jogging	Running	Jump 1	Jump 2	Jump 3	Jump 4	Rowing	E. Bike	Cycling	
15975	0	1	0	0	0	0	0	0	0	Walking
2	**15003**	125	0	0	0	0	0	0	0	Jogging
0	131	**11538**	0	0	0	0	0	0	0	Running
0	0	0	**3594**	7	7	1	0	0	0	Jump 1
0	0	2	16	**3636**	2	0	0	0	0	Jump 2
0	0	1	8	4	**3763**	10	0	0	0	Jump 3
0	0	0	2	1	9	**2011**	0	0	0	Jump 4
0	0	0	0	0	0	0	**8203**	0	0	Rowing
0	0	0	0	0	0	0	0	**10040**	0	E. Bike
0	0	0	0	0	0	0	0	0	**11111**	Cycling

Table 4 describes the confusion matrix of classification results of C4.5 on ideal-placement data. Recognition accuracy on rowing, elliptic bike, and cycling was 100% accurate. In walking activity, only one instance was misclassified as running. Similar to the results of SVM, some instances of jogging were classified as running, while some instances of running were classified as jogging. Some instances in a particular jumping activity was misclassified for other jumping activities.

Table 5. Confusion Matrix (Naïve Bayes)

Walking	Jogging	Running	Jump 1	Jump 2	Jump 3	Jump 4	Rowing	E. Bike	Cycling	
15840	136	0	0	0	0	0	0	0	0	Walking
2	**11612**	3519	0	0	0	0	0	0	0	Jogging
0	3191	**8477**	0	1	0	0	0	0	0	Running
0	0	0	**3541**	65	0	3	0	0	0	Jump 1
0	0	0	62	**3588**	6	0	0	0	0	Jump 2
8	0	0	24	7	**3747**	0	0	0	0	Jump 3
0	0	0	0	9	8	**2006**	0	0	0	Jump 4
0	0	0	0	0	0	0	**8194**	9	0	Rowing
0	0	0	0	0	0	0	0	**10040**	0	E. Bike
0	0	0	0	0	0	0	0	0	**11111**	Cycling

Table 5 describes the confusion matrix of classification results of Naïve Bayes on ideal-placement data. Recognition accuracy on elliptic bike, and cycling was 100% accurate. In walking activity, 136 instances were misclassified as running. In Naïve Bayes, many instances of jogging were classified as running, while many instances of running were classified as jogging. Similar to the results from the other two algorithms, some instances in a particular jumping activity was misclassified for other jumping activity. Misclassification in jogging, running, and all jumping activities was similar to that of C4.5.

Displacement

The recognition performance on displacement data dramatically dropped on every algorithm. The recognition accuracy is described in table 6.

Table 6. Sonsor Displacement Recognition Accuracy

SVM	C4.5	Naïve Bayes
52.10%	60.78%	56.43%

When the constructed models were tested against sensor displacement data, C4.5 was most accurate (60.78%). The accuracy of Naïve Bayes was at 56.43%. The accuracy of SVM was the lowest (52.10%), even it was the highest when recognizing the ideal-placement one.

Table 7 describes the confusion matrix of classification results of SVM on displacement data. There was no 100% recognition accuracy on any of the activities. Although the misclassification percentage was high, the misclassified instances were not widely spread.

Table 8 describes the confusion matrix of classification results of C4.5 on displacement data. There was no 100% recognition accuracy on any of the activities. Misclassification was very disperse on every activities; except for rowing, which was misclassified for only another two activities.

Table 7. Confusion Matrix (SVM)

Walking	Jogging	Running	Jump 1	Jump 2	Jump 3	Jump 4	Rowing	E. Bike	Cycling	
15999	0	0	0	3190	2263	3858	2672	31	4	Walking
0	**15126**	5	0	2945	1670	5203	3082	0	0	Jogging
0	12	**11699**	58	2870	2621	6438	1390	9	0	Running
2	70	165	**3613**	266	578	1849	536	0	256	Jump 1
73	46	494	9	**3968**	196	1942	329	0	457	Jump 2
0	0	517	6	255	**4509**	1778	885	0	5	Jump 3
0	0	0	0	0	953	**2586**	443	0	0	Jump 4
0	0	0	0	0	1721	0	**13405**	0	0	Rowing
0	0	27	77	4236	63	4533	6426	**10040**	0	E. Bike
0	40	0	1263	200	0	5090	10604	0	**11111**	Cycling

Table 8. Confusion Matrix (C4.5)

Walking	Jogging	Running	Jump 1	Jump 2	Jump 3	Jump 4	Rowing	E. Bike	Cycling	
18388	149	1757	963	287	5657	0	7	481	328	Walking
14	**15281**	3014	0	1318	7947	213	0	244	0	Jogging
9	599	**18012**	0	1198	4481	46	87	579	86	Running
287	451	40	**3757**	38	224	1282	13	804	439	Jump 1
293	396	19	4776	**3660**	353	1198	24	977	117	Jump 2
128	288	731	238	492	**5206**	726	39	0	77	Jump 3
598	11	33	89	0	555	**2412**	10	267	7	Jump 4
0	0	0	0	0	0	0	**8246**	4790	2090	Rowing
226	700	1169	0	29	3720	0	68	**19064**	426	E. Bike
3184	190	5832	0	3	50	5	823	4824	**13397**	Cycling

Table 9. Confusion Matrix (Naive Bayes)

Walking	Jogging	Running	Jump 1	Jump 2	Jump 3	Jump 4	Rowing	E. Bike	Cycling	
16225	3839	958	0	0	0	0	91	6904	0	Walking
0	**14249**	9820	0	4	189	8	12	3749	0	Jogging
0	4072	**17299**	0	9	1049	16	17	2635	0	Running
0	251	1457	**3553**	79	596	2	4	1393	0	Jump 1
0	332	1518	74	**3640**	598	0	1	1351	0	Jump 2
8	47	2927	23	15	**4709**	0	12	184	0	Jump 3
0	67	613	1	47	697	**2009**	1	547	0	Jump 4
0	181	2457	0	0	0	0	**11546**	941	1	Rowing
0	2857	7154	0	0	0	0	7	**15384**	0	E. Bike
0	864	7328	0	0	3	0	2726	6275	**11112**	Cycling

Table 9 describes the confusion matrix of classification results of Naïve Bayes on displacement data. There was no 100% recognition accuracy on any of the activities. Misclassification was very disperse on every activities.

4 Conclusion and Future Works

4.1 Conclusions

Activity Recognition (AR) plays a major role in various fields; for instance, pervasive health care, security, and wearable and mobile devices. The approach of using wearable sensors for AR is well-accepted, as it compensates the need to install cameras in image processing approach which can lead to privacy violation. Using wearable sensors for activity recognition can suffer from one disadvantage – sensor displacement. There have been a number of research that studies sensor displacement problem. However, the conclusion cannot be made as which classifier is better than another in recognizing displacement data, as the prior experiments were performed under different conditions and focused on different part of the body. This paper showed the recognition performance evaluation of SVM, C4.5, and Naïve Bayes – on ideal-placement and displacement data on whole body activities, by adopting REALDISP dataset to make such comparison.

Ideal placement data were employed to construct classification models. Displacement data were tested against the models, to examine the effects of sensor displacement on classification accuracy of each algorithm.

Recognition performance of all algorithms on ideal placement data was highly accurate. The accuracy of SVM on this dataset was at 99.95%, which was the highest among the three algorithms. Recognition accuracy of C4.5 was at 99.61%, while that of Naïve Bayes was at 91.73%. On ideal placement dataset, some misclassification patterns can be spotted. On jogging and running data, these two activities were misclassified for one another. There were also some misclassification occurred among the four jumping activities.

When the classification models constructed from ideal placement data were tested against displacement data, recognition performance dropped dramatically. Accuracy of SVM, C4.5, and Naïve Bayes decreased to 52.10%, 60.78%, and 56.43%, respectively. Although the accuracy of SVM was the lowest, the misclassified instances were not widely spread like C4.5 and Naïve Bayes.

In sum, the results of the current work illustrates how recognition accuracy can suffer from sensor displacement, and also which algorithm is the most robust one in handling displacement data. The results can lead to further improvement on recognition algorithm in dealing with sensor displacement data.

4.2 Future Works

Some future works may include the investigation on activities related to specific parts of the body (e.g. trunk, upper extremities, and lower extremities), as this work focused only on whole-body activities. Some other recognition algorithms can be employed to examine the effect of sensor displacement on recognition accuracy. Further investigation can also be made on the activities that were likely to be classified for one another (e.g. jogging and running, and the set of jumping activities).

References

1. Ugulino, W., Cardador, D., Vega, K., Velloso, E., Milidiú, R., Fuks, H.: Wearable Computing: Accelerometers' Data Classification of Body Postures and Movements. In: Barros, L.N., Finger, M., Pozo, A.T., Giménénez-Lugo, G.A., Castilho, M. (eds.) SBIA 2012. LNCS (LNAI), vol. 7589, pp. 52–61. Springer, Heidelberg (2012)
2. Choudhury, T., Consolvo, S., Harrison, B., Hightower, J., LaMarca, A., LeGrand, L., Rahimi, A., Rea, A., Borriello, G., Hemingway, B., Klasnja, P., Koscher, K., Landay, J.A., Lester, J., Wyatt, D., Haehnel, D.: The Mobile Sensing Platform: An Embedded Activity Recognition System. Journal of IEEE Pervasive Computing 7(2), 32–41 (2008)
3. Muncaster, J., Ma, Y.: Activity Recognition using Dynamic Bayesian Networks with Automatic State Selection. In: IEEE Workshop on Motion and Video Computing, pp. 30–37. IEEE Computer Society, Washington, D.C. (2007)
4. Kwapisz, J.R., Weiss, G.M., Moore, S.A.: Activity Recognition using Cell Phone Accelerometers. In: Sensor KDD 2010, pp. 74–82. ACM, New York (2011)
5. Banos, O., Damas, M., Pomares, H., Rojas, I., Toth, M.A., Amft, O.: A Benchmark Dataset to Evaluate Sensor Displacement in Activity Recognition. In: Proceeding of the 14th International Conference on Ubiquitous Computing, pp. 1026–1035. ACM, New York (2012)
6. Chavarriaga, R., Bayati, H., Millan, J.R.: Unsupervised Adaptation for Acceleration-Based Activity Recognition: Robustness to Sensor Displacement and Rotation. Pers. Ubiquit. Comput. 17, 479–490 (2014)
7. Kunze, K., Lukowicz, P.: Dealing with Sensor Displacement in Motion-Based Onbody Activity Recognition Systems. In: Proceeding of the 10th International Conference on Ubiquitous Computing, pp. 20–29. ACM, New York (2008)
8. Banos, O., Tóth, M.A., Amft, O.: REALDISP Activity Recognition Dataset, `http://archive.ics.uci.edu/ml/datasets/REALDISP+Activity+Recognition+Dataset`
9. Banos, O., Tóth, M.A., Damas, M., Pomares, H., Rojas, I.: Dealing with the Effects of Sensor Displacement in Wearable Activity Recognition. Sensors 14(6), 9995–10023 (2014)
10. Banos, O., Toth, M.A.: Realistic Sensor Displacement Benchmark Dataset. Dataset Manual (2014)
11. Han, J., Kamber, M., Pei, J.: Data Ming – Concepts and Techniques. Morgan Kaufmann, Massachusetts (2012)
12. Rokach, L., Maimon, O.: Data Mining with Decision Trees – Theory and Applications. World Scientific, Singapore (2008)
13. Kantardzic, M.: Data Mining – Concepts, Models, Methods, and Algorithms. IEEE Press, US (2003)

Flood Warning and Management Schemes with Drone Emulator Using Ultrasonic and Image Processing

Boonchoo Srikudkao[1], Thanakit Khundate[1], Chakchai So-In[1], Paramate Horkaew[2], Comdet Phaudphut[1], and Kanokmon Rujirakul[1]

[1] Applied Network Technology (ANT) Laboratory
Department of Computer Science, Faculty of Science, Khon Kaen University, Thailand
{Sr.boon,thana.date,comdet.p}@kkumail.com,
chakso@kku.ac.th, kanokmon.r@glive.kku.ac.th

[2] School of Computer Engineering, Suranaree University of Technology, Thailand
phorkeaw@sut.ac.th

Abstract. The objective of this paper is to assess the feasibility of an alternative approach to collect water information relating to flooding crisis by means of a small drone. This information includes aerial images, their geographic locations, water flow velocity and its direction, all of which are normally difficult to obtain and in fact expensive should a conventional helicopter or buoyancy are opted. With a drone, however, these acquisitions can be done by a minimally trained operator and under controlled budget. This paper presents the breakout configuration and integration of various sensors and their data management scheme based on a series of image processing techniques, emulating the tasks required to estimate the key flood related parameters. The experimental results reported herein could provide a basis for determining its potential applications in flood warning and predicting systems, as well as concerns that need to be addressed.

Keywords: Drone, Flooding, Image Processing, Ultrasonic.

1 Introduction

Natural disasters have increasingly carried ever more devastation each year. Flood and inundation, in particular, have immensely disrupted our society and caused damaging consequences, which in many occasions include casualties [1]. Issues involving the disaster arise in many aspects, *e.g.*, rescuing the victims, restoring public amenities and transportation, *etc*. To remedy these issues require devising a systematic plan, taking into account flood related information. Currently, gathering such information remains a challenging task. Despite employing modern data acquisition technologies, most available schemes rely on expensive and relatively constrained technology. Taking aerial images from a helicopter or a satellites for instance involves impedingly high operational and maintenance costs. Moreover, maneuvering helicopter required experienced and highly trained pilots. On collecting water flow parameters by using buoyancies, installations and data acquisition are also complicated and financially

H. Unger et al. (eds.), *Recent Advances in Information and Communication Technology 2015*,
Advances in Intelligent Systems and Computing 361, DOI: 10.1007/978-3-319-19024-2_11

demanding, while their area coverage is too limited to be practical in a real-world scale.

The need for simpler and more economical alternative means for acquiring sufficient data for a rapid and effective preparation of flood warning and management measures has thus been the main motivation of this study. Drones have attracted much broader interests in many applications in the recent years, due to its mobility and economic advantages [2]. Depending on problems at hand, a variety of sensors are attached with this automobile acquiring information underneath for further processing and analysis. To date however, there has not been any implementation of the idea on flood warning and managements. Thus, this paper presents a preliminary drone emulation system to assess the feasibility of its use in gathering information related to flood and inundation, *i.e.*, water level, its flow velocity and direction, by applying image analysis on photographic, ultrasonic and GPS (Global Positioning System) sensors attached to a moving drone. Specifically, mosaic and stitching methods, among others [3], were used to registered unaligned spatial data and related information during the air mobile.

This paper is organized as follows: Section 2 briefly provides an overview of related work on drone implementation and image mosaic. Then, section 3 presents our material and methodology including detailed description. The experimental design including results and discussions will be illustrated in sections 4 and 5, respectively. Finally, section 6 contains our conclusion and possible future work.

2 Related Literatures

2.1 Implementations Based on Drones

Currently there have been a large number of techniques and applications implemented using drones. Mevlana, G. and Yuandong, Z, for instance, proposed automatic face detection system based on skin color and human tracking using an AR. Drone [4]. Pierre-Jean. B., *et. al.* proposed airborne control system of an AR. Drone by using the AR. Drone SDK [5]. Michael, C. K., *et. al.* also constructed another automatic AR. Drone controlling system based on motion capture system [6]. Thanks to these implementations, AR. Drone has been proven to be effective in many areas. Zongjian, L. employed drone in taking aerial photographs and used them for building a model map based on automatic aerial triangulation [7]. Olivares-Mendez, M. A. *et. al.* build a drone which could automatically avoid obstacle objects based on the Miguel Olivares' Fuzzy Software (MOFS) [8]. Nick, D. used a drone incorporated with ultrasonic sensors to capture aerial photographs with corresponding elevations, from which 3D terrain of the respective area were built [9]. Similar fusion from various sensors was also proposed in the work by Abigail, S. R. where a drone equipped with GPS could be deployed to a specific coordinate based on the Coordinated Universal Time [10].

2.2 Image Mosaics

In our design, a drone would follow the flood flow, and as a result images taken along the way would differ as the drone proceeds but fortunately overlapped. To stitch these images to produce the entire coverage, mosaic techniques were considered. Richard, S. compared two misaligned images and merged them using image stitching algorithms [11]. Patrik, N. adopted namely graph cut and watershed in stitching 2 overlapped images [12]. Yingen, X. and Kari, P. also adopted image stitching algorithm in creating a panorama image in a mobile phone by using optimal seam finding and transition smoothing processes [13]. Tejasha, P., *et. al.* proposed an image stitching technique using Scale Invariant Feature Transform (SIFT) [14].

3 Materials and Methods

3.1 Hardware and Components

A drone employed in this study was AR. Drone 2.0 as illustrated in Fig. 1. The drone was equipped with 2 cameras. According to its specification, the frontal one is a HD (1280×720 pixels) with viewing angles of 84.1º×53.8º (92º diagonal viewing angle) while the one attached underneath has a lower resolution of only 480×240 pixels. Both camera can capture a series of images at the rate of 15–30 Hz. The drone has 9 Degrees of Freedom (DoF), where 6 and 3 of which are controlled respectively by gyroscopic and magnetization sensors. In order to measure terrain elevation and water level, an ultrasonic sensor was used. The geographical locations of the taken pictures and heights were annotated by means of integrated GPS. The AR. Drone SDK was adopted to write controlling software for the drone, which communications with the based station over a WiFi connection [2].

Fig. 1. A Parrot AR. Drone 2.0 equipped with 2 cameras and 9 DOF sensors (Source: ref. [2])

More specifically, the drone in this study was controlled by an Arduino Microcontroller, *i.e.*, the Arduino nano 3.0, which has 14 input/ output channels, a USB port for connecting with a host and a reset button [15]. The GPS and WiFi modules integrated with the drone were, respectively, the Ublox NEO-6M GPS 3.1.4 and the NRF24L01 Wireless. Both modules were relatively low cost and powered by 3–5 volt DC supply. The GPS module could transmit the location information with the rate of 9600 baud while the wireless module had the strength of +20dBm, equating approximately 520 meters transmission range. Fig. 2 depicts our system integration including the microcontroller and equipped modules.

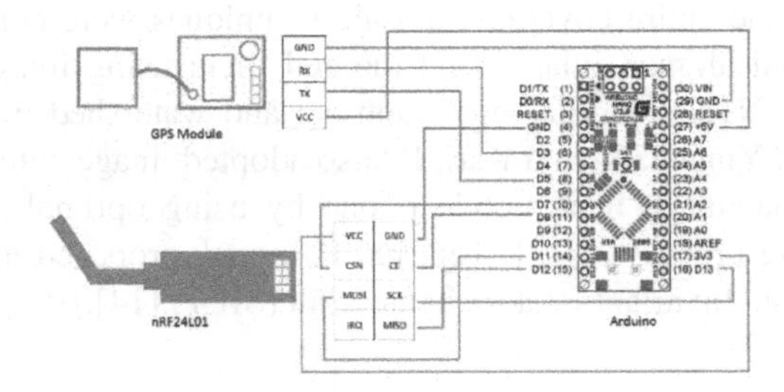

Fig. 2. Overview of System Integration using Arduino nano 3.0, Ublox NEO-6M GPS 3.1.4, and NRF24L01 Wireless

3.2 Overview of the System Integration

Fig. 3 shows the diagram of the proposed system, where the user controls the drone while acquiring the related information, which were published via Google Map API.

The procedures regarding the acquisition of relevant data will be explained in greater detail in the subsequence sections.

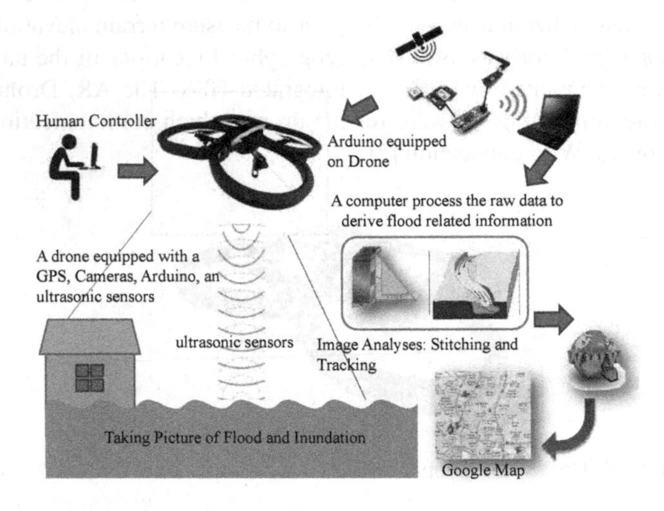

Fig. 3. The diagram showing the overview of the system integration

3.2.1 Image Mosaic and Stitching

As an image sequence taken from the AR, Drone was not yet readily usable since it was nearly impossible if not impractical to ensure steady trajectory of the drone, due to the wind conditions and other environmental factors. The resultant images are thus distorted due to camera perspective and often misaligned. Therefore, to be able to determine flooding condition over the large coverage from this mosaic images, they needed to be firstly registered along overlapping regions, where they were stitched together [14]. Since all of the pictures were taken from nearly the same lighting conditions, the Euclidean distance between pixel intensities were thus sufficiently able to measure the similarity of any pair of registered images, as given in equation (1).

$$d_{f,t}^2(M) = \sum_{(x,y)\in\Omega} \left[f(x,y) - t\big(M(x,y)\big)\right]^2 \tag{1}$$

It should be noted that in this equation, *M* denotes as coordinate transformation and Ω is the overlapping areas of both images.

3.2.2 Determining Water Levels

Integrated with each image taken along its route were underneath water levels, which were measured by an ultrasonic sensor attached with the drone. An ultrasonic sensor exploits sound waves whose frequencies are higher than those audible by human [16]. Depending on health, gender and age of an individual, these values may vary but most ultrasonic sensors use the ones ranging from 20 kHz up to several GHz, called ultrasonic waves. The sensor relies on sending an ultrasonic wave to material (in this case was water) surface, where parts of the incidental wave may propagate through, leaving the other reflect or "echo" back to its origin. The distance between the ultrasonic source and this surface was calculated from the time echo needed to travel back to the source and its velocity. The water levels can then be determined by subtracting this distance from the absolute ceiling of the drone at the respective time.

3.2.3 Moving Object Tracking

On determining the water flow direction and velocity, a plain color object were first dropped onto the water and left afloat. From the first frame in the sequence acquired from the AR. Drone camera, where the object was present, the user marked the object by its color. A Euclidean filter with predefined Euclidean Color (EC) range was then applied to that image and subsequent frames to extract the floating object (shown to the user as a bounding rectangle) at the rate of 25 fps. The centroid (x, y) coordinate of the extracted object together with the location of the drone, given by its GPS, were then used to compute their relative positions and then to control the direction of the drone. This step was implemented using Aforge.Net video processing library.

3.2.4 Determining the Location, Direction, and Respective Velocity

During the flight, the Arduino microcontroller integrated to the drone was fed continuously with latitude (*lat*) and longitude (*lon*) coordinates by the attached GPS. The difference between these consecutives geographical coordinates would be used to compute the displacement and subsequently the drone velocity, while the direction of the drone was obtained from the gyroscope sensor [18], using these formulae.

$$d_{lat} = lat_2 - lat_1;\, d_{lon} = lon_2 - lon_1 \tag{2}$$

$$a = (sin\frac{d_{lat}}{2})^2 + \cos lat_1 \times \cos lat_2 \times (sin\frac{d_{lon}}{2})^2 \quad (3)$$

$$c = 2 \times atan_2\left(\sqrt{a}\right)(\sqrt{1-a}) \quad (4)$$

$$d = R \times c \text{ (where R is the radius of the Earth)} \quad (5)$$

It is worth noting that this information will be transmitted from the Arduino controller back to an emulating computer at the base station over the WiFi connection for further processing.

3.2.5 Connecting with Google Map API

The stitched images associated with their locations and flood related parameters (such as water level, velocity and direction) were then sent to a database and then uploaded to the Google map API per user request. A web application that queried the stored images and relevant information and then displayed them on the Google map API was implemented [19].

4 Experiments

This section describes the method used in our main experiments, namely image stitching, determining water level, object tracking, determining velocity and direction and Google map interface. The first experiment will be the comparison between image stitching and merging algorithms, in terms of their speeds. In this experiment, a medium image size of 480×240 pixels was assumed (as an example shown in Fig. 4 left).

The second experiment was on comparing water level uncertainty between those acquired within the closed and open environment stations (Fig. 5, left and middle). In both environments, their actual water levels, measured form bed to surface, were 40 and 25 centimeters, respectively. The ultrasonic sensor was used to determine the distances twice, *i.e.*, the distance between the drone and the water bed and that between the drone and an obstructing plate floating on the water surface. The water level was subsequently estimated by subtracting these two distances (Fig 5 right).

The third experiment involved finding the appropriate Euclidean range (EC) for object tracking. It was carried out in four places, *i.e.*, a room, a corridor, an open space and a swimming pool as an example shown in Fig. 4 (right). The fourth experiment was to determine the accuracy of direction and velocity by recording the coordinates 3 times at each of the 4 distances, *i.e.*, 30, 50, 70, and 100 meters. The final experiment was to demonstrate the potential use of the emulator by displaying the processed information on a unified Google map web service.

Fig. 4. Examples: image stitching (left) and drone tracking (right)

Fig. 5. Closed (left) and open (middle) stations and the water level estimation (right)

5 Results and Discussions

This section reports the results obtained from our experiments and respective discussions. Table 1 compares the computation times required for registering two images using stitching and merging, respectively. By averaging the registration speeds from 5 trials, the image stitching clearly outperformed the image merging, i.e., 2.08 vs. 2.68 seconds. In addition, Table 2 shows the estimated water level obtained from closed and open environments, whose actual values were 40 and 25 centemeters, respectively.

Table 1. Computation Times Required for Stitching and Merging of Images

	Image Stitching					Image Merging				
No.	1	2	3	4	5	1	2	3	4	5
Time	2.10	1.90	2.20	2.10	2.10	2.70	2.80	2.60	2.70	2.60
Avg.			**2.08**					2.68		
Std.			**0.10**					0.08		

Table 2. Estimated Water Level from Closed and Open Enviroments

	Closed Environment (0.4 m)			**Open Environment (0.25 m)**		
No.	*1st Data*	*2nd Data*	*Level (m)*	*1st Data*	*2nd Data*	*Level (m)*
1	1.283	0.912	0.371	0.723	0.535	0.188
2	1.257	0.921	0.336	0.745	0.541	0.204
3	1.265	0.901	0.364	0.761	0.542	0.219
4	1.274	0.911	0.363	0.789	0.560	0.229
5	1.277	0.935	0.342	0.752	0.525	0.227
6	1.276	0.925	0.351	0.733	0.567	0.166
7	1.262	0.911	0.351	0.729	0.577	0.152
8	1.276	0.928	0.348	0.740	0.561	0.179
9	1.284	0.923	0.361	0.718	0.541	0.177
10	1.273	0.928	0.345	0.736	0.538	0.198
Avg.			**0.353**			**0.194**
Std.			**0.011**			**0.026**

Estimates from boths enviroments resulted in about 5 and 6 cm. errors, respectively. This implies that estimating the water level in controlled environment where less error prone factors, such as wind and water ripple, *etc*., are present yielded better accuracy. Furthermore, Table 3 lists the object tracking results in four different areas, namely, in a room, along a corridor, in an open space, and at a swimming pool.

Table 3. Objects Tracking Success Rate in Four Different Areas at varying E.C. Range

Areas	**EC**	**Success**	**Avg.**	**Std.**	**Areas**	**EC**	**Success**	**Avg.**	**Std**
Room	110	50%	116.75	13.15	Open Space	30	100%	22.00	3.65
	131	70%				41	100%		
	102	**No**				37	100%		
	124	100%				33	100%		
Corridor	147	**No**	158.25	7.80	Swimming Pool	26	100%	35.25	4.79
	164	70%				18	100%		
	159	100%				24	100%		
	163	50%				20	100%		

It can be concluded that the drone can track the object very well at 100% certainty in the outdoor conditions, where the object is sufficiently lit. The result also implies that our framework can be potentially applied in flood warning and managment system, where almost all flood events happend outside. Nontheless, room and corridor results suggest that the drone might as well be succesfully applied under compromised lighting conditions (such as evening or overcast sky) should the EC. ranges are properly tuned. In order to assess the accuracy in terms of direction and velocity, at each of the distances, *i.e.*, 30, 50, 70 and 100 meters, three coordinates were recorded from GPS with elapse times, from which drone velocities were calculated as shown in Table 4..

Table 4. Direction and Veolecity Assement at Differing Distances

Dist.	Rec.	Diff.	T(s.)	Avg.	Std.	Dist.	Rec.	Diff.	T(s.)	Avg.	Std.
30	30	0	38	3.00	2.65	70	65	–5	100	–3.00	1.73
	35	5					68	–2			
	34	4					68	–2			
50	47	–3	75	3.67	3.06	100	101	1	147	2.33	2.89
	49	–1					96	–4			
	43	–7					96	–4			

It could be seen from the table that, when interpreting displacement and computing velocity of the drone from a commercial grade GPS, the error between approximately 2–3 meters could be expected. The final experiment was an implementation of display module for the process result on a Google default map using a web service. Fig. 6 illustrates a screenshot of the web application, including the picture taken from the drone overlaid on the Google map and flood related information, including location name, acquiring date and time, an image taken by the camera, water level, direction and velocity, and coordinates.

- Location Name
- Acquiring Date/ Time
- Images taken by the camera
- Water Level
- Direction and Velocity
- Coordinates

Fig. 6. A screenshot showing the processed result using the Google map web service

6 Conclusions

In this paper an image analysis scheme with ultrasonic and mapping services for economical flood warning and management based on A.R. Drone emulator was presented. The aerial image sequence acquired by the camera attached with the drone as well as corresponding flood related information, estimated by the proposed techniques such as coordinates, water level, direction and velocity, *etc.*, were stored on the database and uploaded onto Google map for display. The preliminary experiments suggested that image stitching was the most effective means of registering misaligned images due to non-steady motion of the camera. Object tracking based on Euclidean Color ranges was employed to determine the water flow. It is worth noting here that the perfect continuous tracking results were achieved outdoor, where lighting conditions ensured reliable object extraction. Nonetheless, under more challenging circumstances, care should be taken in fine tuning the Euclidean parameters for reliable tracking. In our low cost implementation, a commercial grade GPS were used, therefore and error of approximately 2–3 meters are to be expected when determining the drone location and computing its velocity. This flood related information was then ported onto a web based Geographical Information System (GIS) so that it may be used to support flood warning and managements.

References

1. Hallegatte, S., Green, C., Nicholls, R.J., Corfee-Morlot, J.: Future flood losses in major coastal cities. Nature Climate Change 3, 802–806 (2013)
2. Parrot: Parrot AR.Drone Webpage, `http://www.ardrone.parrot.com`
3. Minakshi, K.: Digital Image Processing. In: Satellite Remote Sensing and GIS Applications in Agricultural Meteorology, pp. 81–102 (2003)
4. Mevlana, G., Yuandong, Z.: Autonomous Face Detection and Human Tracking using AR.Drone Quadrotor. Technical Report, Cornell University (2011)
5. Pierre-Jean, B., Francois, C., David, V.: The Navigation and Control Technology inside the AR.Drone micro UAV. In: IFAC World Congress, Milano, Italy, pp. 1477–1488 (2011)
6. Michael, C.K., Christopher, R.M., Michael, L.L.: Autonomous Quadrotor Control with Reinforcement Learning, Technical Report, Rutger University, pp. 1–6 (2011)
7. Zongjian, L.: UAV for Maping-Low Altitude Photogrammetric Survey. In: The International Archives of the Photogrammetry, Remote Sensing and Spatial Information Sciences, China, vol. XXXVII(pt. B1), pp. 1183–1186 (2008)
8. Olivares-Mendez, M.A., Mejias, L., Campoy, P., Mellado-Bataller, I.: Quadcopter see and avoid using a fuzzy controller. In: Proceeding of the International FLINS Conference on Uncertainty Modeling in Knowledge Engineering and Decision Making. World Scientific, Turkey (2012)
9. Nick, D.: Simultaneous localization and mapping with AR.Drone. Master Thesis, Universiteit van Amsterdam (2012)
10. Abigail, S.R.: An Evaluation of a UAV Guidance System with Consumer Grade GPS Receivers. ProQuest, 174 p. (2009)
11. Richard, S.: Image Alignment and Stitching: A Tutorial. Foundations and Trends in Computer Graphics and Vision 2(1), 1–104 (2006)
12. Patrik, N.: Image Stitching using Watersheds and Graph Cuts. Technical Report, Lund University
13. Yingen, X., Kari, P.: Sequential Image Stitching for Mobile Panoramas. In: Proceeding of the International Conference on Information, Communications and Signal Processing, pp. 1–5. IEEE, Macau (2009)
14. Tejasha, P., Mishra, S., Chaudhari, P., Khandale, S.: Image Stitching using MATLAB. International J. of Engineering Trends and Technology 4(3), 302–306 (2013)
15. Arduino: Arduino Webpage, `http://www.arduino.cc`
16. Massa, D.P.: Acoustic/Ultrasound Choosing an Ultrasonic Sensor for Proximity or Distance Measurement Part 1: Acoustic Considerations. Sensors Magazine (1999)
17. AForge.NET: AForge.NET Webpage, `http://www.aforgenet.com`
18. Chamberlain, B.: Finding Distance based on latitude and Longtitude (2002)
19. Google Map API: Google Map API Webpage, `http://developers.google.com`

Concept-Based Text Classification of Thai Medicine Recipes Represented with Ancient Isan Language

Chumsak Sibunruang and Jantima Polpinij

Intellect Laboratory, Faculty of Informatics, Mahasarakham University, Thailand
{chumsak.s,jantima.p}@msu.ac.th

Abstract. This work presents the concept-based text classification for organizing of traditional Thai medicine recipes. These recipes were translated from the Northeastern Thai palm leaf manuscripts. It is noted that each medicine recipe is presented with the ancient Isan language. The proposed method is called '*concept-based text classification*', because we utilize '*concepts*' as document features, where a concept is a surrogate of a word group having a same meaning. The main mechanisms in the method are the *k*-Nearest Neighbor algorithm and an ancient Isan dictionary, called Isan-Thai Markup Language (ITML). The objective of this work is to assign the Thai medicine recipes into predefined 5 groups. They are the groups of medicine recipe for headache and fever, stomachache and abdomen, skin, abscess, and faint and vertigo, respectively. After testing by recall, precision, and *F*-measure, it returns the satisfactory results of automatic text classification.

Keywords: Thai Medicine Recipe, Palm leaf Manuscript, Text classification, Ancient Isan dictionary, Isan-Thai Markup Language.

1 Introduction

Palm leaf manuscript (PLM) is an ancient document form that comprises a significant documentary heritage of the Isan people of Northeastern Thailand [1-2]. The alternative name in a local language is called '*Bailan*'. These materials may contain a vast amount of knowledge, such as traditional Thai medicine recipe. The UNESCO considers the PLMs as a literary heritage that should be preserved [2]. To preserve the PLMs, they are transformed into digital images in order to decrease the problem of PLM damage. Therefore, when an expert wants to read a PLM to obtain the relevant knowledge, the expert can search and read the PLMs in the form of digital images. To do this, the expert can access to the PLMs without directly touching [3].

With a large amount of knowledge contained in the PLMs, over the last few decades, the experts would access and read the ancient documents in order to translate the ancient characters to the present characters. Especially, the knowledge involves to the traditional Thai medicine recipe.

Unfortunately, there is a huge amount of the traditional Thai medicine recipes. So, it is very hard to organize the traditional Thai medicine recipes. As this, text classification is required for handling and organizing the collection of traditional Thai

H. Unger et al. (eds.), *Recent Advances in Information and Communication Technology 2015*,
Advances in Intelligent Systems and Computing 361, DOI: 10.1007/978-3-319-19024-2_12

medicine recipes, where text classification can assign free text documents to one or more predefined classes based on their content. However, it has a problem for Thai medicine recipes classification. This is because, all of these recipes are represented with the ancient Isan language. Therefore, a dictionary of ancient Isan words is also developed, because the developed dictionary will be used for verification and identification the ancient Isan words containing in the Thai medicine recipes.

2 The Northeastern Thai Palm Leaf Manuscript

Palm leaf manuscript (PLM) is an ancient document form that comprises a significant documentary heritage of the Isan people of Northeastern Thailand [1-2]. PLMs contain a vast amount of knowledge that can be classified into eight categories: Buddhism, tradition and beliefs, customary law, economics, traditional medicine, science, liberal arts, and history.

Northeastern Thai PLMs can vary in size. The Isan people used the various sizes in different ways. The longer palm leaf manuscripts are used to record Buddhist stories and doctrine, while the shorter ones are used to record local wisdom related to daily life. In general, the languages written can be Pali, ThaiIsan (dialect Isan), Pali-ThaiIsan, Old Thai, and Khmer. In addition, manuscripts are written in four archaic orthographies (ThamIsan, ThaiNoi, Khmer, and Old Thai).

In order to preserve both knowledge and the manuscripts themselves, there are some projects exploring the most suitable method to digitize and organize the palm leaf manuscripts. For example, the manuscripts are transformed into digital images in order to decrease the problem of PLM damage. Therefore, when an expert wants to read a PLM to obtain the relevant knowledge, the expert can search and read the PLMs in the form of digital images. To do this, the expert can access to the PLMs without directly touching [3].

Furthermore, the some experts would access and read the ancient documents in order to translate the ancient characters to the present characters. To do this, people can have a better understanding for the PLM content. For an example work, some experts of the Conservation Project for Northeastern Palm leaf Manuscript (Mahasarakham University, Thailand) have translated the content of the palm leaf manuscripts in a form that is easier to understand. An example of the original PLM and its result of translation can be presented as Fig.1. The content in PLM is a Thai medicine recipe.

Fig. 1. An example of the original palm leaf manuscript and the result of translation

3 A Dictionary of Ancient Isan Words

The ancient Isan language found in the palm leaf manuscripts is different from the current Thai language, especially a difference of words. Also, some words of the ancient Isan language are also different from the current Isan language. Therefore, A dictionary of ancient Isan words is developed. Consider in Table 1. It shows some of the different words between the ancient Isan language, the current Isan language, and the current Thai language. It is noted that, some ancient Isan words can be similar to the current Isan words. Also, a part of ancient Isan can be illustrated as Fig.2.

Table 1. Examples of different words between the ancient Isan language, the current Isan language, and the current Thai language

The ancient Isan words	The current Isan words	The current Thai words	In English
นางขาม (Nang-kham) ขาม (Kham)	หมากขาม (Mark-kham)	มะขาม (Ma-kham)	Tamarind
เข้ามิ้น (Kao-Min) ขี้มิ้น (Kee-Min) ว่านเหลือง (Wan-Lueng)	ขี้มิ้น (Kee-Min) ขมิ้น (Kha-Min)	ขมิ้น (Kha-Min)	Curcuma
นำแน้ (Nam-Nae)	นำแหน่ (Nam-hnae)	รางจืด (Rang-Chued)	Thunbergia laurifolia

Our dictionary, called *Isan-Thai Markup Language* (*ITML*) [4], is developed manually. It consists of 1,200 ancient Isan words, which are formatted as XSD [5]. It is the controlled vocabulary thesaurus. The words that have similar meaning will be grouped together. Each group will have a concept used as the surrogate of the group. Simply speaking, a group is viewed as a class.

For an example, the words 'ขมิ้น' (Ka-min), 'เข้ามิ้น' (Kao-min), 'ขี้มิ้น' (Kee-min), and 'ว่านเหลือง' (Wan-lueng) have the same meaning. These words refer to '*curcuma*' in English. Therefore, they will be grouped into the same class, and 'เข้ามิ้น' (Kao-min) is used as the concept of the chis class (See in Fig.2).

```
<?xml version="1.0" encoding="UTF-8" ?>
<xs:schema xmlns:xs="http://www.w3.org/2001/XMLSchema">

<xs:element name="เข้ามิ้น" language= "Isan">
   <xs:language>
      <xs:dataType>
         <xs:synonyms >
            <xs:element name="ขมิ้น" langauge = "Thai" type="xs:string"/>
            <xs:element name="ขี้มิ้น" langauge = "Isan" type="xs:string"/>
            <xs:element name="ว่านเหลือง" langauge = "Isan" type="xs:string"/>
            ......
         </xs:synonyms>
      </xs:dataType>
   </xs:language>
</xs:element>
.....
.....
</xs:schema>
```

Fig. 2. An example of the word 'เข้ามิ้น (Kao-Min)' stored in the ITML

4 The Research Method of Thai Medicine Recipe Classification

The proposed methodology consists of two main processes. The overview of the proposed methodology is shown as Fig. 3.

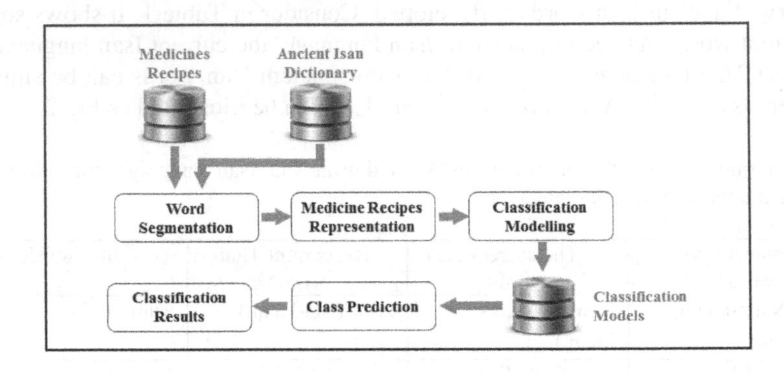

Fig. 3. Overview of the Thai Medicine Recipe Classification Methodology

4.1 Classifier Modelling

Step 1: Word Segmentation

In this work, word segmentation is driven by the longest matching algorithm [6]. The ITML is used as a dictionary to search and detect words from a sentence. After a word is searched and detected by using ITML, the system will provide its concept and uses the concept as an index (surrogate) of a document. Simply speaking, we use the concepts as indexes of each document. It is noted that a document can have many surrogates. Consider a recipe shown as follows.

> "บีงัวใส่เหล้ากิน ๓ ทีหายดีฯ ฮากจันทร์ฝน เอาน้ำท่า กะนาน ๑ น้ำเครือเขาฮอ กะนาน ๑ น้ำหมากนาว กะนาน ๑ เขี่ยว ขี้มิ้น ๑ แง่ง ให้ขูดปั้นเป็นแท่งลูกกอน บ้าหมูหาย"

After segmenting words by the longest matching algorithm and ITML, all indexes of this recipe can be presented as Table 2.

Table 2. The results of word segmentation and transforming them as concepts

Words found	บีงัว	เหล้า	ฮาก จันทร์	เครือ เขาฮอ	น้ำหมาก นาว	ขี้มิ้น	ลูก กอน	บ้าหมู
Concept transformation	บีงัว	เหล้า	ฮาก จันทร์	เครือ เขาฮอ	น้ำหมาก นาว	เข้ามิ้น	ลูก กอน	บ้าหมู

Consider Table 2. After obtaining all significant words, these words will be searched for their concepts through the use of ITML. Finally, these concepts will be used as the surrogate of the medicine recipe. Clearly demonstrated example, after segmenting word, an original word 'ขี้มิ้น (Kee-min)' is resulted, but this word is transformed to the word 'เข้ามิ้น (Kao-min)'. This is because the word 'เข้ามิ้น (Kao-min)' is the concept of 'เข้ามิ้น (Kao-min)'.

Step 2: Medicine Recipe Representation

Traditional medicine recipes are represented by words (features). To transform a text document to a new form that is suitable for text mining, a well-known technique is called vector space model (VSM) [7]). It is also known as a bag of words (BOW). The VSM model is one of information retrieval models (IR model) [7]. It has been developed to retrieve information. This vector representation does not consider the ordering of words in a text. This vector represents the relationship between text documents and words by the number of occurrences of each word in each text document. Each document can be represented in a vector space, $\vec{d} = \{w_{i,1}, w_{i,2}, \ldots, w_{i,t}\}$ called '*bag of words*' (*BOW*) [8]. An example of bag of words can be shown as Fig. 3.

Also, *tf-idf* is applied. This is a weighting technique often used in the areas of information retrieval and text mining [9-10]. The background of this technique is the statistical measure, that is used to evaluate how important a word is to a document in a collection.

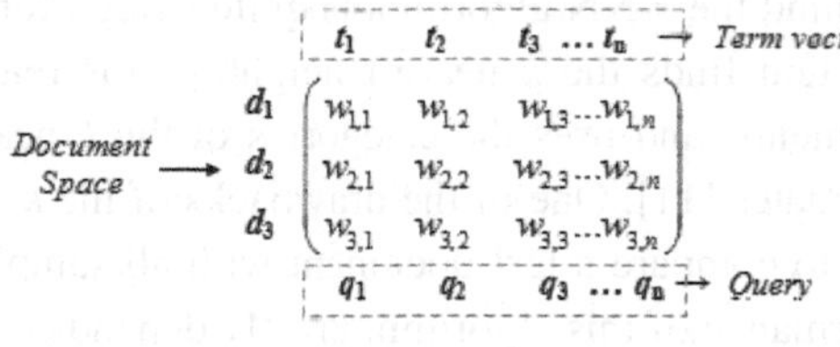

Fig. 4. A Vector Space Model (also known as Bag of Words)

The term frequency $tf_{t,d}$ of term *t* in document *d* is defined as the number of times that *t* occurs in *d*. Relevance does not proportionally increase with term frequency, while the score is *0* if none of the query terms are present in the document. However, the estimate of $tf_{i,j}$ can be calculated by:

$$tf_{i,j} = 1 + \log\ (frequency\ of\ term) \tag{1}$$

Meanwhile, *idf* can be calculated from the document frequency *df*, which is the number of documents in which term *t* occurs. It is described as follows.

$$Idf_t = 1 + \log\ (|D|/df_t) \tag{2}$$

where $|D|$ is the number of documents in the database.

The *tf-idf* weight of a term is the product of its *tf* weight and its *idf* weight and *tf-idf* is defined as follows.

$$tf\text{-}idf = tf \times idf \tag{3}$$

In this work, *tf-idf* is modified a bit. However, we concentrate on '*concept*', not '*term*' or '*word*'. Therefore, *tf* will be changed to *concept frequency* (*cf*). Afterwards, each *cf* value should be greater than or equal to 1. The number of indexes exists in may be equal to the number of unique words in that entire document.

Step 3: Recipe Classifier Modelling

To classify documents into several groups, text classification is applied, where text classification is the process of classifying text documents into one or more predefined categories based on their content. However, by using '*concepts*' as document surrogate, our methodology can be called as the *concept-based text classification*.

This work applies *k*-Nearest Neighbor (*k*-NN) to model the traditional medicine recipe classifiers. This is because the *k*-NN algorithm is one of the most popular algorithms for text classification [11]. It has found that the *k*-NN algorithm achieves very good performance in their experiments on different data sets [12-14]. The *k*-NN classifier is based on the assumption that the classification of an instance is most similar to the classification of other instances that are nearby in the vector space (or bag of words obtained from the step of text representation).

In general, the idea behind the *k*-NN algorithm is quite straightforward. To classify a new document, the system finds the *k* nearest neighbors (as relevant documents) among the training documents, and uses the categories of the *k* nearest neighbors to weight the category candidates [11]. One of the drawbacks of the *k* - NN algorithm is its efficiency, as it needs to compare a test document with all samples in the training set. In addition, the performance of this algorithm greatly depends on two factors, that is, a suitable similarity function and an appropriate value for the parameter *k*.

After *k* nearest neighbors are found, several strategies could be taken to predict the category of a test document based on them. However, a fixed *k* value is usually used for all classes in these methods, regardless of their different distributions. Equation (4) and (5) below are two of the widely used strategies of this kind method.

$$y(d_i) = \arg\max_k \sum_{x_j \in kNN} y(x_j, c_k) \tag{4}$$

$$y(d_i) = \arg\max_k \sum_{x_j \in kNN} sim(d_i, x_j)\, y(x_j, c_k) \tag{5}$$

where d_i is a test document, x_j is one of the neighbors in the training set, $y(x_j, c_k) \in \{0, 1\}$ indicates whether x_j belongs to class c_k, and *similarity* (d_i, x_j) is the similarity function for d_i and x_j. Equation (4) means that the predication will be the class that has the largest number of members in the *k* nearest neighbors; whereas equation (5) means the class with a maximal sum of similarity will be the winner. The latter is thought to be better than the former and used more widely [15].

4.2 Model Usage

After obtaining the traditional medicine recipe classifiers, these text classifiers will be used for automatic classification of the traditional medicine recipe documents. All documents need to pre-processed with the same process described as processing step 1 and 2 in Section 4.2.1, and then these documents will be automatically classified into predefined categories through the use of the medicine recipe classifiers.

5 The Experimental Results

5.1 Dataset

Our dataset contains over 900 Thai medicine recipes. These recipes are translated from the Northeastern Thai palm-leaf manuscripts, done by the staffs of the Conservation Project for Northeastern Palm leaf Manuscript (Mahasarakham University, Thailand). In this study, we have five groups of the Thai medicine recipes: headache and fever, stomachache and abdomen, skin, abscess, and faint and vertigo recipes.

It is noted that this work provides k as 5. To determine the class of a new instance, it will calculate the distance between the new instance and all instances in the training set. Afterwards, it will find the k closest neighbors to the new instance from the training data. Finally, the predicted class label will be set as the most common label among the *k*-nearest neighbors.

For our experiments, we also use 50 documents per group as the test set. It is noted that, the training set and the test set are different.

5.2 The Results and Discussion

We evaluated the results of the experiments by using precision (P) [7], recall (R) [7], and F-measure (F) [7]. The experimental results can be presented in Table 3.

Consider the results in Table 3. It can be seen that the traditional medicine recipe classifiers show the satisfactory results of automatic classification, especially in the term of precision. This is because, with the use of '*concepts*' as document surrogate, this is to support for the semantic analysis during text processing, where the inaccuracy of text classification can be due to the different words, but with a same meaning.

Table 3. The experimental results of the automatic traditional medicine recipe classifiers

Medicine Recipe Classes	Recall	Precision	F-measure
headache and fever	0.78	0.76	0.77
stomachache and abdomen	0.76	0.82	0.78
skin	0.78	0.77	0.77
abscess	0.70	0.80	0.75
faint and vertigo	0.72	0.83	0.77
Average	0.75	0.80	0.77

However, the results of our methodology still show a failure rate. This is because some of concepts are used many Thai medicine recipes. For example, the concept of 'เข้ามิ้น (Kao-Min)' can be found in both of the medicine recipes of heache and fever, skin, abscess, and faint and vertigo. This may lead to poor accuracy of Thai medicine recipe classification.

In addition, by using the *k*-NN algorithm, *k* becomes the most important parameter in a text classification system based on *k*-NN. During the classification process, *k* nearest documents to the test one in the training set are determined firstly. Then, the predication can be made according to the category distribution among these *k* nearest neighbors. Therefore, the system performance is very sensitive to the choice of the parameter *k*. Moreover, it is very likely that a fixed *k* value will result in a bias on large categories.

In another problem, the class distribution in the training set can be uneven, where some classes contain the recipes that may contain more content than others. As a result, the class imbalance problem can be occurred [16]. This is because this has a bias toward the classes with greater number of words, when one of the minority classes can be heavily under-represented in comparison to the other majority class. Finally, it leads to misclassifications.

6 Conclusion

This work proposes a solution to handle and organize the document of traditional Thai medicine recipe translated from the Northeastern Thai palm leaf manuscripts. It is noted that each medicine recipe is represented by the ancient Isan language. Our method is called '*concept-based text classification*', because we use '*concepts*' as document features, where a concept is a surrogate of a group of words that have a same meaning. The main mechanisms in our method are the *k*-Nearest Neighbor algorithm and an ancient Isan dictionary (called Isan-Thai Markup Language). The objective of this work is to assign the Thai medicine recipes into predefined 5 groups. They are the groups of medicine recipe for headache and fever, stomachache and abdomen, skin, abscess, and faint and vertigo, respectively. After testing by recall, precision, and *F*-measure, they return the satisfactory results of automatic text classification.

However, this work may be improved in the future. This is because the *k*-NN shows two disadvantages in this work. Firstly, *k*-NN does not learn anything from the training data. As this, the result of the algorithm may be not well, because it is not being robust to noisy data. In addition, changing *k* can change the resulting predicted class label. Therefore, we may have a new experiment with other algorithms. Furthermore, the class distribution in the training set can be uneven, where some classes contain the recipes that may contain more content than others. This problem is called the class imbalance problem. It should be improved as well, because it can lead to misclassifications.

Acknowledgement. This work is supported by the Faculty of Informatics, Mahasarakham University located in Thailand.

References

1. Iijima, A.: A Historical Approach to the Palm-Leaf Manuscripts Preserved in Wat Mahathat, Yasothon (Thailand), http://www.laomanuscripts.net/downloads/literaryheritageoflaos26_iijima_en.pdf
2. Manmart, L., Chamnongsri, N., Wuwongse, V.: Metadata Development for Palm Leaf Manuscripts in Thailand. In: Proceedings of International Conference on Dublin Core and Metadata Applications (2012)
3. Shi, Z., Setlur, S., Govindaraju, V.: Digital Enhancement of Palm Leaf Manuscript Images using Normalization Techniques. In: Proceeding of the 5th International Conference on Knowledge-based Computer Systems (2004)
4. Polpinij, J.: Concept-Based Cross Language Retrieval for Thai Medicine Recipes. In: Tuamsuk, K., Jatowt, A., Rasmussen, E. (eds.) ICADL 2014. LNCS, vol. 8839, pp. 320–327. Springer, Heidelberg (2014)
5. van der Vlist, E.: XML Schema. O'Reilly (2002)
6. Haruechaiyasak, C., Kongyoung, S., Dailey, M.N.: A comparative study on Thai word segmentation approaches. In: Proceeding of the 5th International Conference on Electrical Engineering/Electronics, Computer, Telecommunications and Information Technology (ECTI-CON) (2008)
7. Baeza-Yates, R., Ribeiro-Neto, B.: Modern Information Retrieval. Addison Wesley (1999)
8. Sivic, J.: Efficient visual search of videos cast as text retrieval. IEEE Transactions on Pattern Analysis ans Machine Intelligence 31, 591–605 (2009)
9. Yang, Y., Pederson, J.O.: A Comparative Study on Features selection in Text Categorization. In: Proceedings of the 14th International Conference on Machine Learning (ICML), Nashville, Tennessee (1997)
10. Soucy, P., Mineau, G.W.: Beyond TFIDF Weighting for Text Categorization in the Vector Space Model. In: Proceedings of the 19th International Joint Conference on Artificial Intelligence (IJCAI) (2005)
11. Manning, C.D., Schutze, H.: Foundations of Statistical Natural Language Processing. MIT Press, Cambridge (1999)
12. Yang, Y., Liu, X.: A Re-examination of Text Categorization Methods. In: Proceedings of the 22nd Annual International ACM SIGIR Conference on Research and Development in Information Retrieval, pp. 42–49 (1999)
13. Joachims, T.: Text Categorization with Support Vector Machines: Learning with Many Relevant Features. In: Proceedings of the European Conference on Machine (1998)

14. Li, B., Chen, Y., Yu, S.: A Comparative Study on Automatic Categorization Methods for Chinese Search Engine. In: Proceedings of the 8th Joint International Computer Conference, pp. 117–120. Zhejiang University Press, Hangzhou (2002)
15. Li, B., Yu, S., Lu, Q.: An Improved k-Nearest Neighbor Algorithm for text categorization. In: Proceedigns of the 20th International Conference on Computer Processing of Oriental Language (2003)
16. García, V., Sánchez, J.S., Mollineda, R.A., Alejo, R., Sotoca, J.M.: The class imbalance problem in pattern classification and learning. In: Congreso Español de Informática (2007)

Vision Inspection with K-Means Clustering on Head Gimbal Assembly Defect

Rawinun Prasertaweelap and Somyot Kiatwanidvilai

Faculty of Engineering, King Mongkut's Institute of Technology Ladkrabang, Bangkok, Thailand
rawinun.prasertaweelap@seagate.com, drsomyotk@gmail.com

Abstract. Head Gimbal Assembly (HGA) is an important feature of read and write process in a Hard Disk Drive (HDD). Currently, HGA circuit inspections are done using human operators under microscope; the vision processing for inspection in automated systems is required. This research work proposes an algorithm for detection the HGA circuit defect by using the blob detection, and then analysis the properties of blob tool. By the measurement properties of the blob tool, the K-Means Clustering can specify the data in each group in 95.45% accuracy with 110 samples.

Keywords: Head Gimbal Assembly, Hard Disk Drive, Blob Detection, K-Means Clustering.

1 Introduction

Currently, image processing is one of the popular choices for inspection process. New technologies with high performance are required for HDD manufacturing process [1-4]. The HGA circuit defect is one of the key factors that impact the read and write performance of HDD. The HGA circuit test has been done by electrical inspection in case of open and short circuit only. It still be a problem when the electrical parametric is within specification from the HGA variation but the vision image found the abnormal connection area.

Nowadays, the image processing has a lot of features for analysis. In HDD manufacturing process, the automatic inspection in each defect type is required and the accuracy and repeatable results are very important. Fig. 1 shows the diagram of a part of HGA process.

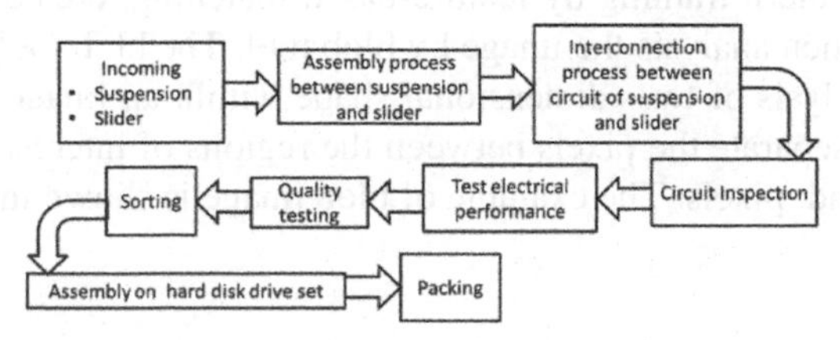

Fig. 1. Diagram of a Hard Disk Drive Process

H. Unger et al. (eds.), *Recent Advances in Information and Communication Technology 2015*, Advances in Intelligent Systems and Computing 361, DOI: 10.1007/978-3-319-19024-2_13

In HDD industry, the new technology in automatic systems is very important. The HDD manufacturing processes always develop and improve for the production speed and product quality. The one important thing in HDD manufacturing is the tool for inspecting the output failure in each type. The different failure has come from the different root cause. Quality improvement will be the high efficiency when specifying the correct root cause. The clustering method is required for classification [2].

2 Vision Processing and Clustering

2.1 Input Image

This paper proposes the using of input image from Thermal Interconnection machine by integrating the industrial camera into the machine. The vision systems composes of the progressive scan CCD with square pixels, VGA resolution of 659 x 494, and LED ring light with red LED. The mechanical setup of machine for vision system uses the reflection concept by fixing the angle of clamp. The reflection device is adopted for capturing the image because the connection area of HGA is not flat. It has an angle on the HGA. The HGA defect images are shown in Fig. 2.

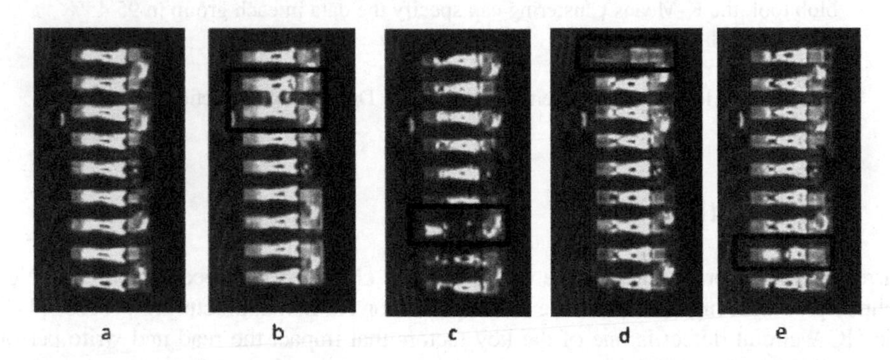

Fig. 2. Image of HGA circuit; (a), (b), (c), (d) and (e) are examples of good, bridge, burn, missing, and no-connect part, respectively

2.2 Image Processing

In this research work, we developed the image processing and clustering on the Cognex VisionPro Quickbuild Version 7.2. The concepts used in the processing are the PatMax for pattern training by feature-based matching, create the fixture coordinate system and then analysis the image by blob tool. The blob tool is a vision tool for detection and analysis of two- dimensional shape within an image. Blob tool concept is the process to separate the pixels between the regions of interest pixels or blob pixels and background. pixels. The example of blob image is shown in Fig. 3.

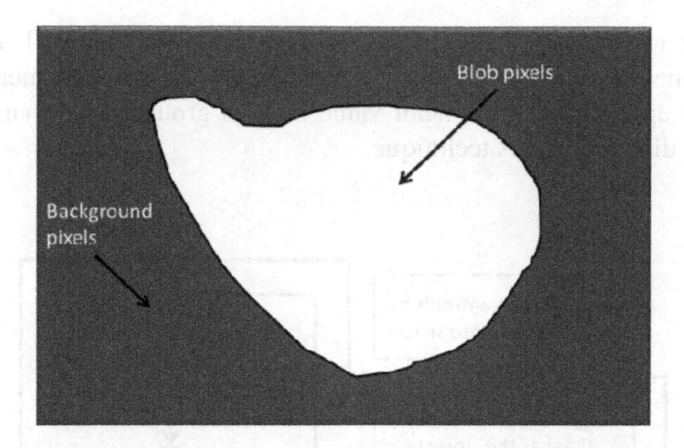

Fig. 3. The example of blob pixels and background pixels on image

The blob tool has ability to measure the properties of an image. Three measurement properties from blob tool used in this research work are the Elongation, Perimeter, and Acircularity; these properties result in different features those are useful in clustering process. The different features show the different characteristics for each input data. The description of each feature is shown in the followings.

The elongation is the dividing of the second moment of inertia between minor axis and major axis. The second moment of inertia is given by (1) which is the calculation of the moment of inertia about the second principal axis of the blob. The elongation can be computed by (2).

$$M_{ijk} = \int I_i(\vec{x}) \cdot x_j \cdot x_k d\vec{x} \tag{1}$$

$$Elongation = \frac{M_{MinorAxis}}{M_{MajorAxis}} \tag{2}$$

The perimeter is the length of the outside edge of the blob that is computed by counting the number of pixel edges.

Acircularity is the measuring index of the lack of circularity of the blob that is computed by the dividing of square of perimeter and area of the blob. Acircularity is given by (3).

$$C = \frac{P^2}{4\pi A} \tag{3}$$

C is the Acircularity, P is the perimeter of the blob and A is the area of the blob [1].

2.3 K-Means Clustering

K-Means Clustering is the technique for classification the data in each small group. The concept of this technique is based on the dividing the group in k groups.

The processing will operate in iterations. In each step, the data will move to some group and the mean value in each group is recalculated for the new mean value. This process will be applied until the mean value in each group is not changed. Fig. 4 shows the brief diagram of this technique.

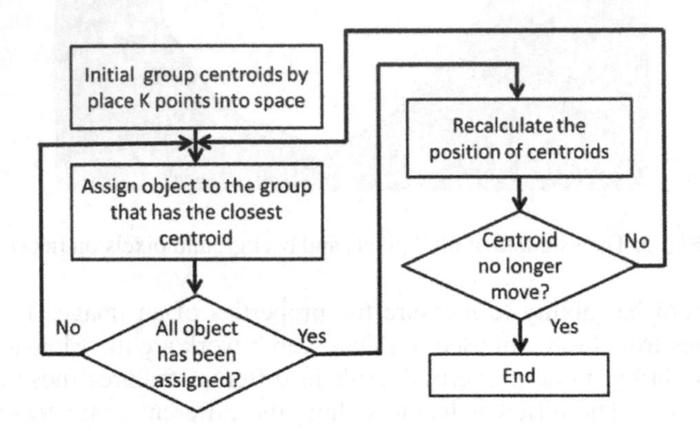

Fig. 4. The process step of K-Means Clustering

K-Means clustering process has the calculation function for clustering data. The distance calculation uses the data point and centroid point to define the minimum distance value which is the Euclidean distance. The Euclidean distance equation is given by (4) when vector x= (x_1, $x2$,....x_n) and another vector y = (y_1, $y2$,....y_n). [5]

$$D(x_1, y_1) = \sqrt{\sum_{i=1}^{n} (x_i - y_i)^2} \quad (4)$$

The Euclidean distance shows that the relationship between data point and centroid on this group is minimized.

3 Experiment Results

3.1 Blob Analysis

The blob analysis of Cognex vision pro program are used for analysis the classification of the HGA image in each defect type. This paper proposes the soft threshold in relative type by 40% low threshold and 60% high threshold. The measurement results in each group of input image are on the Table 1.

Table 1. Blob measurement properties with HGA group

Group	Measurement properties		
	Elongation (-)	**Perimeter (pixel)**	**Acircularity (-)**
Good	106.075	11.416	1.903
Bridge	38.819	14.248	2.544
Burn	29.019	3.898	1.906
Missing	12.020	6.099	1.488
No-connect	28.466	7.145	1.568

3.2 K-Means Clustering Result

This research work utilized the K-Means for clustering the measurement properties from blob detection. In the algorithm, the K-value was specified as 5 and the repeat testing was done in 10 times. The clustering results of all 10 testing results were the same. The center point of each group is changed a little bit in the repeatability testing. The sample size in this experiment is 110; the sample size of good, bridge, burn, missing, and no-connect groups are 30, 27, 11, 22, and 20 samples, respectively. The results in Table 1 show the group properties separated by the final value of centroid. With this result, the good performance from K-Means clustering is achieved with the 95.45% accuracy. The 4.55% errors are from the burn part and no connect part; however, this defect is able to be classified by using the electrical tester in the HGA process.

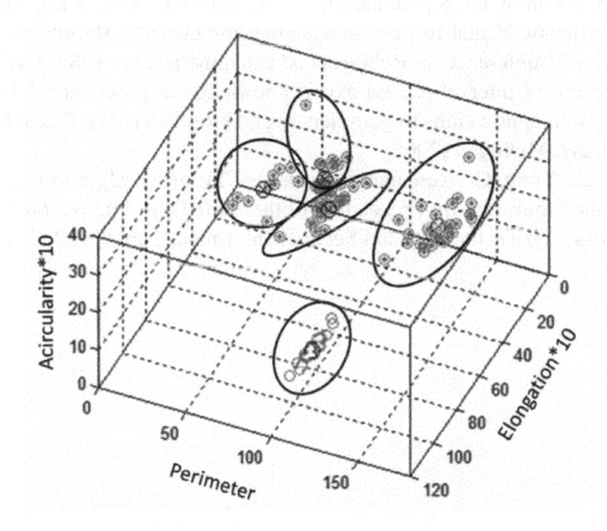

Fig. 5. The raw data and clustering group results

The graph in Fig. 5 shows the measurement data in each image and the clustering groups' results after K-Means clustering. One of the reasons of the error is come from the setting in vision system such as lighting system, docking system. However, the variation in each part is the one thing that impacts the clustering performance. Currently, this error is fixed by using the electrical testing process in HGA production line.

4 Conclusion

The experiment shows the performance of K-Means Clustering when using the blob analysis on the HGA defect type. The results show the accuracy of 95.45% with 110 samples. In this experiment, the incorrect clustering is the burn and no connect defect which is able to be classified by the electrical tool in the process line. The effective measurement properties are adequate for clustering and this technique was applied to the production line in the HGA manufacturing.

Acknowledgement. This work was supported by Seagate Technology (Thailand) Co., Ltd. and DSTAR, KMITL under grant no. SOW#152980.

References

1. Cognex: Image Processing & Machine Vision Training. Cognex Corporation, Massachusetts (2014)
2. Chee, W.M., Afzulpurkar, N.V., Dailey, M.N., Saram, P.B.: A Bayesian Approach to Automated Optical Inspection for Solder Jet Ball Joint Defects in the Head Gimbal Assembly Process. IEEE Transactions on Automation Science and Engineering 11, 1155–1162 (2014)
3. Saenthon, A., Kaitwanidvilai, S., Kunakorn, A., Ngamroo, I.: A New Edge Detection Technique for an Automatic Visual Inspection System using Genetic Algorithms. In: Proceeding of the International Conference on Robotics and Biomimetics, pp. 1882–1887 (2008)
4. Araiza, R.: The use of interval-related expert knowledge in processing 2-D and 3-D data, with an emphasis on applications to geosciences and biosciences. ProQuest Information and Learning Company, Michigan (2008)
5. Na, S., Xumin, L., Yong, G.: Research on k-means Clustering Algorithm: An Improved K-means Clustering Algorithm. In: Proceeding of the 2010 Third International Symposium on Intelligent Information Technology and Security Informatics, pp. 63–67 (2010)

Vision-Based Site Selection for Emergency Landing of UAVs

Mohammad Mahdi Dehshibi[1], Mohammad Saeed Fahimi[2], and Mohsen Mashhadi[2]

[1] Pattern Research Center (PRC), Iran
dehshibi@iranprc.org
[2] Department of Computer Engineering,
Islamic Azad University-Saveh Branch, Saveh, Iran
ms.fahimi69@gmail.com, mashhadimohsen@yahoo.com

Abstract. In this paper, a new method is proposed for finding the suitable forced landing sites for UAVs. This approach does not have any limitations of the previous few researches done in this area. For finding the suitable landing sites, we first segment the aerial images based on classification using both color and texture features. Classification is performed based on k-nearest neighbor algorithm by incorporation of Gabor filters in HSV color space. Then, a geometric test is carried out for finding appropriately sized and shaped landing sites. Output images highlight the selected safe landing locations. Experimental results show the effectiveness of the proposed method.

Keywords: Aerial image segmentation, Feature extraction, Surface type classification, k-nearest neighbor (KNN), Unmanned Arial Vehicles (UAVs).

1 Introduction

Nowadays, applications of machine vision can be found in every aspect of life [10-20]. Using the unmanned systems in the war is not new, but what will be new in the future is how such systems are used in the civilian space. Unmanned Aerial Vehicles (UAVs) are going to be used in the civil and commercial applications extensively, and are receiving noticeable attention by industry and research community. For performing majority of civilian tasks, Unmanned Aerial Vehicles are restricted to flying only in distinct spaces, which are commonly not above populated areas [1]. Current UAV technologies have not an acceptable level of safety, especially when an engine failure happens and so an emergency or forced landing in the civil areas is required.

Piloted aircraft in the same situation have a pilot on board who is able to do a complex decision making process for choosing a suitable landing site. If UAVs fly usually in civilian airspace, then an important unresolved problem is finding a safe landing location for a forced landing which must be dealt with [2].

The main purpose of this paper is designing a system for choosing autonomously the "safe" landing sites for a UAV by using machine vision and image processing

H. Unger et al. (eds.), *Recent Advances in Information and Communication Technology 2015*,
Advances in Intelligent Systems and Computing 361, DOI: 10.1007/978-3-319-19024-2_14

techniques. A "safe" landing site is a place which has three properties including (i) Does not cause any injury to a person; (ii) Does not damage the environment; (iii) Minimize damage to the UAV [2]. These three properties are listed in order of priority. It means minimizing damage to the UAV itself has the lowest priority. For example a UAV forced landing system should choose a lake instead of a busy road for landing. The criteria of landing site selection for UAVs are based on the criteria that a human pilot considers in a forced landing scenario. These include:

- Size
- Shape
- Slope
- Surface
- Surroundings
- S(c)ivilisation

These factors are known as the six S' and many of them are still important for selecting a landing site in a UAV forced landing situation. To date, there are very few publications on landing site selection for UAVs forced landing based on image processing and machine vision techniques [3, 4]. One of the best researches has tackled the specific problem of a UAV forced landing, has been done by Fitzgerald [2]. In [2] based on the "size", "shape", "surface" and "slope" criteria, for finding the safe landing sites these steps were proposed: (1) Segmenting the image, (2) Finding sites with suitable **size** and **shape**, (3) Classifying the **surface** type, (4) Estimating the slope.

In spite of testing different methods, Fitzgerald didn't get acceptable segmentation results. He proposed a simple method for extracting regions from aerial images which have similar texture and also are free of obstacles. But he used some assumptions in his method which are not valid under every condition. For instance, he used the edge detection measure in his algorithm for objects identification in the image based on the assumption that distinct edges situate between boundaries of objects. This assumption is valid only under the condition that the contrast between objects or regions in the image is enough and that the spatial resolution is high sufficient. Another drawback is that he used the intensity measure to eliminate some of the manmade objects in the image which usually are the white building or roof tops, as these areas are most likely to reflect the sun. But on more cloudy days or at soon or late times of the day, it is possible that some objects do not be detected. In this paper, by considering the "size", "shape" and "surface" criteria for landing site selection.

The remainder of the paper is organized as follows. Section 2 describes the segmentation step including the feature extraction and the k-nearest neighbor (KNN) classification. Then, section 3 explains the method of finding suitably sized and shaped landing areas. After that, section 4 represents the final results of landing site selection. Finally, conclusions are given in section 5.

2 Image Segmentation Based on Surface Type Classification

Image segmentation is one of the most difficult problems in image processing and computer vision which has a lot of useful applications. A segmentation algorithm is performed to semantically divide the image into some regions, or objects, to be used by the next processing steps for interpretation. The segmentation of different land cover regions in aerial images is known as a complicated problem. The natural scene typically has many regions including grass, water, tree, building, etc. It is really a challenging task to separate these regions correctly [5]. Some of the segmentation methods in the literature work well, but they have different parameters which need manual accurate tuning for every image to reach the optimal segmentation performance and this is not suitable for the purpose of automatic (unsupervised) segmentation. In this paper, we use a method for partitioning aerial images into different regions based on pixel level classification.

2.1 Feature Extraction

The features that we tested for aerial image segmentation include (1) Color features and (2) Texture features. We consider the features of RGB, HSV and LAB color spaces for the segmentation task. The analysis of texture is an important step for aerial image segmentation. However, many existing texture segmentation methods are orientation dependent and therefore cannot correctly classify textures after rotation [5]. In this research, we use the orientation independent textures. So the algorithms are independent from the direction that UAV approaches to the area.

Gray Level Co-occurrence Matrix

Gray Level Co-occurrence Matrix (GLCM) features are based on statistical properties of the GLCM. The GLCM is a matrix of relative frequencies which describe how often two gray level pixels appear in a specific distance and at a certain orientation in an image area.

The features of normalized GLCM at each pixel are computed in a $w \times w$ window with that pixel in the center. For deriving rotation invariant features, 4 orientations (0°, 45°, 90°, and 135°) are considered. For the distance measure, one of the values of 1, 2 and 3 and for number of quantization levels, one of the values of 8 and 16 are used after comparison. Selecting the suitable size of window is also important. The features we consider to extract from GLCM are: "contrast, homogeneity and energy" or just "contrast and homogeneity". Also we test different combination of Haar like features. Assume we extract contrast and homogeneity features in 4 orientations and 1 given distance in every window. As a result, the feature vector length will be 8.

Wavelet Transform

Manthalkar et al. in [6] introduced a method for extracting rotation and scale invariant texture features different type of wavelet filters. Cao et al in [5] used this method for extracting features of aerial images and got the good results. In this method, a multi-level wavelet decomposition of a small area of the image is computed. Then, by calculating the energy of each decomposed image, the rotation invariant features for a pixel are derived. If the decomposed image is $x(m, n)$, where $1 \leq m \leq M$ and $1 \leq n \leq N$, and i denotes the decomposition level, the energy features are:

$$en_i = \frac{1}{MN}\sum_{m=1}^{M}\left(\sum_{n=1}^{N}|x(m,n)|\right) \tag{1}$$

$$en_{istd} = \frac{1}{MN}\sqrt{\sum_{m=1}^{M}\left(\sum_{n=1}^{N}(|x(m,n)| - en_i)^2\right)} \tag{2}$$

To get rotation invariant features, the energy features in LH and HL channels in each level of decomposition are grouped. The feature vector is a vector of mean and standard deviation of all HL and LH channel in the proposed decomposition. This feature vector is given as Eq. 3 and 4.

$$EN_i = 0.5 \times [en_{iHL} + en_{iLH}] \tag{3}$$

$$EN_{istd} = 0.5 \times [en_{istdHL} + en_{istdLH}] \tag{4}$$

In this paper, 3-level wavelet decomposition is used and in every decomposition level, both mean and standard deviation features is derived. So the feature vector length will be 6 ($[EN_1, EN_{1std}, EN_2, EN_{2std}, EN_3, EN_{3std}]$) and we use it to characterize each class. Note that all the features are normalized from 0 to 255. To this end, every elements of feature vector for all pixels is mapped to 0-255 range individually and independently of other elements. Therefore, maximum and minimum of each element should be calculated through the image and then using ($\frac{value\ of\ every\ element - \min(value)}{\max(value) - \min(value)} \times 255$) formula and considering the fix part of answer, normalized value of desired element will be derived.

To choose the wavelet filter in [6], one Daubechies wavelet (Db4) and three Biorthogonal wavelet (Bior5.5, Bior4.4, and Bior3.3) have been tested and compared. The best result has been gain by Db4 and Bior4.4. In this paper different type of Orthogonal and Biorthogonal wavelet filters including Daubechies, Coiflets, Symlets, Discrete Meyer, Biorthogonal and Reverse Biorthogonal for choosing the best one are used. For defining the window size, different amounts are tested and the results are compared.

Local Binary Patterns

One of simple methods to prepare high accurate texture features of an image is Local Binary pattern (LBP). We compute the normalized histogram of LBP features at each pixel in a $w \times w$ window with that pixel in the center. The uniform rotation-invariant LBP ($LBP_{P,R}^{riu_2}$) is computed by selecting (P, R) parameters, once as $(8, 1)$ and at the second time as $(16, 2)$ which result the feature vector length of 10 and 18, respectively. Selecting the suitable size for the window is also important. We also test Local Binary Pattern (normalized) Histogram Fourier Features. In this case, considering $(8, 1)$ and $(16, 2)$ as (P, R) parameters, respectively result the feature vector length of 38 and 138.

Gabor Filters

Gabor features can be used to infer texture of an image region. There are a large number of publications which show that Gabor features can successfully discriminate between textures [2]. Chang et al. [7] claimed that Gabor filters are the best textural features out of the methods considered. A Gabor filter is a linear and local filter that is defined by a certain orientation and spatial frequency. It acts as a band-pass filter with optimal joint localization properties in both the spatial domain and the frequency domain [8]. Gabor filters are popular because the human vision system uses similar banks of directional band-pass filters with similar frequency and orientation representations [9].

We convolve the gray-scale image with two-dimensional 3×3 Gabor filters with various orientations (rotations) and frequencies (scales). Considering different orientations cause independency of image rotation. The output is a set of Gabor filtered images (one for each filter) that retain spatial information and can therefore be used for segmentation purposes.

2.2 Classification

There are different regions including water, grass, tree, road and building in aerial images; but in different places, the color of waters is different; also the shape of trees and their color are different and are affected by season changing; also buildings have different shapes, some of them have flat roofs and others sloped ones. Such issues affect surface type classification and make the problem more complicated.

There are different methods for image classification with a number of advantages and disadvantages. However, good results are usually obtained by careful selection of features and appropriate training practices [2]. In all of the classification methods, the features of test samples are extracted and compared with features of training data set. Then one of the output classes is assigned to each of the test samples. In pattern recognition, the kNN algorithm is a method for classifying objects based on closest training examples in the feature space. In this method, an object is assigned to the most common class amongst its *k* nearest neighbors. The neighbors are taken from a set of objects for which the correct classification is known. This can be thought of as the training set for the algorithm.

The classification process in kNN method like other classifiers has two steps of training phase and classification phase. The training phase of the algorithm consists only of storing the feature vectors and class labels of the training samples. In the classification phase, *k* is a user-defined constant, and an unlabeled vector is classified by assigning the label which is most frequent among the *k* training samples nearest to that query point.

Choice of the classification classes is an important component of classifier designing. The classes must include the different surface types that may be encountered by the classifier. The classifier would have to be able to distinguish between these classes correctly, so that the UAV is able to land on the appropriate target. These classes are (1) Grass, (2) Tree, (3) Water, (4) Road, and (5) Building.

Appropriate training data are also important to the operation of any classifier. We trained our kNN classifier on 150 sample images for each of the classes of grass, tree, water, road and building. For assessing the performance of the kNN classifier (trained on 150 sample images), 280 test images were manually selected and categorized. These images were then used as inputs to the kNN classifier. The best kNN classification results are obtained by considering one of the values of 1, 2, 3 or 4 for parameter k and L1 norm (Sum of absolute differences) or L2 norm (Euclidean distance, Sum of square differences) as the distance metric (the distance from the test point to each of its k nearest neighbors). We consider parameter k equal to 3 and use L2 norm as distance metric in our tests.

For any classification problem, a suitable set of features must be chosen. Good features are ones that allow discrimination between the output classification classes [2]. The best results of classification are obtained by combination of HSV color feature and Gabor texture feature. So for extracting the features of test samples or the features of training data set, we consider the given color image in HSV color space and separate it into three H, S, and V channels. Then we filter each channel of the image with Gabor filters in different orientations and frequencies. The mean of filtered images is calculated in every channel. The result is 3 images for 3 channels, that by calculating the average of each of them, we obtain 3 values finally. These 3 values will be used as feature vectors related to the considered image. This classifier performed extremely well on the test sample set, achieving a classification accuracy of 97%.

2.3 Image Segmentation Based on Classification

Our proposed method for image segmentation consists of identifying the objects present in an aerial image given a set of known patterns. In aerial images, the image contains several regions of different patterns and we label each pixel with one of the given classes based on specified features. Evidently, the labeling process subsumes image segmentation but besides segmenting the image to different regions, it assigns each region to one of the objects patterns.

We perform a per-pixel classification task and define the class for each pixel. For this purpose, we consider a window around each pixel and classify the area inside it using k-nearest neighbors algorithm (kNN). Then we assign the label of classification result to the central pixel of window. After computing all pixels in the image we obtain a segmented image which surface type of each segment is also defined.

3 Finding Sites with Suitable Size and Shape

In the previous step, aerial images have been segmented into a number of homogenous areas and simultaneously the surface type of each area has been defined by classification. In the final stage, a geometric test for finding appropriately sized and shaped landing sites should be performed. All areas that are too small or the incorrect geometric shape would be rejected, leaving only areas large enough for a UAV landing.

The algorithm in this phase involves the use of a mask, which is circular in shape and also is scalable. We have chosen circular shape for some reasons including possibility of approaching to the candidate landing site from different directions, wings of UAV and minimizing the processing time.

Size of the mask is determined by the pixel resolution calculation and is dependent on the category of UAV (small, medium or large) and the current height above ground. For example, a small UAV may have a landing site requirement of 15×60 meters, as opposed to a larger UAV requiring a landing site of 30×200 meters. Fig. 1 shows some example of these masks with different dimensions.

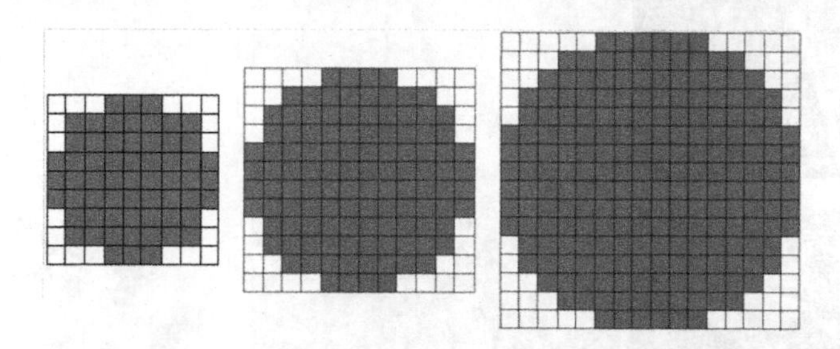

Fig. 1. Mask definitions of Landing Site matrix

For geometric test, we choose an arbitrary landing mask size (12x12 pixels). Note that the value of landing mask dimension depends on the height from which the image is taken. So when the height above ground decreases, bigger value for the mask dimension should be considered. However, as the differences between the heights of our test images are not very large, we use the same value for all of them.

The mask moves over the output image from the previous steps and is compared with the proper segments of the image. If the mask can be fitted in an area, that area would be a candidate for landing of UAV.

4 Experimental Results

Fig. 2 shows the final results of landing site selection. In this figure, firstly aerial images are segmented; then by a geometric test, areas with suitable size and shape are located. White color areas in the final results (output of the geometric test) are the areas which are not suitable for landing because of non-suitable size and shape. Other areas are candidate landing sites and are shown with a special color based on their surface type (water with blue color, grass with yellow color, tree with green color, building with red color and road with pink color).

Based on the assumption that landing on the natural objects has a lower chance of injury to people, natural areas are more suitable for landing than man-made areas. So the priority for landing is in the order of:

(1) Grass
(2) Water
(3) Tree
(4) Building
(5) Road

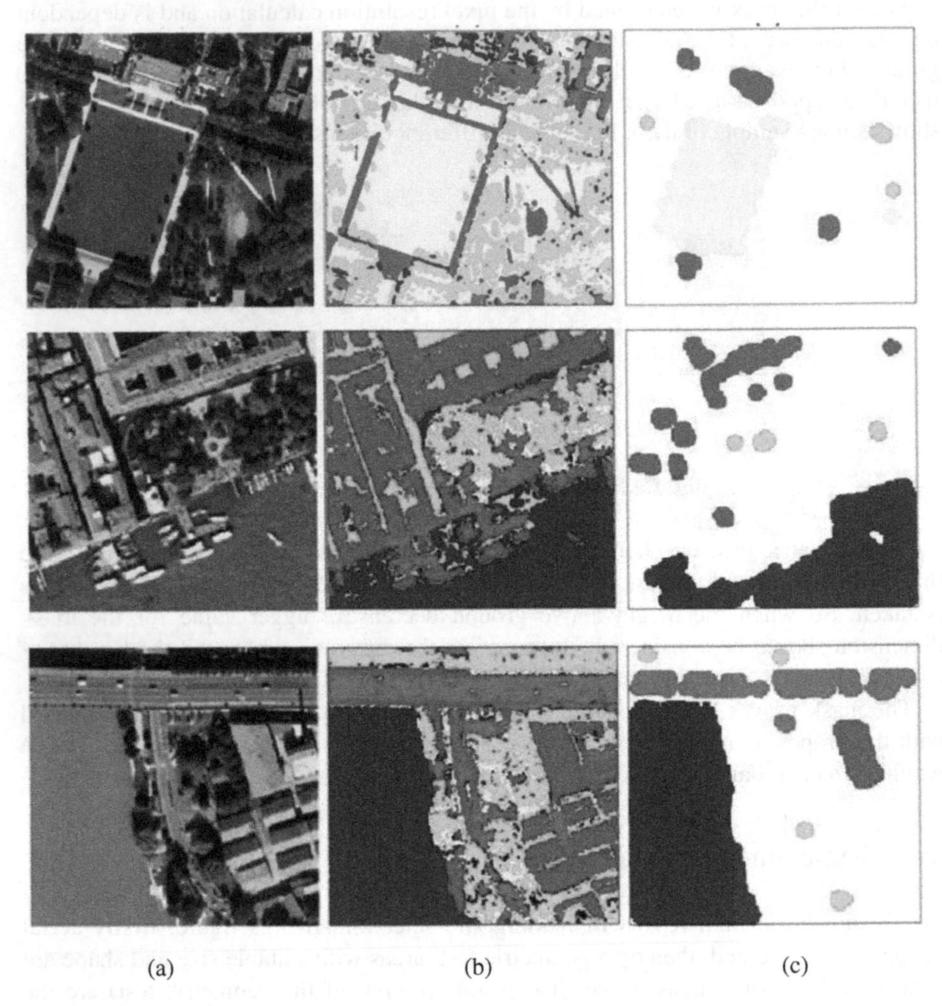

Fig. 2. Experimental results of landing site selection: (a) Original image, (b) Segmentation result, (c) Final result

5 Conclusion

In this paper, we proposed a process for landing site selection for UAVs forced landing which has not any limited assumption of the previous work. We segmented and determined the proper areas for forced landing of UAVs. A method based on k-nearest neighbor (kNN) classification was used for segmentation considering both color and texture features of the areas. We evaluated the classification performance over a variety of 'color' and 'texture' features and combination of them and found the most appropriate ones. Experimental results indicated that the best performance is obtained by using Gabor filters in HSV color space. The segmentation method produces consistent results and also it is easy to implement and able to segment aerial images automatically without any supervision i.e. without a priori knowledge of image content. After segmenting the image into a number of regions, in the final step, the algorithm locates areas of a given size and shape suitable for a UAV forced landing.

References

1. Mejias, L., Fitzgerald, D.L., Eng, P.C., Xi, L.: Forced landing technologies for unmanned aerial vehicles: towards safer operations. Aerial Vehicles, 415–442 (2009)
2. Fitzgerald, D.L.: Landing site selection for UAV forced landings using machine vision (2007)
3. Shen, Y.-F., Rahman, Z., Krusienski, D., Li, J.: A vision-based automatic safe landing-site detection system. IEEE Transactions on Aerospace and Electronic Systems 49, 294–311 (2013)
4. Lu, A., Ding, W., Li, H.: Multi-information Based Safe Area Step Selection Algorithm for UAV's Emergency Forced Landing. Journal of Software 8, 995–1002 (2013)
5. Cao, G., Mao, Z., Yang, X., Xia, D.: Optical aerial image partitioning using level sets based on modified Chan–Vese model. Pattern Recognition Letters 29, 457–464 (2008)
6. Manthalkar, R., Biswas, P.K., Chatterji, B.N.: Rotation and scale invariant texture features using discrete wavelet packet transform. Pattern Recognition Letters 24, 2455–2462 (2003)
7. Chang, K.I., Bowyer, K.W., Sivagurunath, M.: Evaluation of texture segmentation algorithms. In: Proceeding of the IEEE Computer Society Conference on Computer Vision and Pattern Recognition, pp. 294–299. IEEE (1999)
8. Daugman, J.G.: Uncertainty relation for resolution in space, spatial frequency, and orientation optimized by two-dimensional visual cortical filters. JOSA A 2, 1160–1169 (1985)
9. Jain, A.K., Farrokhnia, F.: Unsupervised texture segmentation using Gabor filters. In: Proceeding of the IEEE International Conference on Systems, Man and Cybernetics, pp. 14–19 (1990)
10. Aghaahmadi, M., Dehshibi, M.M., Bastanfard, A., Fazlali, M.: Clustering Persian viseme using phoneme subspace for developing visual speech application. Multimedia Tools and Applications 65, 521–541 (2013)
11. Bastanfard, A., Nik, M.A., Dehshibi, M.M.: Iranian face database with age, pose and expression. In: International Conference on Machine Vision, ICMV 2007, pp. 50–55. IEEE (2007)
12. Dehshibi, M.M., Alavi, S.M.: Generic Visual Recognition on Non-Uniform Distributions Based on AdaBoost Codebooks. In: Proceeding of the International Conference on Image Processing, Computer Vision, and Pattern Recognition, pp. 1046–1051 (2011)
13. Dehshibi, M.M., Allahverdi, R.: Persian Vehicle License Plate Recognition Using Multiclass Adaboost. International Journal of Computer and Electrical Engineering 4, 355–358 (2012)
14. Dehshibi, M.M., Bastanfard, A.: A new algorithm for age recognition from facial images. Signal Processing 90, 2431–2444 (2010)
15. Dehshibi, M.M., Bastanfard, A., Kelishami, A.A.: LPT: Eye Features Localizer in an N-Dimensional Image Space. In: IPCV, pp. 347–352 (2010)

16. Dehshibi, M.M., Fazlali, M., Shanbehzadeh, J.: Linear principal transformation: toward locating features in N-dimensional image space. Multimedia Tools and Applications 72, 2249–2273 (2014)
17. Dehshibi, M.M., Shanbehzadeh, J.: Persian Viseme Classification Using Interlaced Derivative Patterns and Support Vector Machine. Journal of Information Assurance and Security 9, 148–156 (2014)
18. Dehshibi, M.M., Shanbehzadeh, J., Alavi, M.: Facial family similarity recognition using Local Gabor Binary Pattern Histogram Sequence. In: Proceeding of the 12th International Conference on Hybrid Intelligent Systems, pp. 219–224 (2012)
19. Dehshibi, M.M., Vafanezhad, A., Shanbehzadeh, J.: Kernel-Based Object Tracking Using Particle Filter with Incremental Bhattacharyya Similarity. In: Proceeding of the 13th International Conference on Hybrid Intelligent Systems (HIS), pp. 50–54 (2013)
20. Yazdani, D., Arabshahi, A., Sepas-Moghaddam, A., Dehshibi, M.M.: A multilevel thresholding method for image segmentation using a novel hybrid intelligent approach. In: Proceeding of the 12th International Conference on Hybrid Intelligent Systems (HIS), pp. 137–142 (2012)

Watermarking Capacity Control for Dynamic Payload Embedding

Hussain Nyeem[1], Wageeh Boles[2], and Colin Boyd[2,3]

[1] Khulna University of Engineering and Technology (KUET), Khulna, Bangladesh
[2] Queensland University of Technology (QUT), Brisbane, Australia
[3] Norwegian University of Science and Technology (NTNU), Trondheim, Norway
h.nyeem@kuet.ac.bd, w.boles@qut.edu.au,
colin.boyd@item.ntnu.no

Abstract. Despite significant improvements in capacity-distortion performance, a computationally efficient capacity control is still lacking in the recent watermarking schemes. In this paper, we propose an efficient capacity control framework to substantiate the notion of watermarking capacity control to be the process of maintaining "acceptable" distortion and running time, while attaining the required capacity. The necessary analysis and experimental results on the capacity control are reported to address practical aspects of the watermarking capacity problem, in dynamic (size) payload embedding.

Keywords: Capacity control, dynamic payload, embedding capacity, fragile watermarking.

1 Introduction

Digital image watermarking has drawn much attention for improving the embedding capacity-distortion performance [1–5, 9–18, 25, 27–29]. Particularly, recent fragile watermarking schemes aim at achieving high capacity with the lowest possible distortion. (Fragile watermarks, by definition, become invalid for any possible modification in a watermarked image.) In some fragile watermarking applications (*e.g.*, annotation) the payload size (*i.e.*, watermark plus any side information) significantly varies [21]. We call such a payload *dynamic*, which requires varying capacity with a relatively high upper bound. Achieving this requirement, however, is challenging (specially under the perceptual constraints of images) and requires capacity control.

Watermarking *capacity control* is the process to achieve the required capacity while maintaining a low level of distortion. Ideally the process should minimise the distortion. However, when the payload is dynamic this may lead to an unacceptable running time and so in this paper we study the notion of capacity control to be the process of maintaining "acceptable" *distortion* and *running (or computational) time*. (The term "acceptable" means to be minimum, but its level may vary with the applications.) Here, distortion is the degree of perceptual degradation incurred by the embedding function, and running time [6] is measured by the number of machine-independent operations executed.

In watermarking research, the best capacity-distortion performance are shown by the reversible schemes [1–5, 9–15, 18, 25, 27–29]. These fragile schemes introduce an

H. Unger et al. (eds.), *Recent Advances in Information and Communication Technology 2015*,
Advances in Intelligent Systems and Computing 361, DOI: 10.1007/978-3-319-19024-2_15

invertible distortion in a watermarked image. Tian [28] pioneered the DE scheme addressing the low capacity and/or noticeable visual artefact problems of earlier feature compression based reversible schemes [2, 8, 9]. Tian's scheme is later generalised by Alattar [1], and then improved using the sorting of pixel-pairs by Kamstra and Heijmans [12] for higher embedding capacity. Kim *et al.* [13] improved Tian's scheme using the simplified location map to reduce the overhead data.

Additionally, Ni *et al.* [18] introduced the histogram shifting (HS) scheme, which does not require any location map (but a pair of peak/zero points in the histogram). Ni *et al.*'s scheme is later improved by using the difference-histogram [14], rhombus predictor and sorting [25], adaptive and multilevel embedding [15, 24] for better perceptual quality and higher embedding capacity. Reversible contrast matching (RCM), another invertible transform, -based scheme is proposed by Coltuc and Chassery [5] and extended by Chen *et al.* [3]. Thodi and Rodríguez [27] combined the HS and DE techniques and introduced the prediction error expansion (PEE) scheme, which was later improved by Hu *et al.* [11] for better capacity-distortion performance.

However, the above (and many others) DE-, HS-, RCM- and PEE-based reversible schemes usually have an inefficient capacity control and thus may not be suitable for dynamic payload embedding. They often require a *recursive* (or *multi-level*) embedding in support of capacity control. A recursive embedding re-embeds any remaining part of a payload recursively in a watermarked image until the required capacity is achieved. The repeated alterations of pixels may not always incur more distortion if the same bit-plane(s) is used for re-embedding, but they are more likely to significantly grow the running time of the schemes. To demonstrate the need for an efficient capacity control, and to thus address a novel aspect of the embedding capacity problem, form the motivation of the research reported in this paper.

As the main contribution, this paper presents an efficient capacity-control framework. It is more than challenging to develop a unified capacity control framework for watermarking due to the variety of its techniques and applications. A case of fragile watermarking schemes is therefore considered for dynamic payload embedding, where we often face the dilemma of limiting the size of payloads or sacrificing the performance of an embedding scheme. In this paper, we determine the impact of the inefficient (*e.g.*, recursive) capacity control of watermarking schemes on their overall performance. Thereby, we validate the proposed framework with the asymptotic analysis and necessary experiments of watermarking schemes having different capacity control.

The rest of the paper is organised as follows. A new capacity control framework is proposed in Sec. 2. In Sec. 3, asymptotic analysis of capacity control of watermarking schemes are given in light of the proposed framework. Experimental results are discussed in Sec. 4 followed by the conclusions in Sec. 5.

2 A New Capacity Control Framework

In this section, we present a capacity control framework for efficient embedding of dynamic payload. (We adopt necessary notations from [20].) As discussed in Sec. 1 and shown in Fig. 1 (*existing scenario*), current capacity control ideally aims at minimizing the distortion only and thus lacks consideration of the running time. Here, the Hu *et al.* scheme [11] (that we will analyse in Sec. 3.2 to validate our framework) is a prominent example that fits the existing scenario. Therefore, our proposed framework (incorporating the *extended scenario* in Fig. 1) aims to ensure the attainment of the required capacity with both the least possible distortion and running time simultaneously. For an input image and payload, the framework seeks a suitable capacity parameter setting with possible user intervention to update the predefined thresholds (or to reconsider the inputs/embedding scheme, in a worst case scenario). The general steps are discussed below.

Let an embedding function, $E(\cdot)$, embeds a watermark, W and side information, S_{info}, in an input image, I such that $\bar{I} \leftarrow E(I, payload)$. Here, $\bar{I}$ is the watermarked image and the *payload* is computed by a concatenation function, $Concat(\cdot)$, *i.e.*, $payload \leftarrow Concat(W, S_{info})$. To find the least possible distortion, *Dist* and running time, E_t, the set of capacity parameters *par* and the thresholds (T_1, T_2) for (*Dist*, E_t) are initialized with their minimum possible values. With that setting, a capacity estimation function, $Est(\cdot)$ computes the total (available) capacity, C_t. The required capacity, C_p is determined using $Size(\cdot)$ that returns the bit-length of its input such that $C_p \leftarrow Size(payload)$. Until the C_p is achieved, *par*, T_1 and T_2 are updated as shown in Fig.1.

The influence of increasing C_p on *Dist* and E_t. can be controlled by the capacity parameter, *par*, and the thresholds T_1 and T_2. An efficient updating of *par* is here crucial for the capacity control to minimize E_t. For example, based on the difference between C_t and C_p, an adaptive update of *par* may significantly minimize the time needed to reach the required capacity level. Bearing this in mind, we define the capacity control efficacy in Def.1 below. Here, we consider only C_p, since C_t accounts for C_p (possibly with an increasing *Dist* and E_t for $C_t \geq C_p$).

Definition 1. (capacity control efficacy). *An embedding function, $E(\cdot)$ is said to have an efficient* capacity control, *if it ensures the attainment of the required capacity, C_p with minimum possible distortion, Dist and computation time, E_t, where Dist and E_t grow with the minimum possible amount/step as C_p grows.*

In order to demonstrate the viability of Def. 1, and thus to determine the capacity control efficacy of $E(\cdot)$ for dynamic payload embedding, we pose the following questions: (*i*) what are the *Dist* and E_t values of $E(\cdot)$ for the lower bound of the dynamic payload? and (*ii*) at what rate should *Dist* and E_t of $E(\cdot)$ be changed for the increasing C_p? Analysing watermarking schemes in light of these questions would lead to the conclusion that the lower the (quantitative) values of the said parameters, the higher the efficacy of the schemes.

3 Capacity Control Analysis

We analyse and experiment with the Nyeem, Boles, and Boyd (or NBB) scheme [22,23] and Hu, Lee, and Li (or HLL) scheme [11] below as to determine the performance of their capacity control for dynamic payload embedding. The choice of the HLL scheme is made as it is a prominent watermarking scheme having capacity control that closely represents the existing capacity control scenario, as mentioned in Sec. 2. We presented the NBB scheme in [22, 23] that follows the capacity control framework proposed in

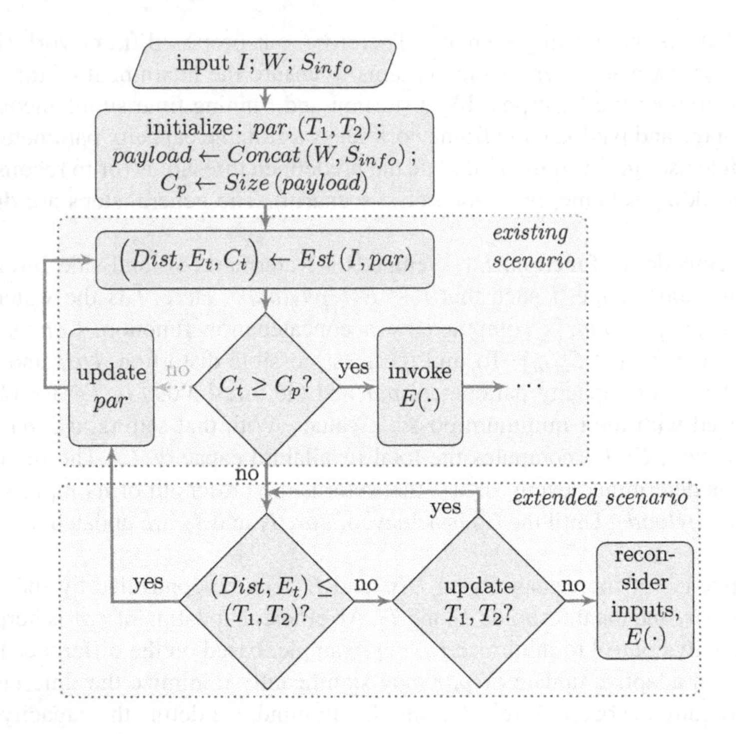

Fig. 1. Flow-chart of the proposed capacity control framework for dynamic payload embedding [19]

this paper. Since, both schemes have capacity control, they can be used for dynamic payload embedding.

3.1 The NBB Scheme Analysis

NBB scheme [22, 23] is proposed to provide continuous security protection and to minimize the legal-ethical issues of medical images. That scheme embeds payloads in the LSB (least significant bit) planes of the border pixels of input images. A greater capacity control is targeted in terms of N_{BW} and N_{LSB} with their thresholds T_{BW} and T_{LSB} respectively, as shown in Fig. 2. Here, N_{BW} is the number of pixels in a given border width, N_{LSB} is the number of LSB-planes, and c_t is calculated using $C_{total} = 2N_{BW} \times (r + c - 2N_{BW}) \times N_{LSB}$. We note that Fig. 2 does not explicitly show any consideration for E_t as shown in the proposed framework in Fig. 1. Because the NBB scheme does not consider recursive embedding, its running time always remains in $O(n)$. The capacity control running time of NBB embedding function is $n \times (c_5 + c_6 + c_7 + c_8)$, where c_5 to c_8 are time constants for the steps shown in Fig. 2.

3.2 The HLL Scheme Analysis

The HLL scheme [11], on the other hands, expands the (median) prediction errors (*i.e.*, $p_e = x - \hat{x}$) using classical DE rule (*i.e.*, $p'_e = 2p_e + b$) and its variant (*i.e.*, $p'_e = 2p_e - b$). (Where, x and $\hat{x}$ are original and predicted versions of the pixels, p_e and p'_e are original and expanded versions of the errors, respectively and b is the watermark bit.) Thereby,

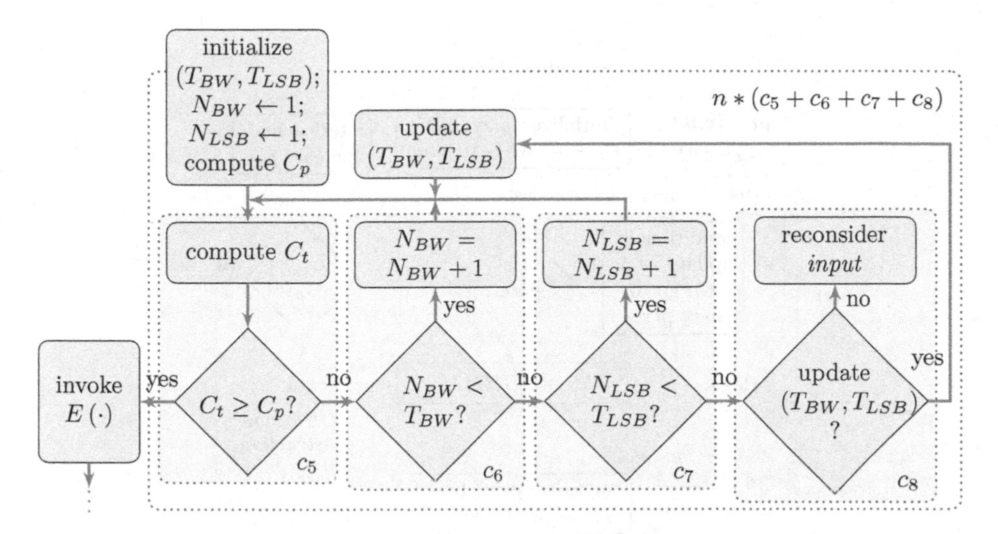

Fig. 2. Capacity control flow-chart of NBB scheme [19]

an interleaving approach (*e.g.*, adding a bin first rightward, then leftward, and so on or *vice-versa* for an embeddable region) is introduced for capacity control, considering a first round embedding. This consideration, however, is also suggestive of possible successive embedding rounds. To demonstrate the consequences of such embedding, we perform an asymptotic analysis of the HLL scheme. to determine its rate of growth of running time.

HLL capacity control has two main parts: *OUF (over-/under-flow) map construction* and *scanning* as shown Fig. 3. For the given I and C_p, the OUF location map, M is constructed recursively for each *pixel*—a vector for the current pixel and the $flag$—a flag-bit for HS direction (*e.g.*, 0 for the left). The JBIG (Joint Bi-level Image Experts Group) compression is then used: $\hat{M} \leftarrow JBIG(M)$, where $\hat{M}$ is JBIG compressed version of M. (For more details, see [11].)

The OUF map construction and scanning have the running time of $c_4 n \times (c_1 n + c_2 \log n)$ and $c_3 n$, respectively, where n is input size—the total number of pixels, and c_1 to c_4 are time constants—the fixed time period taken by the set of operations. So, the capacity control running time becomes $c_1 c_4 n^2 + c_2 c_4 n \log n + c_3 n$ leading to an asymptotic upper bound $O(n^2)$ (considering the JBIG running time is $O(\log n)$ as being arithmetic coding based [26]). With the running time of $E(\cdot)$ in $O(n)$, the HLL scheme's embedding performance depends on its capacity control, leading to an overall upper bound of $O(n^2)$. So, for a k-round embedding, the running time will be in $O(n^{2k})$, where even with a small value, k will severely impact on the HLL scheme's overall performance.

4 Experimental Results and Discussion

We examined the performance of the schemes in question [11,22] with varying payload size. We performed several experiments using 150 test images (from [7]). An example of a few test-set images are shown in Fig. 4. As shown in Fig. 5 (1st row), unlike NBB scheme, where the capacity control running time remains steady and much lower,

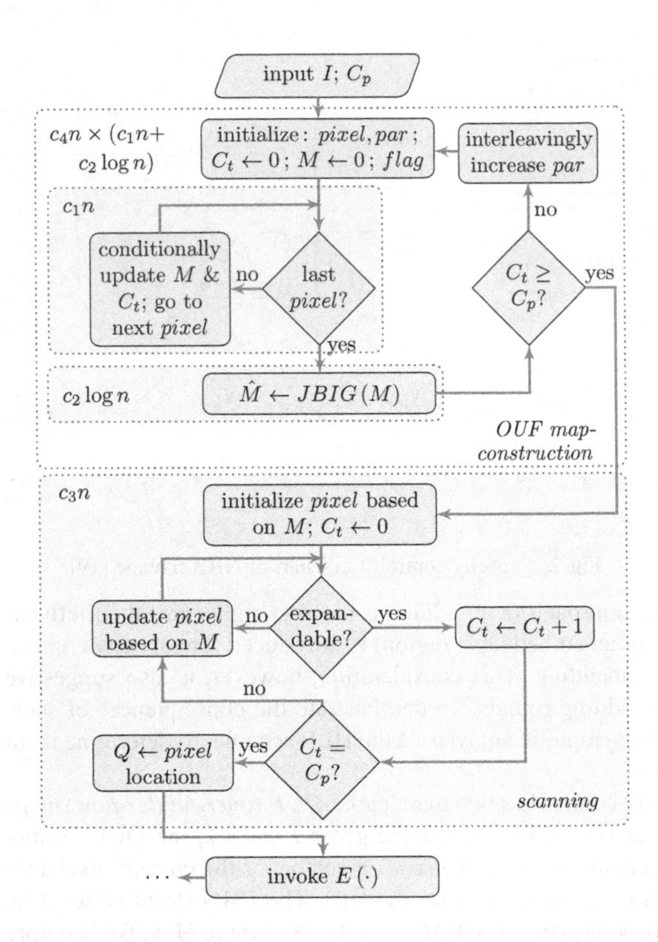

Fig. 3. Capacity control flow-chart of the HLL scheme. (Dotted-blocks indicate their approximate worst-case running time [19].)

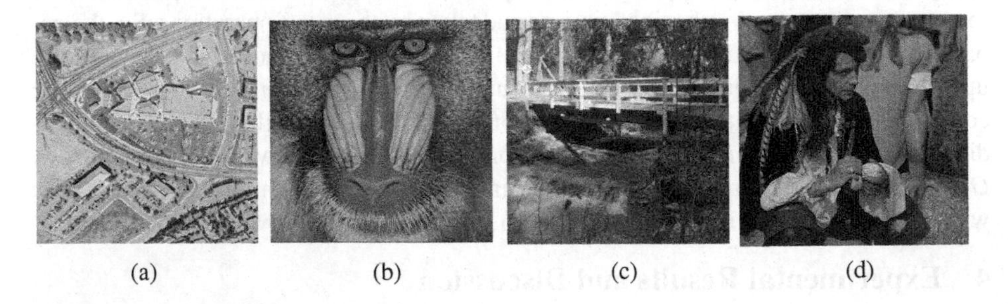

Fig. 4. Sample test images of size $512 \times 512 \times 8$: (a) Aerial, (b) Mandrill, (c) Stream and bridge, and (d) Man. (Available here [7]).

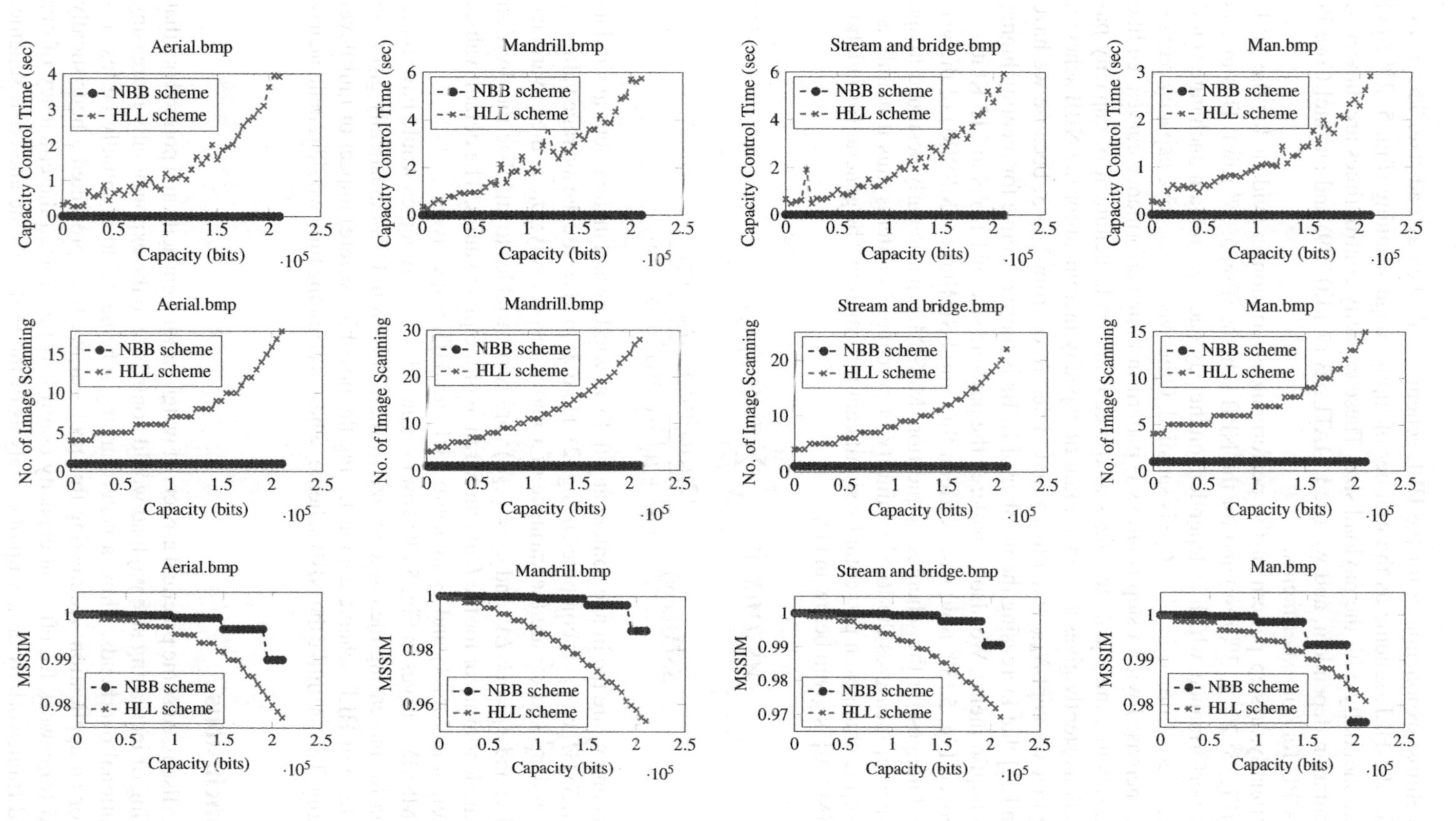

Fig. 5. Evaluation and comparison of capacity control efficacy for *Aerial*, *Mandrill*, *Stream and bridge*, and *Man*. From top, capacity control time (1st row), number of scanning the input image (2nd row), and MSSIM (3rd row).

it grows almost exponentially for the HLL scheme, with the payload size. This is expected for the HLL scheme as the number of input image scanning (Fig. 5: 2nd row) grows exponentially with the payload size. These capacity control times are, however, implementation dependent, and we used MATLAB (7.14.0.739) and an Intel Core i5 3.2GHz CPU for our experiments.

Additionally, a step pattern in the performance variation is evident for the HLL scheme (Fig. 5: 2nd & 3rd row) and for the NBB scheme (Fig. 5: 3rd row). This means that their performance, which is dependent on the payload size, remain unchanged until the capacity condition (*i.e.*, $C_t \geq C_p$) is satisfied. Otherwise, respective capacity parameters are increased with a step (up/down) pattern in their performance curves. Unlike the HLL scheme, the equal step-sizes also mean that each increment of capacity parameters consistently gives a fixed amount of capacity increment for the NBB scheme. We note that we kept N_{BW} fixed to 25 and varied N_{LSB} from 1 to 5, because we have shown in [22] that increasing the number of LSBs is more effective for meeting higher capacity requirements. We also considered the payload size of 1 Kbits to 215 Kbits.

Moreover, Fig. 5 (3rd row) shows Mean Structural SIMilarity (MSSIM) of the watermarked images for the schemes in question. MSSIM, a particularly designed image quality metric, measures the local similarity of perceptual contents thus it mainly accounts for the changes in perceptually significant information.The general formulation of MSSIM [30] is given below in (1).

$$MSSIM(X,Y) = \frac{1}{M}\sum_{j=1}^{M} SSIM(x_j,y_j) \tag{1}$$

$$SSIM(x,y) = \frac{(2\mu_x\mu_y + c_1)(2\sigma_{xy} + c_2)}{(\mu_x^2 + \mu_y^2 + c_1)(\sigma_x^2 + \sigma_y^2 + c_2)} \tag{2}$$

where, x_j and y_j are the image content at j-th local window and their structural similarity index, $SSIM(x_j,y_j)$ is computed using (2). Here, μ_x and μ_y are the average values of x and y, and σ_x^2 and σ_y^2 are the variance of x and y, respectively; σ_{xy} is the covariance of x and y; and $c_1 = (k_1L)^2$ and $c_2 = (k_2L)^2$ are two variables to stabilize the division with weak denominator for the L dynamic range of the pixel values. The default values of the weight factors, k_1 and k_2 are set to 0.01 and 0.03, respectively.

This MSSIM curves in Fig. 5 (3rd row) suggest that both schemes would have more distortion for increasing the payload size further. This would also drastically grow the running time of HLL scheme demonstrating the possible (severe) impact of multilevel embedding. This is unlike the NBB scheme, where its running time would remain nearly steady.

5 Conclusions

We have discussed some practical aspects of watermarking capacity and pointed out that embedding of increasing size payload would contribute to the exponentially increasing computational overheads. Thus, a watermarking scheme could eventually be less efficient for an application. Addressing this problem, we have presented a heuristically designed framework for efficient capacity control. We examined the efficiency of our proposed framework by an asymptotic analysis of the HLL scheme and NBB scheme, where these schemes closely represent the existing and proposed frameworks of capacity control, respectively. We also verified the efficiency of the proposed framework by analysing the performance variations (resulting from the dynamic payload embedding) of HLL and NBB schemes.

We argue that the existing capacity control is more interpreted as a trade-off between capacity and distortion requirements. However, failure to also consider the running time may render a scheme less practicable for an application, especially if dynamic payload is a requirement. Therefore, addressing the questions posed in Sec. 2, we have shown that the capacity control of NBB scheme outperforms that of the HLL scheme. The asymptotic analysis and experimental results demonstrate further the consequences of an inefficient capacity control and thus validate the efficiency of the proposed capacity control framework. We note that the given results are implementation dependent, and possible optimized implementation could give better results than what we found. However, the trends shown in the graphs (in Fig. 5) will apply for any implementation.

Moreover, since the HS-, DE-, RCM- and PEE-based schemes usually have a similar capacity control principle like the HLL scheme, they should have more or less similar effect on the running time, for increasing payload size as well as for recursive embedding. Although different watermarking principles and application requirements make it a more challenging task, the proposed capacity control framework may be reduced to a generalized form in future.

Acknowledgement. The authors would like to Thank to Lance De Vine (QUT) for JBIG help.

This paper presents a part of the work of the first author's PhD project and thus contains some of the material available in that thesis [19].

References

1. Alattar, A.M.: Reversible watermark using difference expansion of quads. In: Proceedings of IEEE International Conference on Acoustics, Speech, and Signal Processing, vol. 3, pp. 377–380. IEEE (2004)
2. Celik, M.U., Sharma, G., Tekalp, A.M., Saber, E.: Lossless generalized-lsb data embedding. IEEE Transactions on Image Processing 14, 253–266 (2005)
3. Chen, X., Li, X., Yang, B., Tang, Y.: Reversible image watermarking based on a generalized integer transform. In: Proceedings of IEEE International Conference on Acoustics, Speech, and Signal Processing, pp. 2382–2385. IEEE (2010)
4. Coatrieux, G., Pan, W., Cuppens, B.N., Cuppens, F., Roux, C.: Reversible watermarking based on invariant image classification and dynamic histogram shifting. IEEE Transactions on Information Forensics and Security 8, 111–120 (2013)
5. Coltuc, D., Chassery, J.M.: Very fast watermarking by reversible contrast mapping. IEEE Signal Processing Letters 14(4), 255–258 (2007)
6. Cormen, T.H., Leiserson, C.E., Rivest, R.L., Stein, C.: Introduction to Algorithms, ch. 2, 3rd edn. MIT Press (2011)
7. The USC-SIPI image database, `http://sipi.usc.edu/database`
8. De Vleeschouwer, C., Delaigle, J.F., Macq, B.: Circular interpretation of bijective transformations in lossless watermarking for media asset management. IEEE Transactions on Multimedia 5, 97–105 (2003)
9. Fridrich, J., Goljan, M., Du, R.: Lossless data embedding-new paradigm in digital watermarking. EURASIP Journal on Applied Signal Processing 2002(1), 185–196 (2002)

10. Guo, X., Zhuang, T.G.: A region-based lossless watermarking scheme for enhancing security of medical data. Journal of Digital Imaging 22, 53–64 (2009)
11. Hu, Y., Lee, H.K., Li, J.: DE-based reversible data hiding with improved overflow location map. IEEE Transactions on Circuits and Systems for Video Technology 19, 250–260 (2009)
12. Kamstra, L., Heijmans, H.J.: Reversible data embedding into images using wavelet techniques and sorting. IEEE Transactions on Image Processing 14, 2082–2090 (2005)
13. Kim, H.J., Sachnev, V., Shi, Y.Q., Nam, J., Choo, H.G.: A novel difference expansion transform for reversible data embedding. IEEE Transactions on Information Forensics and Security 3, 456–465 (2008)
14. Kim, K.S., Lee, M.J., Lee, H.Y., Lee, H.K.: Reversible data hiding exploiting spatial correlation between sub-sampled images. Pattern Recognition 42, 3083–3096 (2009)
15. Lee, S., Yoo, C.D., Kalker, T.: Reversible image watermarking based on integer-to-integer wavelet transform. IEEE Transactions on Information Forensics and Security 2, 321–330 (2007)
16. Lin, C.C., Tai, W.L., Chang, C.C.: Multilevel reversible data hiding based on histogram modification of difference images. Pattern Recognition 41, 3582–3591 (2008)
17. Ni, Z., Shi, Y.Q., Ansari, N., Su, W., Sun, Q., Lin, X.: Robust lossless image data hiding designed for semi-fragile image authentication. IEEE Transactions on Circuits and Systems for Video Technology 18, 497–509 (2008)
18. Ni, Z., Shi, Y.Q., Ansari, N., Su, W.: Reversible data hiding. IEEE Transactions on Circuits and Systems for Video Technology 16(3), 354–362 (2006)
19. Nyeem, H.: A digital watermarking framework with application to medical image security. Ph.D. thesis, QUT, School of Electrical Eng. and Computer Science, Australia (July 2014)
20. Nyeem, H., Boles, W., Boyd, C.: Developing a digital image watermarking model. In: Proceedings of 13th International Conference on Digital Image Computing: Techniques and Applications, pp. 468–473. IEEE, Piscataway (2011)
21. Nyeem, H., Boles, W., Boyd, C.: A review of medical image watermarking requirements for teleradiology. Journal Digital Imaging 26, 326–343 (2013)
22. Nyeem, H., Boles, W., Boyd, C.: Utilizing least significant bit-planes of roni pixels for medical image watermarking. In: Proceedings of 15th International Conference on Digital Image Computing: Techniques and Applications, pp. 1–8. IEEE, Piscataway (2013)
23. Nyeem, H., Boles, W., Boyd, C.: Content-independent embedding scheme for multi-modal medical image watermarking. BioMedical Engineering Online 14(7) (2015)
24. Peng, F., Li, X., Yang, B.: Adaptive reversible data hiding scheme based on integer transform. Signal Processing 92, 54–62 (2012)
25. Sachnev, V., Kim, H.J., Nam, J., Suresh, S., Shi, Y.Q.: Reversible watermarking algorithm using sorting and prediction. IEEE Transactions on Circuits and Systems for Video Technology 19, 989–999 (2009)
26. Simpson, M., Biswas, S., Barua, R.: Analysis of compression algorithms for program data. University of Maryland (2003)
27. Thodi, D.M., Rodríguez, J.: Expansion embedding techniques for reversible watermarking. IEEE Transactions on Image Processing 16, 721–730 (2007)
28. Tian, J.: Reversible data embedding using a difference expansion. IEEE Transactions on Circuits and Systems for Video Technology 13(8), 890–896 (2003)
29. Tsai, P., Hu, Y.C., Yeh, H.L.: Reversible image hiding scheme using predictive coding and histogram shifting. Signal Processing 89, 1129–1143 (2009)
30. Wang, Z., Bovik, A.C., Sheikh, H.R., Simoncelli, E.P.: Image quality assessment: From error visibility to structural similarity. IEEE Transactions on Image Processing 13(4), 600–612 (2004)

Two-Stage Method for Information-Theoretically Secure Data Encryption

Wolfgang A. Halang[1], Li Ping[1], Maytiyanin Komkhao[2], and Sunantha Sodsee[3]

[1] Chair of Computer Engineering, Fernuniversität in Hagen, Germany
wolfgang.halang@fernuni-hagen.de
[2] Faculty of Science and Technology,
Rajamangala University of Technology Phra Nakhon, Bangkok, Thailand
maytiyanin.k@rmutp.ac.th
[3] Faculty of Information Technology,
King Mongkut's University of Technology North Bangkok, Thailand
sunanthas@kmutnb.ac.th

Abstract. It is known from information theory that eavesdropping and gaining unauthorised access to computers can be prevented by cryptography employing one-time keys. Regarding this fact, a practically feasible data encryption and user authentication scheme is presented, whose two stages are already information-theoretically secure each on its own. The first stage generates packet-long one-time keys by a chaos-theoretical method, and the second one adds redundancy in form of allowing to encrypt any plaintext by a randomly selected element out of a large set of possible ciphertexts. Obliterating the symbol boundaries in transmission units, this cryptosystem removes a toehold for cryptanalysis not addressed before.

Keywords: Unbreakable encryption, eavesdropping, secure communication, authentication, chaos theory.

1 Introduction

In information and communication technology, increasingly datasets of any size are exchanged between computers in form of streams via data networks. To guarantee the confidentiality of such messages' contents, a plentitude of methods to encrypt the data streams was developed [7]. Currently used encryption methods usually employ the same keys during longer periods of time, lending themselves to cryptanalytic attacks. It was shown, for instance, that the rather widespread asymmetrical RSA-cipher with keys 768 bits long has at least theoretically been broken. The symmetrical cryptosystem DES is already regarded as unsafe, too. Other ciphers such as 3DES or AES are still being considered safe, but only because the presently available computing power is insufficient to carry out simple brute-force attacks. In some countries law requires to deposit the keys used with certain agencies. Thus, these countries' secret services do not need any cryptanalysis whatsoever to spy out encrypted data.

H. Unger et al. (eds.), *Recent Advances in Information and Communication Technology 2015*,
Advances in Intelligent Systems and Computing 361, DOI: 10.1007/978-3-319-19024-2_16

In consequence, only perfectly secure one-time encryption appears to be feasible in the long run. Perfect security is achieved, if encryption of a plaintext yields with equal probability any possible ciphertext, and if it is absolutely impossible to conclude from the ciphertext to the plaintext in a systematic way. According to the theorem of Shannon [8] fundamental for information theory, a cryptosystem is regarded as perfectly safe only then, if the number of possible keys is at least as large as the number of possible messages. Hence, also the number of keys is at least as large as the one of possible ciphertexts which, in turn, must be at least as large as the number of possible plaintexts. Based on these considerations, in the sequel a novel system for data encryption and user authentication is presented, which works with one-time keys as long as packets to be stored or transmitted.

The method employed to generate one-time keys is a chaos-theoretical one. Mathematical chaos [9] is one of the most well-known and potentially useful classes of non-linear dynamics. The dynamical behaviour of non-linear systems has gained strong interest in recent decades. As a result, non-linearity has become a major topic in mathematics and engineering sciences. Although chaotic systems are governed by simple and low-order deterministic rules, their dynamics are random-like and complex. These characteristics let chaotic systems become potential candidates for sources of pseudo-randomness such as building blocks in cryptographical applications. Summaries of various corresponding research activities and designs can be found in [1,5]. Here, more complicated chaotic systems, viz. spatiotemporal ones, are utilised as sources of pseudo-randomness due to their good performance. In particular, coupled map lattices are adopted as such spatiotemporal systems, which are more complex than other chaotic systems and can serve as multiple sources of pseudo-randomness [6].

To yield perfectly safe encryption, one-time keys need to be truly random. Since deterministic methods as the one described in Section 2 are only able to produce sequences of pseudo-random bits, however, in Section 3 we propose to make up for this deficiency by adding a second encryption stage. This stage features ample redundancy by allowing freely selectable plaintext segments to be encrypted by elements randomly chosen out of large sets of possible ciphertexts, and it blurs the boundaries between data symbols encrypted together. Thus, it is made impossible to conclude from boundaries between data items in ciphertexts to the boundaries of data items in the resulting plaintexts, removing a toehold for cryptanalysis which has been neglected so far. This paper ends with stating the two-stage encryption and decryption algorithms and some considerations pointing to a wide range of options for implementation and for modifications during communication processes.

2 Spatiotemporal Chaos Yielding Pseudo-random Bits

In order to produce one-time keys for encryption purposes, a novel Multiple Pseudo-Random Bits Generator (MPRBG) based on spatiotemporal chaos was proposed and comprehensively studied [6]. By their very nature, spatiotemporal chaotic systems are dynamical systems, which are often described by partial differential equations, coupled ordinary differential equations or Coupled Map Lattices (CMLs). These dynamical systems exhibit chaotic properties in both time and space. Among them, the ones described by CMLs are most widely used, due to their digital nature and favourable combination of computational complexity and representation of the original systems.

Spatiotemporal chaos is created in CMLs by local non-linear dynamics and spatial diffusion. By adopting various non-linear mappings for local chaos, and various discretised diffusion processes, which are also regarded as coupling, different forms of CMLs can be constructed. Commonly used are the logistic map as local map and nearest-neighbour coupling.

A general nearest-neighbour CML can be described as

$$x_{i+1,j} = (1-\epsilon)f(x_{i,j}) + \frac{\epsilon}{2}[f(x_{i,j+1}) + f(x_{i,j-1})], \tag{1}$$

where $i = 1, 2, ...$ is the time index, $j = 1, 2, ..., L$ with $L \geq 2$ is the lattice site index with a periodic boundary condition, f is a local chaotic map in the interval I, and $\epsilon \in (0, 1)$ is a coupling constant. Here, the logistic map is taken as local map, which is described by

$$f(x) = rx(1-x), \tag{2}$$

where $r \in (0, 4]$ is a constant. An example of the spatiotemporal chaos generated by Eqs. (1) and (2) is shown on the left side of Fig. 1.

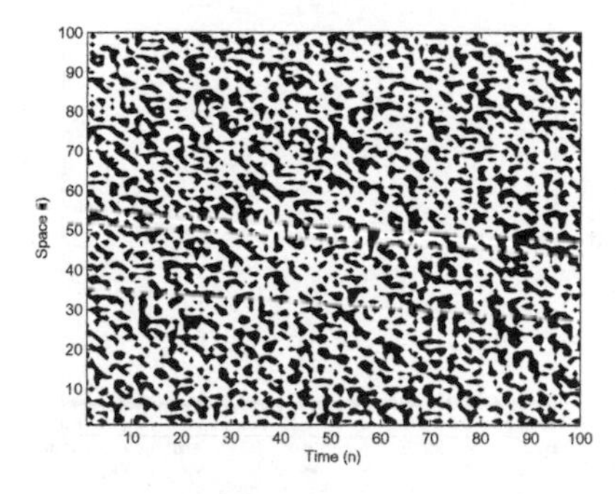

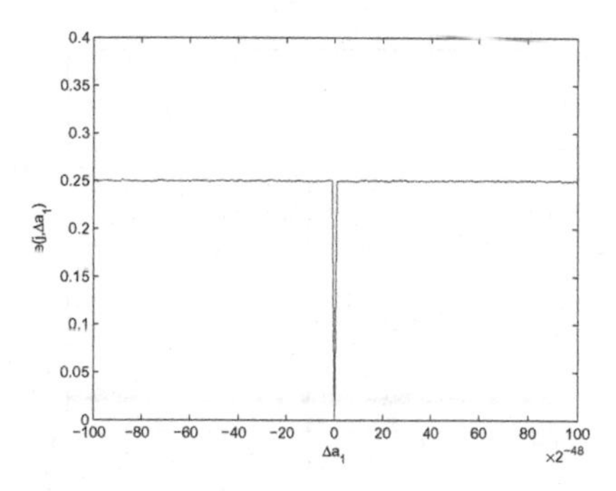

Fig. 1. Left: pattern of a CML with $\epsilon = 0.9$, $r = 4$ and $L = 100$; right: error function

Thus, a CML with L lattice sites can simultaneously generate L pseudo-random bit sequences by digitising the chaotic outputs of the lattice sites. The state variable $x_{i,j}$ of the j-th site can be regarded as a pseudo-random number, which means that $\{x_{i,j}\}_{i=1}^{\infty}$ is a Pseudo-Random Number sequence (PRNS), denoted by PRNS_j. Therefore, L PRNSs can simultaneously be generated from a CML of size L. Further, by digitising the PRNSs, i.e. by transforming the sequence of real numbers to a binary sequence, Pseudo-Random Bit Sequences (PRBSs) can be obtained. Here, a PRBS is generated by concatenating the mantissae of the numbers $x_{i,j}$ in a certain floating-point representation [6].

In order to prevent the lattice sites from falling into synchrony, a criterion was derived by analysing the Lyapunov exponents λ_j, $j = 1, 2, \ldots, L$, of the CML, namely that λ_2 must be bigger than 0, i.e.

$$\lambda_1 + \ln[1 - \epsilon + \epsilon \cdot \cos(2\pi/L)] > 0, \tag{3}$$

or

$$r\left[1 - \epsilon\left(1 - \cos\frac{2\pi}{L}\right)\right] > 2. \tag{4}$$

When satisfying Condition (4), the CML with r, ϵ and L can exhibit chaotic behaviour without its sites falling into a state of synchronisation.

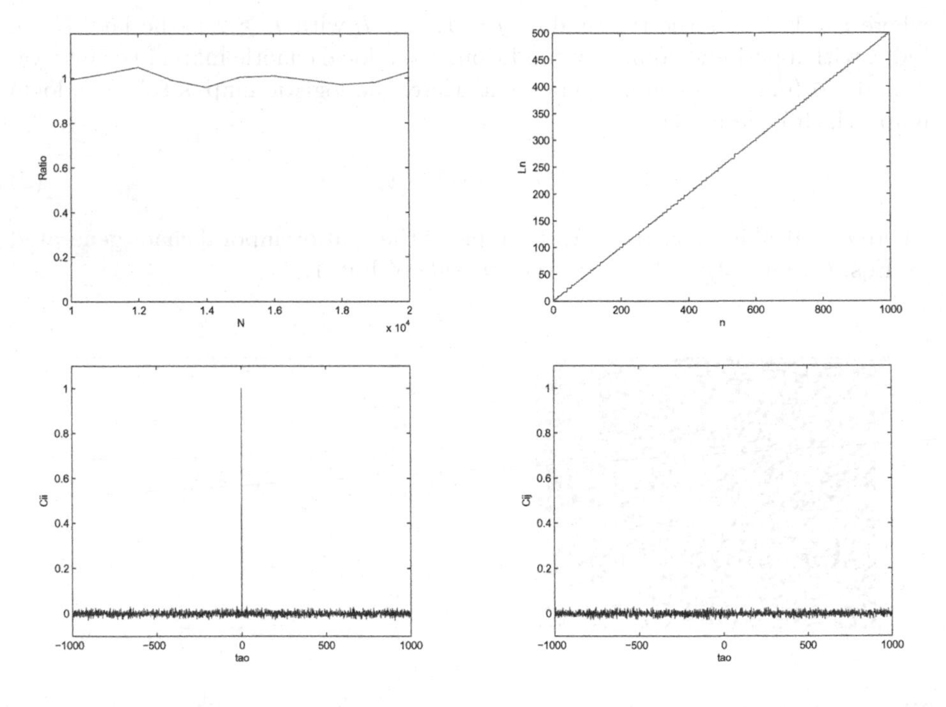

Fig. 2. Cryptographic properties of a pseudo-random bit sequence: 0:1 ratio, linear complexity, auto- and cross-correlation

The cryptographic properties of this approach were studied. The first one is a long period, which is ensured by spatiotemporal chaotic systems, although there exists a problem of short periods along with the chaotic orbits, when chaotic maps are realised in computers with finite precision. The period of CMLs with L lattices is about $10^{-0.4L} \cdot 2^{52\times0.47L} \approx 10^{7L}$, and the period of the PRBSs generated by CML-MPRBGs is also about 10^{7L}. Therefore, when $L > 5$, the period provided by CML-MPRBGs satisfies the basic cryptographic requirement, since a length of order $O(2^{100})$ is cryptographically long. The second property is balance, which means that a PRBS has a uniform distribution, i.e. with about the same numbers of 0's and 1's in the binary sequence. The third one is strong linear complexity. The forth property, that a PRBS has a cross-correlation close to zero, can be used to encrypt several plaintexts at one time. The last one is that a PRBS has a δ-like auto-correlation, which measures the extent of similarity between the PRBS and a shift of itself by some positions. These five properties have been investigated numerically by calculations as depicted in Fig. 2. Based on them, the novel scheme to generate multiple streams of pseudo-random bits promises to be advantageous for fast and secure encryption.

A one-way cyclically coupled logistic-map lattice with certain parameters has the best cryptographic properties among the six most simple pseudo-random bits generators [6]. Based on this, a stream cipher is designed, which carries out encryption as follows:

$$\begin{aligned} f(x_{i,j}, a_j) &= (3.9 + 0.1a_j)x_{i,j}(1 - x_{i,j}), \\ x_{i+1,j} &= (1 - \epsilon)f(x_{i,j}, a_j) + \epsilon f(x_{i,j-1}, a_{j-1}), \\ K_{i,j} &= \text{int}[x_{i,j} \times 2^u] \bmod 2^v, \\ C_{i,j} &= M_{i,j} \oplus K_{i,j}, \qquad j = 1, ..., L,\ i \in \mathbb{N} \end{aligned} \tag{5}$$

where the index 0 is to be replaced by L for the cyclic coupling of lattice sites, $u, v \in \mathbb{N}$ suitably chosen, $K_{i,j}$, $M_{i,j}$ and $C_{i,j}$ are keystream, plaintext and ciphertext, respectively, and $\oplus$ denotes bitwise antivalence. Actually, the CML serves as a PRBG to produce L keystreams by applying the algebraic operations *int* and *mod* on the outputs of the CML. Plaintexts are subjected to bitwise exclusive or with keystreams to produce ciphertext. Encryption parameters are assumed as $a_j \in [0, 1]$, denoted in vector form as $\mathbf{a} = \{a_1, a_2, ..., a_L\}$.

The configuration and parameters of decryption are the same as those of encryption, which is described as

$$\begin{aligned} f(y_{i,j}, a'_j) &= (3.9 + 0.1a'_j)y_{i,j}(1 - y_{i,j}), \\ y_{i+1,j} &= (1 - \epsilon)f(y_{i,j}, a'_j) + \epsilon f(y_{i,j-1}, a'_{j-1}), \\ K'_{i,j} &= \text{int}[y_{i,j} \times 2^u] \bmod 2^v, \\ M'_{i,j} &= C_{i,j} \oplus K'_{i,j}, \qquad j = 1, ..., L,\ i \in \mathbb{N} \end{aligned} \tag{6}$$

where $a'_j \in [0, 1]$ are decryption parameters, denoted as $\mathbf{a}' = \{a'_1, a'_2, ..., a'_L\}$. When $\mathbf{a}' = \mathbf{a}$ and $y_{0,j} = x_{0,j}$ holds for the seed values, these two CMLs are synchronised, i.e. $y_{i,j} = x_{i,j}$, $i \in \mathbb{N}$, thus producing identical keystreams, $K'_{i,j} = K_{i,j}$. As a result, plaintext is decrypted, $M'_{i,j} = M_{i,j}$.

For the keystreams in this cipher to have proper statistical properties, the parameters a_j, $j = 1, 2, ..., L$, are set to guarantee that the logistic map's coefficient r in Eq. (2) falls into the range $[3.9, 4.0]$, and ϵ is fixed as 0.95. The quantity v is assumed as 32 for the following reasons, and u is selected accordingly, e.g. as 52 when double-precision floating-point arithmetic is used. First, the leading 4 bits are discarded for their bad statistical properties. Then, the smaller v is, the harder it is to break the cipher with known-plaintext attacks. Finally, from the implementation point of view, the larger v is, the more efficient the cipher will be. Therefore, a trade-off between efficiency and security leads to fix v as 32 by considering that common computers work with numbers 32 bits or 64 bits wide. When determining L, the following considerations are important. There is no evident influence of L on the cryptographic properties of the keystream except for its period equal to about 10^{7L}, there is no influence on the encryption speed either, and the cost of breaking the cipher is about 2^{40L}. Therefore, in investigating a concrete cipher thereafter, L is assumed as 4 in order to let the keystream's period be 10^{28} and the cost of breaking the cipher be up to 2^{160}, which are suitable choices from the cryptographic point of view.

A keyspace is defined as a set of all possible keys, which should be studied in depth when designing a cipher. An error function is used here to determine the size of the cipher's keyspace. When $\mathbf{a}' \neq \mathbf{a}$, the decrypted plaintext, $M'_{i,j}$, can deviate from the original one, $M_{i,j}$. The error function is defined as

$$\begin{aligned} e(j, \Delta\mathbf{a}_t) &= \tfrac{1}{T}\sum_{i=1}^{T} |m'_{i,j} - m_{i,j}|, j = 1, 2, ..., L, \\ m'_{i,j} &= 2^{-32} \cdot M'_{i,j}, \quad m_{i,j} = 2^{-32} \cdot M_{i,j}, \end{aligned} \tag{7}$$

where $\Delta\mathbf{a}_t = \{\Delta a_1, \Delta a_2, ..., \Delta a_t\}$ ($\Delta a_j = a'_j - a_j,\ j = 1, 2, ..., t,\ t \leq L$), and T is the length of encryption. The error function vs. Δa_1 with $T = 10^5$ is plotted on the right side of Fig. 1. It is shown that the error function is not equal to zero but 0.25, even if Δa_1 takes on an extremely small value 2^{-47}. In other words, the parameter a'_1 is sensitive to any differences equal to or larger than 2^{-47}. Similarly, the error functions of Δa_j ($j = 2, 3, ..., L$) were computed, indicating that the parameters a'_j ($j = 2, 3, ..., L$) are also sensitive to any differences equal to or larger than 2^{-47}. Therefore, the keyspace is 2^{47L}.

Since ciphertext is generated by direct bitwise application of the antivalence operator between plaintext and keystream, the cryptographic properties of the keystream have significant effects on the security of the cipher. Owing to the symmetric configuration of the CML, all keystreams have similar cryptographic properties. Some cryptographic properties of a keystream among the L ones, such as probability distribution, auto-correlation and run probability, were investigated numerically. In summary, L keystreams have satisfactory random-like statistic properties. Moreover, the security of the cipher was evaluated by investigating its confusion and diffusion properties and using various typical attacks, such as the error function attack, the differential attack, the known-plaintext attack, the brute-force attack and the chosen-plaintext/ciphertext attack.

3 Most General Form of Encrypting Bit Patterns

All known cryptographic methods subject the data elements to be transmitted, may that be bits, alphanumerical characters or bytes containing binary data, may they be single or in groups, always as unchanged entities to encryption. Shannon's [8] information-theoretical model of cryptosystems is founded on this restrictive basic assumption as well. Consequently, information such as the boundaries between data elements and their number perpetuates observably and not encrypted into the ciphertext: as a rule, to any plaintext symbol there corresponds exactly one ciphertext symbol. Since even block ciphers seldom work with data entities exceeding 256 bits, the symbols in plaintexts and in ciphertexts are ordered in the same sequences or, at least, their positions lie very close together. Thus, corresponding symbols in plaintext and ciphertext can rather easily be associated with one another.

As this feature facilitates code-breaking, a counter-acting enhancement for cryptographic systems was devised [3]. The method's fundamental idea is based on the observation that, ultimately, technical realisations represent all symbols in binary encodings. Correspondingly, for encryption the most general among all possible forms of replacing one bit pattern by another one is employed. This allows to blur the boundaries between the plaintext symbols, and to use several, randomly selected encryptions for a single bit pattern, having more bit positions in the ciphertext than in the plaintext.

In a state t of a communication process, the number m_t of bit positions to be encrypted according to [3] is determined by an arbitrarily selectable method. It is decisive that the parameter m_t is different from the number of bit positions k encoding the plaintext alphabet. Thereby the boundaries between the plaintext symbols are annihilated. Then, for m_t bits each in a stream, an encryption with n bit positions is determined by means of a state-dependent *relation*

$$R_t \subset \{0,1\}^{m_t} \times \{0,1\}^n. \tag{8}$$

Here, the parameter n may not be smaller than m_t, as information would get lost otherwise, and it should not be equal to m_t either, in order to prevent the disadvantages mentioned above. Choosing $m_t \neq k$ and $n > m_t$ inherently ensures that it is not easily possible anymore to conclude from the boundaries between the ciphertext symbols on the ones between the plaintext symbols.

Contrary to the conventional cryptographic methods, the relation R_t does not need to be a mapping: it is even desirable that with any element in $\{0,1\}^{m_t}$ as many elements of $\{0,1\}^n$ as possible are related by R_t, allowing to randomly select among them one as encryption. For $n > m_t$, the set of possible encryption elements is embedded in a considerably larger image set, significantly impeding code analysis for an attacker. Moreover, every element in $\{0,1\}^n$ should be a valid cipher of an element in $\{0,1\}^{m_t}$, to completely exhaust the encryption possibilities available. Then, unique decipherability is given, if and only if the inverse relation is a surjective (onto) mapping:

$$R_t^{-1} : \{0,1\}^n \longrightarrow \{0,1\}^{m_t}. \tag{9}$$

Different from Kerckhoffs' [4] principle, this decryption function is not known publicly – and the relation R_t used for encryption is not only publicly unknown, but no function either. Publicly known is only, that R_t^{-1} is a totally arbitrary mapping among all possible ones mapping the finite set $\{0,1\}^n$ onto another finite set $\{0,1\}^{m_t}$.

The number of all possible relations $R_t \subset \{0,1\}^{m_t} \times \{0,1\}^n$, for which R_t^{-1} is a surjective mapping, amounts to $\frac{2^n!}{(2^n-2^{m_t})!}$. The set of these relations comprises, among others, all possibilities to permutate bits in their respective positions, to insert $n-m_t$ redundant bits, each of which may have either one of both possible values 0 or 1, at $\binom{2^n}{2^{n-m_t}}$ positions in the output bit patterns as well as to link the bit positions of the encryption elements by the most general computations.

4 Implementation of Two-Stage Data Encryption

A system for encrypting data packets to be transmitted over wired or wireless communication networks is to be devised now, which should be both information-theoretically secure and practically feasible. Since the PRBG described in Section 2 does not produce truly random bit sequences, we combine it with the method of Section 3 in defining the following two-stage encryption algorithm.

1. As many iterations of the PRBG in Eq. (5), i.e. recurrent floating-point calculations, are carried out as required for the concatenated resulting bit sequences to match or exceed the length of a data packet to be encrypted. Then, bitwise antivalence is formed between the packet and the bit sequence serving as one-time key.
2. The bit string resulting from step (1) and having the same length as the original packet is further processed by the method of Section 3, i.e. the bit string is partitioned into segments of length m_t (cp. Fig. 3), and for each segment an image is randomly selected among the images the segment relates to by relation R_t. The concatenation of all images thus determined, and each being n bits long, constitutes the packet's final enciphering ready for transmission.

To decrypt such a data packet, a receiver carries out the inverse operations expected to be applicable to the packet in the current state of a communication.

1. A packet received is partitioned into n bits long segments, each of which is subjected to the mapping R_t^{-1}. The results are concatenated.
2. As many iterations of the PRBG in Eq. (6) are carried out as required for the concatenated resulting bit sequences to match or exceed the length of the data packet to be decrypted. Bitwise antivalence is finally formed between the packet and the bit sequence.

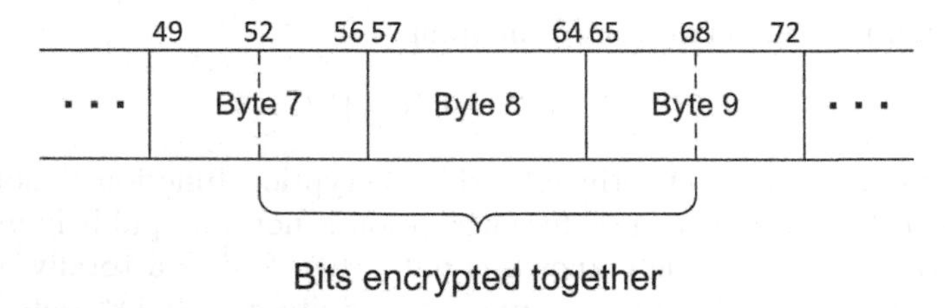

Fig. 3. Extracting bit strings from a data packet

The condition for the above algorithms to work correctly is that a sender's and a receiver's PRBG run in synchrony, i.e. their parameters have the same values and their iteration counts are equal. Since a communicating device usually acts in turn as sender and receiver, it needs to maintain both the relation R_t and the inverse R_t^{-1} for each of its communication partners.

Breaking this cipher were only possible, if an eavesdropper had such an amount of ciphertexts available as required by pertaining analyses – totally disregarding the necessary computational power. Among the following – non-exhaustive – variety of implementation options there are some further measures preventing sufficiently long encipherings, generated with certain choices of parameter sets and encryption relations, to arise in the first place.

- The order of the two steps applying one-time keys and the encryption relation may be reversed in the algorithms above.
- The seed and parameter values of the PRBG running in sender and receiver may be modified frequently and irregularly. To implement this, one of the communication partners may serve as master and may employ a physical phenomenon with truly random behaviour, e.g. measuring white noise. At randomly determined points in time, either new seed and parameter values or just array indices may be included into the data packets transmitted. The latter case resembles the iTAN procedure of on-line banking. The indices would identify locations in read-only memory modules, whose production, transport and installation would typically represent a confidential and authentic channel to transfer secret information.
- The selection of images among the ones provided by relation R_t may be based on a physical phenomenon with truly random behaviour.
- Communicating units may, during operation at randomly selected points in time, sufficiently often vary the parameter m_t between 1 and an installation-dependent upper bound, thus modifying the relation R_t correspondingly.
- Similarly, communicating units may turn to use a completely different encryption relation R_t provided in real-only memory, which only requires to transmit its identifier and some parameter values.
- A simplified version of the relation-based encryption may entail the contents of R_t and other parameters to be supplied in form of pseudo-random bit sequences as well. A pertaining protocol may co-ordinate that both sender and receiver proceed, rather frequently at random instants, from one state to the next. In the course of a state transition, the relation R_t and the parameter m_t are re-defined. For co-ordination, as few details as possible should be transmitted between the communicating units for reasons of confidentiality.

Owing to its high complexity, the encryption procedure lends itself for authentication purposes in a straightforward way. To authenticate a packet, the receiver just needs to check for expected values in certain data fields. The bit patterns found there will be different if the packet does not come from the correct source, or anything has gone wrong.

5 Conclusion

Currently applied cryptosystems to secure confidential data transmission have either already been broken or are expected to be broken soon. Moreover, in certain countries their keys need to be escrowed with government agencies. In order to prevent eavesdropping and gaining unauthorised access to computers, the method of choice is, therefore, information-theoretically secure one-time encryption. Since providing a very large number of rather long one-time keys is practically infeasible, an efficient chaos-theoretical approach for the continuous generation of pseudo-random bit strings to be employed as one-time keys was presented. To compensate for this method's deficiency, viz. the lack of genuine randomness, it was combined with encrypting a bit pattern by a longer one selected truly at random within a larger set of possible ciphers.

To encrypt not by means of a bijective, i.e. invertible, function, but by a relation with a surjective mapping as inverse, is already known from the patent [2]. There the symbols of a plaintext alphabet are bijectively mapped onto equivalence classes of symbols in an image set of generally higher cardinality. To encrypt a plaintext symbol, out of the equivalence class corresponding to it in the image set an image symbol is selected, and that either randomly or so that the enciphering becomes as invulnerable by statistical methods as possible. Whereas according to [2] plaintext symbols are replaced one-to-one by image symbols in ciphertexts, the second stage of the method presented here blurs the boundaries between data items encrypted together, rendering it impossible to conclude from boundaries between data items in ciphertexts to the boundaries of data items in plaintexts. Thus, a toehold for cryptanalysis left open by a silent assumption in Shannon's communication theory was eliminated.

References

1. Álvarez, G., Li, S.J.: Some Basic Cryptographic Requirements for Chaos-based Cryptosystems. International Journal of Bifurcation and Chaos 16(8), 2129–2151 (2006)
2. Günther, C.-G.: Ein universelles homophones Codierverfahren, German patent 39 04 831 (1989)
3. Halang, W.A., Komkhao, M., Sodsee, S.: A Stream Cipher Obliterating Data Element Boundaries. Thai patent registration 140-100-1271 (2014)
4. Kerckhoffs, A.: La cryptographie militaire. Journal des Sciences Militaires 9. Serie (1883)
5. Kocarev, L.: Chaos-based Cryptography: A Brief Overview. IEEE Circuits and Systems Magazine 1, 6–21 (2001)
6. Li, P.: Spatiotemporal Chaos-based Multimedia Cryptosystems. Fortschr.-Ber. VDI Reihe 10 Nr. 777. VDI-Verlag, Düsseldorf (2007)
7. Menezes, A.J., van Oorschot, P.C., Vanstone, S.A.: Handbook of Applied Cryptography. CRC Press, Boca Raton (1997)
8. Shannon, C.E.: Communication Theory of Secrecy Systems. Bell System Technical Journal 28, 656–715 (1949)
9. Silva, C.P.: A Survey of Chaos and its Applications. In: IEEE MTT-S International Microwave Symposium Digest, vol. 3, pp. 1871–1874 (1996)

A Concept for Trust Derivation from User Activities

Doan Trung Son and Mario Kubek

University in Hagen,
Faculty of Mathematics and Computer Science, Hagen, Germany
{doan.trung.son.vn,dr.mario.kubek}@gmail.com

Abstract. Business activities are usually based on trust and reputation of the participating actors. Online social networks present to their members manifold possibilities to meet new business partners, while an evaluation of their trustworthiness is still a quite unsafe and risky matter. Basing on communication activities, a new concept to obtain reliable trust values for any two users is introduced and its generalisation to a global, network wide trust system will be proposed. Last but not least, a fully decentralised processing for those trust values is proposed.

Keywords: trust, social networks, user activity, random walk, pagerank.

1 Introduction

In online businesses, security and trust are the most important factors for merchants and customers to protect their goods, money and transaction data from any unwanted loss. Recently, online social networks (ONS) became a huge marketplace [1], where people meet, negotiate, buy and sell any kind of products. Usually, those people never met before and rely on honesty and trustworthiness of the respective business partners. Of course, those media also attract people that intend to use them in an unhonest and unlawful manner. Consequently, the problem of distinguishing honest business partners from others such as cheating people appears [2]. While the problem of secure communications and transactions is quite well addressed by a series of cryptographic methods and protocols [3], the problem of giving trust to somebody is still an open problem, especially if people have never met in real life.

Trust can be understood as the reputation of people, i.e. the overall quality of character seen or judged by people in general [4]. It becomes clear that it will be quite hard to measure this by any quantitative values.

In [5] was figured out that a reliable trust estimation can be derived over a longer observation period, only, i.e. it requires a longer time of mutual communication and activities involving interactions in the social networks as well as in reality. Hereby, trust can be mostly understood a the predictability of activities of the other users in the respective environment. While it is relatively easy to determine the predictability of a limited number of activities of a user and measure

H. Unger et al. (eds.), *Recent Advances in Information and Communication Technology 2015*,
Advances in Intelligent Systems and Computing 361, DOI: 10.1007/978-3-319-19024-2_17

it by a percentage value over a longer period, it is quite difficult to generate such a value at the first short term contact merchants and customer usually have. Recently, most merchants rely on customer evaluation on their web pages or evaluation activities of third parties, which still presents a lot of possibilities for manipulations [6]. The evaluation of the customer remains usually hidden and is done in the form of (secret) black lists of the merchants or groups of them. Consequently, it is intended to generalise the concept of *trust chains* such that from the pairwise trust values and the structure of the whole network a more or less objective trust value which is also protected against manipulation for each user can be derived.

In the following sections, a new concept for trust derivation shall be introduced, basing on the frequent use of online social networks. First, the generation of mutual trust between any two users is described in sec. 2. In sec. 3 shall be shown, how that pairwise trust estimation can be combined with the estimations of other users to a global trust value for all participants using a random walker approach. Last but not least, a concept for an implementation and a simulation setup is given.

2 Pairwise Trust of OSN-Users

Several psychological and sociological publications deal with the problem of understanding trust [7] [8][9][10]. It becomes clear that trust is not a fixed value but a parameter changing over time depending on very subjective rules and also feelings. Normally, user are carefully in the beginning, slowly gain trust until they fully believe in each other. Of course, this growth process may be durable and suddenly disturbed, if one of the partners occurs unreliable, e.g. by a single lie. From our point of view and following [5], trust might be quantified. Differing from the human approach of trust building, a technical system must be based on exact measurements of suitable parameters and algorithms as well as on how to combine them to a reliable trust value. For the communication of users and the exchange of contents, users of OSNs may use a limited set of activities. A user usually can or has to

- register and establish a profile showing his interests (respectively content or information he offers or is looking for);
- establish, add or eventually remove friends (from the set of other OSN users), eventually divided into groups (note that the friendship relation is not in all systems a symmetric one, i.e. in some systems, A can be a friend of B without B being a friend of A);
- read, write (post) and redistribute content from other users while sometimes the system adds any new items to this set;
- communicate with other users by like (FaceBook), +1 (Google+) or other operations, comment their contributions or mail with them;
- establish groups or communities, which are a broadcast possibility for their members to all other members.

While it is hard to analyse the content of a OSN, the appearance of communication activities is a strong instrument to evaluate the relation among two users. Moreover, posting or communicating some content is published with the expectation to obtain some reward, i.e. like's of those content, comments, mail etc. In such a manner, some relations of cause and impact appear with every activity in a OSN, which can be measured, predicted and evaluated condering their frequency by real numbers. Those numbers later will be referred to as cumulative trust value between two users, i.e., user u_x trusts user u_y to a certain extent, denoted as $T(u_x, u_y)$. Note at this point that trust is not a symmetric relation, i.e. $T(u_x, u_y) \neq T(u_y, u_x)$.

Summarizing our understanding means that the trust $T(u_x, u_y)$ between any two users u_x and u_y mostly depends on:

1. the time the two users know each other,
2. the similarity of their interests,
3. the mutual predictability of their activities and last but not least,
4. some (often initially given) mutual sympathy (which is of course hard to model).

Being a friend (note that we use the term in business and private matters, although it is usually called a partner in business) is in both real world and social networks a special expression of trust and subject to a permanent evaluation.

Trust non-monotonically and dynamically changes and is adapted to the changing conditions of contexts, in which user activities take place. Of course, also external, real-world influences effect users' trust and may result in a rapid increase or decrease of the trust value among any two users.

In the described approach, (only) the above activities will be used to cause a (periodical) increase or decrease of trust between any pair of users starting from an initially given trust $T_0(u_x, u_y)$, which depends on hardly predictable personal circumstances and preferences.

In detail, the following rules apply to generate a cumulative trust value $T(u_x, u_y)$ over the continuous interactions and activities with other users for a longer period of time.

For the special example of *Google+* the following rules were derived.

1. Being liked from a user u_x will increase $T(u_x, u_y)$
2. Positive comments have a more intense, increasing effect as likes.
3. If u_y posts interesting (i.e. usually similar) content, which u_x reads, it will increase $T(u_x, u_y)$. Frequent, consecutive like activities may increase the trust value stronger over time.
4. Posting uninteresting, offending content will decrease $T(u_x, u_y)$, especially (and therefore in an exponential manner) if it happens in an uninterrupted series.
5. Reaching a given trust value $T_f(u_x, u_y)$ will result in adding u_y as friend by u_x;
6. In the same manner a much lower value of $T_{uf}(u_x, u_y)$ may result in an 'unfriend'-activity.

7. Finding a triadic closure i.e. recognizing that u_y and u_z are friends, may increase the trust $T(u_x, u_z)$. A differentiation using strong and weak ties may be also useful.
8. Also, a new friend may be randomly added with a small probability, representing a new friend from the real world.

Currently, a concrete quantitative analysis of trust alterations of $T(u_x, u_y)$ is not given in this article. These values, however, must be later empirically derived and be confirmed in a simulation process.

From the human psychology it is clear, that the transition between no and full trust is definitely not a linear function, but more or less a sigmoid dependency, if few exceptional events resulting in an immediate loss of trust are not considered at the moment[1].

1. In the beginning, the first activities of a users are not adequately recognised.
2. After some time of doubt, positive activities result in a significant increase of trust.
3. When a time of probation is over, full trust is given.
4. This process, however, is reversible.
5. Some activities may result in an immediate loss of trust, this may be modeled again with a small probability $p_{lie}(u_x, u_y)$ representing that u_x is cheated by u_y such that any trust is destroyed and $T(u_x, u_y) = 0$.
6. It must be discussed whether activities shall be considered over all time using e.g. a (sliding) window approach or for a specific time slice only.

As the linear combination of activities influencing the trust value of user u_x for user u_y is aggregated in $T(u_x, u_y)$, which can vary in a big range, normalisation should be introduced to map $T(u_x, u_y)$ to $t(u_x, u_y)$ with values in an interval of [0,1] following [11], which preserves the underlying trust semantics. The question is now how $t(u_x, u_y)$ can be suitably derived from $T(u_x, u_y)$?

A sigmoid function is often used [12] and the suggested solution for our propose:

$$t(u_x, u_y) = \frac{1}{2} + \frac{T(u_x, u_y) - T_{off}}{2\sqrt{1 + (T(u_x, u_y) - T_{off})^2}}, \quad (1)$$

whereby T_{off} describes the user characteristics, i.e. how much initial trust is given and how much positive activities must be performed in order to obtain an increased trust value.

Now, the pairwise trust functions must be used to generate a (global) trust value for each user, which shall not solely depend on a special pairwise business relation but be an overall trust evaluation of this user in his (complex) network of relations.

[1] Lies are an important strategic possibility of individuals in society to reach their goals.

3 Random Walks-Based Trust Calculation

From the above said, it becomes clear that the (global) trust value of a user depends on on the trust of all users knowing him as well as the trustworthiness of those users. E.g. if a user A trust a user B with 100 percent and has an own trust estimation of 10 percent only, this user probably cannot convince the community that B is reliable.

By considering those relations, the similarity to the calculation of PageRank [13] is highly visible. Indeed, the results of [14] show that we can use and specify the PageRank calculation for our needs. Another advantage is that it is known that the PageRank of a node can be obtained by a fully decentralised working, random walker based method.

While PageRank reflects — as intended in the before cited original publication — only topological aspects of nodes embedded into (web-) graphs, [14] includes other factors generating graphs with weighted edges in other words continuous-valued networks, which may influence the role of a node in a system. Originally, the transition probability of a random walker from a node u_x to a node u_y $p(v_x, v_y) = \frac{1}{|N_u|}$ is the only parameter influencing the PageRank besides the topological properties of the underlying graph.

In order to obtain a TrustRank TR, the global trust value for each node in a complex OSN, the trust values $t(u_x, u_y)$ can be used, i.e. the trust a user u_x has in another user u_y. With this assumption, the transition probability of a random walker to move from v_x to v_y can be defined as

$$p(v_x, v_y) = \frac{t(u_x, u_y)}{\sum_{\forall u_a \in |N_{u_x}|} t(u_x, u_a)}, \tag{2}$$

where $\sum_{u_a \in N_{u_x}} p(v_x, v_a) = 1$. It is easy to see that now the random walker will prefer links with a higher trust.

The TrustRank is now easy to calculate and can be obtained faster by using k random walkers.

If $k \in \mathbf{N}$ random walkers are used, then the TrustRank can be calculated by

$$TR_{u_x}(t) = \frac{\sum_{\forall k} f_{u_{x_k}}(t)}{\sum_{\forall k} step_k(t)}, \tag{3}$$

where $f_{u_{x_k}}(t)$ is the number of all visits of the k-th random walker on u_x so far in all its $step_k(t)$ steps until time t.

It is clear that the counted trust value $TR(u_x)$ is still a value, which depends on the network size, i.e. the bigger the network is, the smaller all values are. In order to make these values comparable, a normalization must be carried out using the size of the network. For centralized OSN, this value is known to the provider.

For any other cases, [14] suggests a small trick using a property of mean values which helps to cope with this situation, viz. the mean value of a small number of samples already approximates the real mean value normally quite well. Based on the knowledge above and some basic mathematics it is known that the average TrustRank of all nodes in a community $\overline{TR}$ is given by

$$\overline{TR} = \frac{\sum_i TR_i}{n} = \frac{1}{n}. \quad (4)$$

Hence, to calculate the average TrustRank, n can be estimated from a smaller number of samples than by considering the entire number of nodes by

$$n = \frac{\sum_{i=1(1)K} TR_i}{\overline{TR}} = \frac{1}{\overline{TR}}, \quad (5)$$

with $K < |V|$. In other words, the network size is estimated from a sample of TR values whose mean value will converge to $\frac{1}{n}$. Now, only a good estimation for K is needed. This can be replaced, however, by considering the deviation of the calculated mean value. The calculation can be stopped when the deviation is small enough and/or the mean value is stable enough.

With the above method, a trustworthiness of a node u_y can be counted from the trust, any user u_x gives to that node by $t(u_x, u_y)$. Since this value will be kept on u_y, the question on possibilities of its manipulation comes up. In [5], a protocol is introduced to check the validity of an (electronic) coin by keeping it on a set of previous machines and checking those history information.

A similar approach can be used for the trust values. Hereby, a random walker carries the just counted trust value $TR(u_x)$ with him and distributes it on the s next nodes on his way. After some time, all possible successors will have an (almost) correct value of $TR(u_x)$. This value can therefore be obtained from any (doubting) node accessing u_x by visiting those nodes that are reachable within s steps from u_x. This method works correctly as long as u_x cannot allocate and manipulate a larger number of nodes (how much depends on s and the out-degree of its successor nodes).

Last but not least, it shall be mentioned that the security mechanism from [5] may be applied to avoid any manipulation of the trust values by the user. It is mainly based on the propagation of the trust values along a randomly chosen trail through the network and the selection of a (smaller) group of nodes as witnesses for the confirmation of the respective locally stored value.

4 Simulation and Implementation

So far, only a limited, small simulation has been setup to prove the described concept. It contains the simulation of just 200 nodes in a small-world network generated by the algorithm introduced by Watts and Strogatz [15].

Since we do not have an ONS to obtain realistic user data in the first version of simulation, the initialisation has to also include the generation of trust weights for the edges of the network. Therefore the Richardson technique [16] is applied to uniformly choose a continuous value for the directed edges between two nodes u_x and u_y in scaled intervals $[max(\gamma_{u_y} - \varphi_{u_x u_y}, 0), min(\gamma_{u_y} + \varphi_{u_x u_y}, 1)]$, while the quality parameter γ_{u_y} was chosen from Gaussian distribution with $\mu = 0.5$ and $\delta = 0.25$; the noise parameter $\varphi_{u_x u_y}$ has been set to $1 - \gamma_{u_y}$.

The first experiments had the goal to prove that

1. random walkers are a suitable tool to calculate the PageRank of nodes in complex networks and to obtain an idea of the speed of convergence and
2. a realistic distribution of trust values can be obtained.

For both goals, satisfying results could be achieved. Fig. 1 shows once more that the euclidean distance between the values obtained from PageRank and the Random walks-based method converges to zero indicating that both algorithms are quite strongly correlated. If the number of iterations is big enough (i.e. $\geq$ 2.000 iterations), the result reveals an acceptably small distance of around 0.008. The results, however, still do not cover real-world conditions with a large number of nodes involved. Also, the number of needed steps to achieve convergence is quite high such that mechanisms are needed to improve the convergence speed.

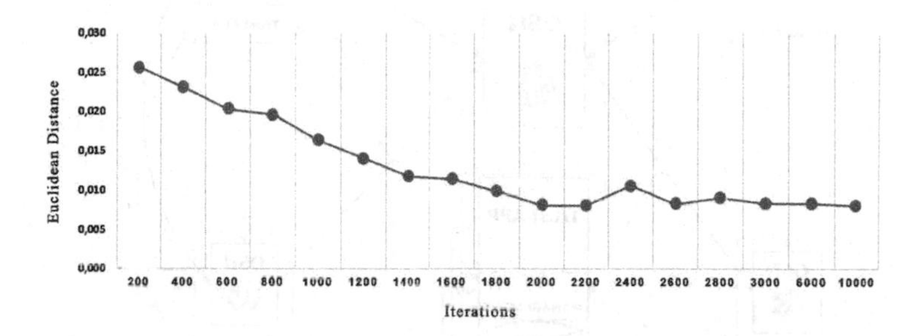

Fig. 1. Euclidean distance correlation between Page Rank and Random walks-based method on binary-valued networks

Good results could also be obtained from the statistics of the obtained trust rank in Fig. 2. It is worth to point out that the TrustRank values follow a gaussian distribution. Additionally, the simulation result of TrustRank shows the mean value of users' (not normalised) TrustRank $\overline{TR}$ of 0.0049855 after 10.000 iterations. It also shows that the summation of all values amounts to TR 0.9971, which is approximately equal to 1, which is fairly suitable according to the theoretic statements given in the previous section.

In Fig. 3 the intended implementation of the developed trust management system is shown in the context of any OSN, in our case *Google+*.

It is to be seen that the suggested system mostly consists of an application running in parallel to the OSN. It is able to collect data from the social network, in particular it can access any neighbourhood (friendship) information of a particular user and record all its activities (since this happens locally on the user's computer, no security concerns may arise). From those activities, the respective pairwise, local trust values can be calculated.

In addition, the application contains a management system for a population of random walkers, controlled as suggested in the literature by [5], [14]. The random

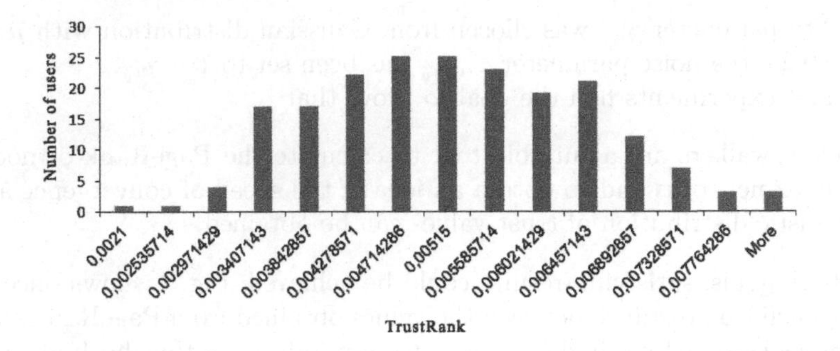

Fig. 2. Distribution of *TrustRank* values of two hundreds of users

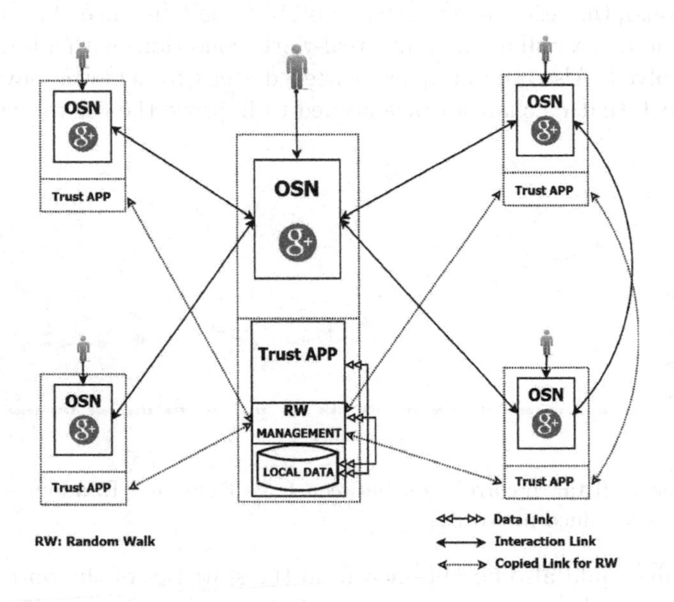

Fig. 3. Blockscheme of the intended trust management system

walkers may follow the copied links of the OSN and generate the global trust values for the users. In addition, those trust values are copied along a random trail in order to avoid unwanted manipulations as described above.

5 Conclusion and Outlook

A fully decentralised concept to calculate global trust in an OSN was introduced. It is based on the evaluation of the predictability of user activities and uses random walkers for all communication and calculation processes. After a short startup time, trust values can be derived for every user, even when the information available on a particular user (e.g. when the user just joined the network) is sparse. In such a manner, the concept may contribute to endeavours to make online trading more safe. First experiments have been conducted to prove the technical soundness of this concept.

In future works, larger simulations will be carried out and an implementation embedded in a real OSN environment along with its test results will be provided to obtain more detailed information about the dynamics and reliability of the proposed mechanisms including a study of its practicability in the daily use.

References

1. Easley, D., Kleinberg, J.: Networks, Crowds, and Markets: Reasoning About a Highly Connected World. Cambridge University Press (2010)
2. Mazar, N., Amir, O., Ariely, D.: The Dishonesty of Honest People: A Theory of Self-Concept Maintenance Social Science Research Network Working Paper Series (2007)
3. Liu, C., Marchewka, J.T., Lu, J., Yu, C.S.: Beyond concern: a privacy-trust-behavioral intention model of electronic commerce. Information and Management 42(1), 127–142 (2004)
4. Jøsang, A.: Trust and Reputation Systems. In: Aldini, A., Gorrieri, R. (eds.) FOSAD 2007. LNCS, vol. 4677, pp. 209–245. Springer, Heidelberg (2007)
5. Unger, H., Bohme, T.: A decentralized, probabilistic money system for P2P network communities. In: Proceedings of the Virtual Goods Workshop, Ilmenau, pp. 60–69 (2003)
6. Jøsang, A., Golbeck, J.: Challenges for robust of trust and reputation systems. In: Proceeding of the 5th International Workshop on Security and Trust Management (2009)
7. Deutsch, M.: The resolution of conflict. Yale University Press, New Haven, London (1973)
8. Grabner-Kräuter, S., Kaluscha, E.A.: Empirical research in on-line trust: a review and critical assessment. Int. J. Hum.-Comput. Stud. 58(6), 783–812 (2003)
9. McKnight, D.H., Choudhury, V., Kacmar, C.J.: Developing and Validating Trust Measures for e-Commerce: An Integrative Typology. Information Systems Research 13(3), 334–359 (2002)
10. McKnight, D.H., Chervany, N.L.: The Meanings of Trust. UMN university report (2003)
11. Marsh, S.P.: Formalising Trust as a Computational Concept. PhD thesis, University of Stirling (1994)
12. Gibbs, M.N., MacKay, D.J.C.: Variational Gaussian process classifiers. IEEE Trans. Neural Netw. Learning Syst. 11(6), 1458–1464 (2000)
13. Page, L., Brin, S., Motwani, R., Winograd, T.: The pagerank citation ranking: Bringing order to the web. Technical Report, Stanford Digital Library Technologies Project (1998)
14. Sodsee, S.: Placing Files on the Nodes of Perr-to-Peer Systems. PhD thesis, Fernuniversitat in Hagen, published in VDI Fortschrittsberichte Informatik, Dusseldorf, vol. 218 (2012) ISBN: 978-3-18-381610-1
15. Watts, D., Strogatz, S.: Collective dynamics of small-world networks. Nature (393), 440–442 (1998)
16. Richardson, M., Agrawal, R., Domingos, P.: Trust Management for the Semantic Web. In: Fensel, D., Sycara, K., Mylopoulos, J. (eds.) ISWC 2003. LNCS, vol. 2870, pp. 351–368. Springer, Heidelberg (2003)

Web Attack Detection Using Chromatography-Like Entropy Analysis

Akkradach Watcharapupong and Thanunchai Threepak

Department of Computer Engineering, Faculty of Engineering
King Mongkut's Institute of Technology Ladkrabang, Bangkok, Thailand
akkra_watch@yahoo.com, ktthanun@kmitl.ac.th

Abstract. Web services are mostly attacked in various ways directly and indirectly. We calculate the Shannon entropy from web server log files, especially access logs, and then estimate the entropy distance to detect intrusions and identified them by distinct attack word lists as general, cross-site script, and SQL injection attacks. The experiment shows that our proposed chromatography-like entropy analysis method can detect and identify these behaviors.

Keywords: Web Security, Database Security, Anomaly Detection, Entropy Analysis.

1 Introduction

Many techniques are used to implement anomaly detection systems. Fuzzy logic, genetic algorithm, hidden Markov model, Bayesian methodology, artificial neural network, self organizing map, support vector machine, principle component analysis, singular value decomposition, data mining, and statistical methods [1], [2] are examples of techniques which are used to discover system profiles and detect anomalies.

Entropy based detection is the one of interesting techniques that is applied in many researches. Entropy analysis is the process which converts data to entropy domain and make decision by some specified criteria. This scheme is applied in many research topics such as anomaly detection in network applications [3], anomaly detection in network traffic [4-7], detecting computer worm using entropy analysis [8], [9], anomaly detection in stream [10], in space shuttle engine [11], in computer network behavior [12], and also web attacks [13] and anomaly SQL statement [14] detections.

In this paper, we propose the chromatography-like entropy analysis method to detect some web attacks. Our approach uses the Shannon entropy analysis to analyze data in web server log files by assuming that intrusive requests change the entropy and the entropy distance greater than conventional requests.

Details of each sequence are described in next sections. Section 2 describes principle of entropy. Section 3 exhibits the variations of entropy calculation. Section 4 explains the proposed chromatography-like entropy analysis. Section 5 shows our experimental results and Section 6 is our conclusion and future works.

H. Unger et al. (eds.), *Recent Advances in Information and Communication Technology 2015*,
Advances in Intelligent Systems and Computing 361, DOI: 10.1007/978-3-319-19024-2_18

2 Web Intrusion Techniques

Log files of services are the sources of any attacking evidences. Most web servers collect all details of request information to log files. We focus the open-source Apache web server project in our research. Legitimate complicated dynamic requests (to call web application code with parameter passing) are usually as follows:

```
GET /file?var1=value&var2=value&var3=value
```

However most critical web application attacks [15], e.g. code injection and directory traversal techniques, will insert too many special characters and/or have unorthodox word pattern in the resource strings. These will raise client requests' complexity as follows:

```
A: GET /../../../../../../some/file
B: GET /..%%35%63..%%35%63/file.exe?/val1+val2
C: GET /file.cgi?var=<SCRIPT>bogus()</SCRIPT>
```

In A, generally intruder tries to traverse beyond proper directory structure to get the server's internal stuff unexpectedly.

In B, intruder may attempt to launch server's internal executable command (file.exe) with arguments (*/var1 var2*) and uses evasive character encoding (%%35%63) to cross to a target.

In C, intruder tries to embed client-side script as a bogus routine, a JavaScript code, into variable (*var*). It may be transferred and run on other web client victims after all. This is also known as the Cross-Site Scripting (XSS) technique.

And more sophisticated attacks along with more novel web programming techniques, it is hard to distinguish between natural and unnatural ones responsively such as web application programs access information in databases using function calls. They set up connections at first then pass commands which mostly are the SQL SELECT-statements as those function's arguments to database services. These are also known as the SQL injection attacks. Attacks mostly try to find something (e.g., raw information, database schemata, and DBMS configurations) or try to perform some negative actions (e.g., data manipulation, privilege escalation, and server-side command execution).

3 Entropy Calculations

Shannon entropy [16] is the widely used method to measure uncertainty value in the system. In principle, the entropy, H(X), of continuous random variable X with probability density function p(x) is given by equation (1).

$$\mathrm{H}(X) = -\int \mathrm{p}(x)\log_2(\mathrm{p}(x))\,dx \tag{1}$$

For discrete domain, the entropy value of discrete variable X with density probability $\mathrm{p}(x)$ is specified by equation (2).

$$\mathrm{H}(X) = -\sum_{x\in X} \mathrm{p}(x)\log(\mathrm{p}(x)) \tag{2}$$

Entropy method is applied to measure complexity in various areas including text data. Entropy text analyzer [12] modifies original Shannon entropy to measure complexity in text data instead of random variables. Entropy of text T are calculated by using equation (3) where text T contains λ words with n different ones, and, p_i (i = 1 to n), is number of time that the ith-word happen in the text T.

$$E_t(p_1, \ldots, p_n) = \frac{1}{\lambda}\sum_{i=1}^{n} p_i[\log_{10}(\lambda) - \log_{10}(p_i)] \tag{3}$$

$$E_{Rel} = \frac{E_t}{E_{max}} \times 100 \tag{4}$$

$$E_{max} = \frac{1}{\lambda}\sum_{i=1}^{\lambda} 1[\log_{10}(\lambda) - \log_{10}(1)] = \log_{10}(\lambda) \tag{5}$$

The relative entropy (E_{Rel}), in equation (4), is formulated to normalize entropy value. This function calculates the percentage of entropy value by its maximum entropy value. The relative entropy value is usually uses to compare the complexity in different data domain.

The maximum entropy (E_{max}), in equation (5), is the entropy of text with all words which occur exactly once. This value is calculated to measure the highest complexity in the system.

4 Chromatography-Like Entropy Analysis

Chromatography is a scientific method to separate mixture material. In paper chromatography, small dots of samples are place into chromatography papers and put the papers into solvent liquid. When the solvent rises through the paper, it carries the samples going upward. Different samples are move in different distance.

Entropy-based chromatography is the detection technique that applied from paper chromatography and text entropy. Separation method in Entropy-based chromatography processes by using relation that entropy value of different combined texts changes unequally. Following definition of chromatography principle is defined for entropy-based chromatography technique. Mobile phase is the relative entropy value of examined text. Solute is a group of inspected incident's words. Solvent is a group of words in log record that needs to identify normal or abnormal behavior. And, detector is the entropy distance value.

Principle behinds entropy-based chromatography method is the difference between the entropy value of related texts and entropy value of unrelated texts. To illustrated this principle, we define *W* is text of *N* words with different *M* ones and *PW* is list of frequency of each identical word in *W*

$$W = (w_1, w_2, \dots, w_N) \tag{6}$$

$$PW = (p_1, p_2, \dots, p_M) \tag{7}$$

From above definition, entropy of text *W* equals to

$$E_w(p_1, \dots, p_M) = \frac{1}{N}\sum_{i=1}^{M} p_i[\log_{10}(\lambda) - \log_{10}(p_i)] \tag{8}$$

While p_i ($i = 1$ to M), is frequency of the ith-word happened in the text *W*. When another word appends into text *W*, entropy of new text is vary into one of following cases.

Case 1: Word *x* is a single word appending into text *W* and word *x* is already happened in text *W*. Text *WX* is text *W* after appends with word *x*. And, *PWX* is list of frequency of each word in *WX*.

$$WX = (w_1, w_2, \dots, w_N, x) \tag{9}$$

$$PWX = (p_1, p_2, \dots, p_M + 1) \tag{10}$$

From previous description, entropy of text *WX* equals to

$$\begin{aligned} E_{wx}(p_1, \dots, p_M + 1) = {} & \frac{1}{N+1}\sum_{i=1}^{M-1} p_i[\log_{10}(\lambda) - \log_{10}(p_i)] \\ & + \left(\frac{p_M + 1}{N+1}\right)[\log_{10}\lambda - \log_{10}(p_M + 1)] \end{aligned} \tag{11}$$

Case 2: Word y is a single word appending into text W and word y is not happened in text W. Text WY is text W after appended with word y. And, PWY is list of frequency of each word in WY.

$$WY = (w_1, w_2, \ldots, w_N, y) \tag{12}$$

$$PWY = (p_1, p_2, \ldots, p_M, 1) \tag{13}$$

From previous description, entropy of text WY equals to

$$\begin{aligned} E_{wy}(p_1, \ldots, p_M, 1) &= \frac{1}{N+1} \sum_{i=1}^{M} p_i[\log_{10}(\lambda) - \log_{10}(p_i)] \\ &+ \left(\frac{1}{N+1}\right)[\log_{10}\lambda - \log_{10}(1)] \end{aligned} \tag{14}$$

From equations (11) and (14), entropy value of text when append with new word (E_{wy}) is always greater than entropy value of the same text that appends with recurring word (E_{wx}). Consequently, the entropy value of attack signature content appended in log text that has related attack content are less than entropy value of attack signature appended in log text that does not has related attack content. Entropy-based chromatography technique applies this relation to detect intrusive events.

Entropy-based chromatography is a process to measure the relation of log records with any attack signatures. Following definitions are used for illustrated this technique.

- A = intrusion type A word list
- B = intrusion type B word list
- C = intrusion type C word list
- Ref = reference word list
- X = word list of log records

Size of word lists A, B, C and Ref must be equal. Each word in word list Ref are identical and not appear in A, B and C. While compare with paper chromatography, X operates as solvent. A, B, C, and Ref word lists operate as solute. And entropy of combined text is mobile phase.

After identify word lists, following entropy values of combined text are calculated.

- E(A+X) is a relative entropy of combined text A and X
- E(Ref+X) is a relative entropy of combined text Ref and X

Entropy value of combined text Ref and X is used as normal event reference point. Entropy value of other combined text that equals to *E*(Ref+X) is identified as common event. In contrast, entropy of combined text that is less than *E*(Ref+X) is identified as anomaly event with related to solute text. The Entropy Distance (*ED*) on combination of Ref and X with other text is defined for detecting anomaly criteria. *ED* value of any text Z is defined as follows

$$ED_Z = E(Ref + X) - E(Z + X) \tag{15}$$

ED_Z is used to identify text X that relates to inspected event Z. If ED_Z is greater than 0, X relates to inspected event Z. But if ED_Z is equal to zero, X does not. However ED_Z cannot be negative because *E*(Ref+X) is always greater than *E*(Z+X).

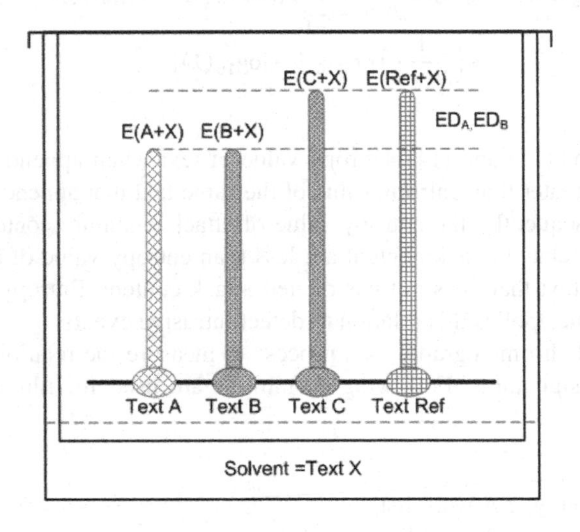

Fig. 1. Entropy-based chromatography model

Figure 1 is a model of entropy-based chromatography illustrated as paper chromatography. In this figure, ED_A and ED_B are greater than zero and ED_C is equal to zero, which means text X relates to intrusion type A and B but does not relate to intrusion type C. Text X may be an information of combined attack using A and B techniques.

5 Web Attack Detection Using Chromatography-Like Entropy Analysis

In our research, detection process that uses to identify complicated anomaly behaviors in web technology such as general attacks, cross site script attacks, SQL injection attacks is summarized as follows:

```
Procedure : Chromatography-like Entropy Analysis
Notation :
  [GEN] = General web attack word list
  [XSS] = Cross site script attack word list
  [SQL] = SQL injection attack word list
  [REF] = reference word list
  [LOG] = log parameters word list
Algorithm:
   1. Define [GEN], [XSS], [SQL], and [REF]
   2. For each line in www log
   3. Filter the parameters of GET method and decode any
        encoded URL
   4. Split into [LOG] using specified delimiter
   5. Append [LOG] into [GEN], [XSS], [SQL] and [REF]
   6. Calculate ED_GEN, ED_XSS, ED_SQL and classify anomaly
        events
```

1. Specified the general attack, cross site script attack, SQL injection attack, and reference word lists. Sample of words in each word list are as follows:

Table 1. Sample of attack word lists

Type	Words
General attack	['passwd', 'htaccess', 'win.ini', 'winnt', 'uff0e', ...]
Cross site script attack	['javascript', 'cookie', 'documentURI', 'referer', 'getElementByID', 'script', 'src', 'embeded'..],
SQL injection attack	['select', 'from', 'where', 'insert', 'into', 'value', 'update', 'delete', ...]

2. Read each line in log file and process following detection method.
3. Screen only parameter values in a GET method string and decode an encoded URL content. For example, a log record of GET method is "/directory /method ?var1=val1 &var2=val2&var3=val3%20val4". After screen parameters and change '%20' to spacebar. We get word list as ['val1', 'val2', 'val3 val4'].
4. Split each parameter word list using delimiter [/ <>?=]. The previous word list become ['val1', 'val2', 'val3', 'val4'].
5. Append all into general attack, cross site script, SQL injection, and reference word lists
6. Calculate the Entropy Distance values of each attack word list (ED_{GEN}, ED_{XSS}, and ED_{SQL}) and process detection method as follows:

- General attacks are detected when ED_{GEN} greater than 0
- Cross site script attacks are detected when ED_{XSS} greater than 0
- SQL injection attacks are detected when ED_{SQL} greater than 0

6 Experimental Results

The first web server log file from an anonymous commercial website that operates 24-hours a day is used in our experiment. It contains 3,526,362 query records of 197 operation days. After choosing table names to inspect anomaly signals with our proposed method, the SELECT-statements are screened then there are 52,837 log records which accesses targeted tables. Another log file, from web server of our internal university office, is used to examine our proposed process. This log file contains 73,124 records of HTTP requests from various sources. After combine them and omit static requests then we get 16,278 dynamic requests. By manual checking, all requests in log file are normal.

Our experiment has two phases. First, we calculate both relative entropy and entropy distance values of each attack word lists to detect anomaly records in normal log file. This process measures a false positive rate of the proposed method. Second, we add simulated attack records into log file and use our method to detect anomaly records. This process measures a false negative rate.

To measure false positive rate, log parameter word lists are generated from log file of all normal requests. After we appended log parameter word list into each attack word list, Relative entropy of each word list and entropy distance value of each attack will be calculated. Relative entropy of each word lists of normal log file shows in figure 2. In this figure, relative entropy values of all word lists are equal. (X-axis is the relative entropy and Y-axis is the line number of 16,278).

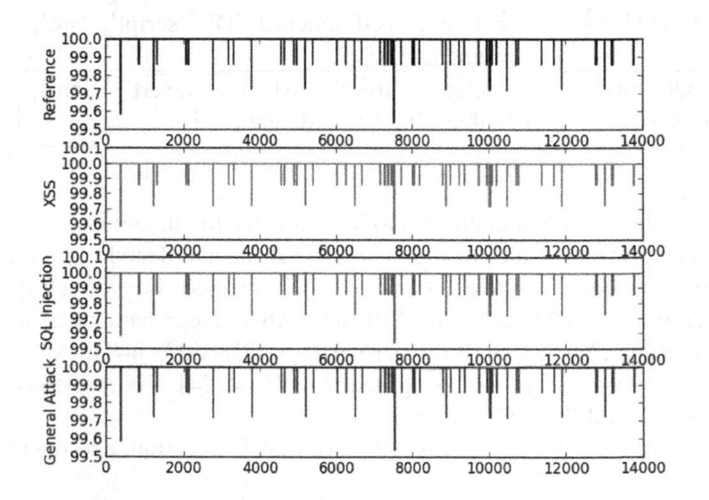

Fig. 2. Relative entropy value of all combined word lists

To measure false negative rate, we add attack requests information using NESSUS vulnerability scanner program. Then it has more 100 records which are 35 cross site script attacks, 42 SQL injection attacks, and 23 general web attacks. After that, log parameter word lists are generated from combined log file and append into each attack word list. Relative entropy of each wordlist are calculated and shown in figure 3. (X-axis is the relative entropy and Y-axis is the line number of 16,278).

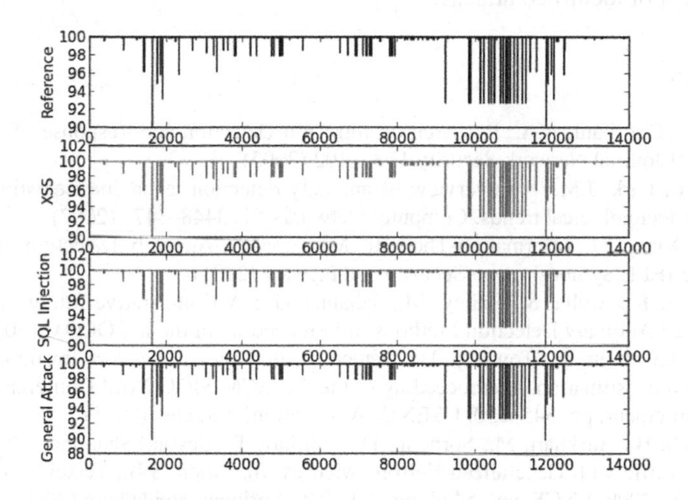

Fig. 3. Relative entropy of combined texts

Entropy distance of each attack word list of anomaly log file is shown in figure 4. In this figure, 92 records of attack are detected, i.e., 19 general attacks, 31 cross site script attacks, and 42 SQL injection attacks. From this experiment, false negative rate is equal to 8%. (X-axis is the entropy distance and Y-axis is the line number of 16,278).

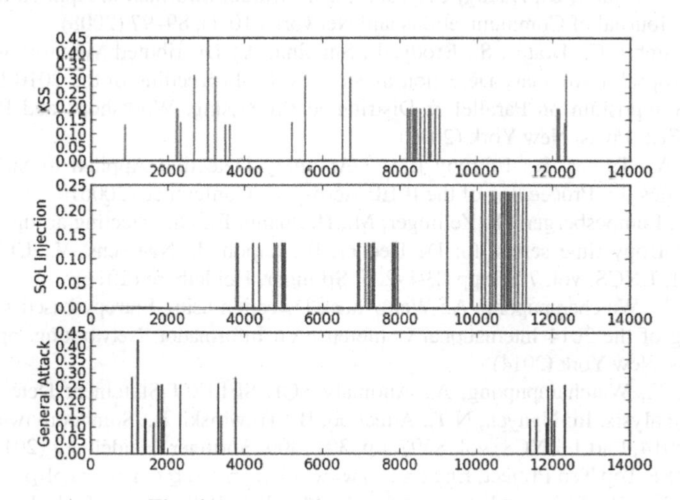

Fig. 4. Entropy distance of combined text

7 Conclusion and Future Works

As shown, our proposed chromatography-like entropy analysis method can detect intrusions because they changed the entropy significantly. And it can identify some web attacks which are classified as in distinct attack word lists with likelihood of each attack types, because of the entropy distance. In future work, we need to classified confident level of identified attacks.

References

1. Kabiri, P., Ghorbani, A.A.: Research on Intrusion Detection and Response: A Survey. International Journal Network Security 1, 84–102 (2005)
2. Patcha, A., Park, J.M.: An overview of anomaly detection techniques: existing solutions and latest technological trends. Computer Networks 51, 3448–3470 (2007)
3. Lee, W., Xiang, D.: Information-Theoretic Measures for Anomaly Detection. In: Proceeding of the IEEE Symposium on Security and Privacy (2001)
4. Nyalkalkar, K., Sinha, S., Bailey, M., Jahanian, F.: A Comparative Study of Two Network-based Anomaly Detection Methods. In: Proceeding of the INFOCOM (2011)
5. Gu, Y., McCallum, A., Towsley, D.: Detecting anomalies in network traffic using maximum entropy estimation. In: Proceeding of the 5th ACM SIGCOMM Conference on Internet Measurement, pp. 345–350. USENIX Association, Berkeley (2005)
6. Tellenbach, B., Burkhart, M., Sornette, D., Maillart, T.: Beyond Shannon: Characterizing Internet Traffic with Generalized Entropy Metrics. In: Moon, S.B., Teixeira, R., Uhlig, S. (eds.) PAM 2009. LNCS, vol. 5448, pp. 239–248. Springer, Heidelberg (2009)
7. Nychis, G., Sekar, V., Andersen, D.G., Kim, H., Zhang, H.: An empirical evaluation of entropy-based traffic anomaly detection. In: Proceeding of the 8th ACM SIGCOMM Conference on Internet Measurement, pp. 151–156. ACM, New York (2008)
8. Wagner, A., Plattner, B.: Entropy based worm and anomaly detection in fast IP networks. In: Proceeding of the 14th IEEE International Workshops on Enabling Technologies: Infrastructure for Collaborative Enterprise, pp. 172–177. IEEE Press, New York (2005)
9. Kopylova, Y., Buell, D., Huang, C.T., Janies, J.: Mutual Information Applied to Anomaly Detection. Journal of Communications and Networks 10(1), 89–97 (2008)
10. Arackaparambil, C., Bratus, S., Brody, J., Shubina, A.: Distributed Monitoring of Conditional Entropy for Anomaly Detection in Streams. In: Proceeding of the 2010 IEEE International Symposium on Parallel & Distributed Processing, Workshops and Phd Forum, pp. 1–8. IEEE Press, New York (2010)
11. Agogino, A., Tumer, K.: Entropy Based Anomaly Detection Applied to Space Shuttle Main Engines. In: Proceeding of the IEEE Aerospace Conference (2006)
12. Winter, P., Lampesberger, H., Zeilinger, M., Hermann, E.: On detecting abrupt changes in network entropy time series. In: De Decker, B., Lapon, J., Naessens, V., Uhl, A. (eds.) CMS 2011. LNCS, vol. 7025, pp. 194–205. Springer, Heidelberg (2011)
13. Threepak, T., Watcharapupong, A.: Web Attack Detection using Entropy-based Analysis. In: Proceeding of the 2014 International Conference on Information Networking, pp. 244–247. IEEE Press, New York (2014)
14. Threepak, T., Watcharapupong, A.: Anomaly SQL SELECT-Statement Detection Using Entropy Analysis. In: Nguyen, N.T., Attachoo, B., Trawiński, B., Somboonviwat, K. (eds.) ACIIDS 2014, Part I. LNCS, vol. 8397, pp. 301–309. Springer, Heidelberg (2014)
15. The OWASP Top Ten Project, `https://www.owasp.org/index.php/Top10`
16. Shannon, C.E.: Prediction and entropy of printed English. Bell Systems Tech. Jour. 30, 50–64 (1951)

Context-Aware Tourism Recommender System Using Temporal Ontology and Naïve Bayes

Chakkrit Snae Namahoot[1,*], Michael Brückner[2], and Naruepon Panawong[3]

[1] Department of Computer Science and Information Technology, Faculty of Science, Naresuan University, Phitsanulok, Thailand
chakkrits@nu.ac.th

[2] Department of Educational Technology and Communication, Faculty of Education, Naresuan University, Phitsanulok,Thailand
michaelb@nu.ac.th

[3] Department of Applied Science, Faculty of Science and Technology, Nakhon Sawan Rajabhat University, Nakhon Sawan, Thailand
jnaruepon.p@gmail.com

Abstract. In this paper, we present a Context Aware Thai Tourism Recommender System (CAT-TOURS) that applies a complex Naïve Bayes Model with boundary values, tourism ontology for Thailand and a temporal ontology to support decision making in tourism. Promising results are presented in the form of precision, recall and F measure for Websites related to Thailand's tourism industry. We compare the results with those gained with Latent Semantic Indexing (LSI).

This research was guided by the following aims: (1) find a simple method to classify Thai tourism Web documents that contain information on more than one topic, and (2) take into account time constraints in the process of making recommendations.

Keywords: Recommender System, Temporal ontology, Tourism, Naïve Bayes, LSI.

1 Introduction and Related Work

Currently, a vast number of Web sites provide services for finding travel information. However, tourists need to know more about touristic places and areas, e.g. relating attractions, hotels, dining, One Tambon One Product (OTOP, a government program to stimulate markets for local products) shops, and events. Most tourists search for interesting areas with a search engine, but the set of search results is often difficult to consume and confusing to understand because of the overload of information, mixed unwanted information, and uncategorized incoherent presentation. This leads to waste of time when extracting all relevant information and leads to inconvenient information gathering, even from a single information source.

* Corresponding author.

H. Unger et al. (eds.), *Recent Advances in Information and Communication Technology 2015*,
Advances in Intelligent Systems and Computing 361, DOI: 10.1007/978-3-319-19024-2_19

Automatic document clustering has gained increasing interest in the research community, especially in the course of the ever accelerating increase in the amount of information presented on Web pages in the World Wide Web. A review of machine learning for online document clustering of online news, blogs, e-mail, and digital library [1] refers to Rocchio's Algorithm [2], K-Nearest Neighbor [3], Naïve Bayes Algorithm (see, for example, [4]), and Support Vector Machine and states that Naïve Bayes is the best for e-mail, numeric and document clustering. Naïve Bayes also needs less data for the learning step and is not too complex. An additional advantage is that Naïve Bayes is convenient to develop when compared to other algorithms and is very fast [5].

When tourists seek travel recommendations, appropriate solutions have to take into account such contexts as time. An ontology-based recommender system should, therefore, support a temporal ontology. There are many representations of temporal ontologies, and some are listed in Table 1.

Table 1. Temporal ontology representations

Name	Specification	Source
SWRL Temporal Ontology	OWL GUI by Protégé	http://protege.cim3.net/cgi-bin/wiki.pl?SWRLTemporalOntology
W3C Temporal Ontology	OWL	http://www.w3.org/TR/owl-time/
RETR	Based on Allen's notation of events	[6]
CHRONOS	Handling of temporal ontologies with Protégé	[7]

However, not so many temporal ontologies deal with terms and appellations of non-Western calendars used in countries of Southeast Asia, among other regions in the world. There are some approaches based on the work of Eade [8], e.g. by Brückner [9], which can be applied not only to historic documents but also to contemporary events in the tourism market.

The tourism domain is not only well represented on the World Wide Web but has also gained increasing attention by semantic modeling experts. A number of ontologies have been created and published with different approaches and content (see [7], for a recent overview of tourism ontologies and their ranking following an Analytical Hierarchy Process). Some of these ontologies can be used to classify Web documents. This research was guided by the following aims: (1) to find a simple method to classify Thai tourism Web documents that contain information on more than one topic, and (2) to take into account time constraints in the process of making recommendations.

In this paper, we reduce the error in Web page clustering and apply sets of words, thereby introducing an improvement of the Naïve Bayes algorithm - not only by adding synonyms and related words into the clustering technique but also by allowing the categorization of a Web page into more than one group by adjusted boundary values of the Naïve Bayes algorithm. As an example, a Web page may contain (1) เที่ยว (*travel*), ภู (*hill*), เขา (*mountain*) categorized as Attraction; (2) อาหาร (*food*), ร้านอาหาร (*restaurant*), ทาน(*dine*), and กิน (*eat*) categorized as Dining; and (3) ที่พัก (*accommodation*), รีสอร์ท (*resort*) , นอน (*sleep*) categorized as Accommodation. Consequently, this Web page should be categorized into three categories: Attraction, Dining and Accommodation. In the following, we lay out the methodology of this research, report on the testing and results, and finally draw conclusions and indicate some directions of future work.

2 Methodology

The Context-Aware Thai Tourism Recommender System (CAT-TOURS) is divided into four parts: the novel (1) Naïve Bayes algorithm with Boundary values (NBB), (2) a tourism ontology focusing on Thailand, (3) a temporal ontology, and (4) the Tourism Recommender System. These parts will be elaborated on in the following.

2.1 Naïve Bayes with Boundary Values (NBB)

A Naïve Bayes classifier is a simple probabilistic classifier based on applying Bayes' Theorem with strong independence assumptions. These assumptions make the computation of Bayesian classification approach more efficient but this assumption severely limits its applicability. Depending on the precise nature of the probability model, the Naïve Bayes classifiers can be trained very efficiently by requiring a relatively small amount of training data to estimate the parameters that are necessary for classification [10].

In this paper, we propose an improvement of the Naïve Bayes algorithm called Naïve Bayes with Boundary values (NBB) for the classification of Web documents. We concentrated on touristic Web sites linked from Truehits, a DMOZ-like Web site in Thai language, and implemented the system as follows:

1. Apply Naïve Bayes algorithm for learning with 1,048 tourism Web pages from Truehits and six categories including Attraction (233 Web pages), Accommodation (200 Web pages), Dining (318 Web pages), Souvenir (54 Web pages), OTOP (88 Web pages), and Events (155 Web pages).
2. Improve the Naïve Bayes algorithm for a Thai tourism Web classification with increased efficiency.
3. Test of the Web document classification by applying 500 Web pages from Google search results with the NBB algorithm and measure the efficiency using F-Measure.
4. Apply NBB to CAT-TOURS.

The NBB algorithm is based on Eq. (1). Since the results of the calculation of the NB were negative in all categories and the frequency of queries to the group with the most negative results of calculations, we used the absolute value (positive values) to determine the Web category classification in Eq. (2).

$$C_{map} = \operatorname*{argmin}_{c \in C} \left(P(c) \textstyle\sum_{k=1}^{n} \log\left(P(t_k|c)\right)\right) \tag{1}$$

$$C_{map} = \operatorname*{argmax}_{c \in C} \left(P(c) \textstyle\sum_{k=1}^{n} \log\left(P(t_k|c)\right)\right) \tag{2}$$

$$C_{iMax} <= \text{threshold} >= C_{iMin},$$
$$C_{iMax} = \max(Cmap) \text{ and } C_{iMin} = \min(Cmap), \text{with } 1 <= i => 6 \tag{3}$$

where Cmap is the Web classification result and means the probability of the multiplied result by the probability P (c) with P (t_k | c) or briefly called probability of Naïve Bayes (P(NB)), Ci is probability of Attraction (At), Accommodation (Ac), Dining (D), Souvenir (S), OTOP (O) and Events (E). t_k is the word frequency, C is a category, P(c) is the probability of each of category and i is category counter 1 to 6 (i.e., Attraction, Accommodation, Dining, Souvenir, OTOP and Event). The improved Naïve Bayes algorithm with Boundary values (NBB) can be described as follows:

1. Initialize algorithm parameters
2. Used Web pages from test data set 1,048 Web pages from Truehits Web directory (by removing HTML Tags and choosing only web contents) and divide into six categories (Attractions, Accommodation, Dining, Souvenir, OTOP, and Events).
3. Calculate P (NB) of each category with Naïve Bayes algorithm (equation 1) using terms from tourism ontology([11], [12]) and many words from variation word and dictionary for tourism Web classification.
4. Define minimum (CiMin) and maximum (CiMax) of P (NB) of each category as threshold values using equation 2 and for Web classification automatically (Table 2).
5. Test 500 websites for classification by calculating P(NB)and compare P(NB) with CiMin and CiMax in Table 2 of every category using following condition:
   If P(NB) is between CiMin and CiMax of each category, then categorize; if not, then do not categorize
6. Repeat for all categories

Terminate after 500 websites and summarize Web classification results.

The NBB algorithm for Web classification can classify a Web page into more than one category, which is more in line with the diverse content on tourism information Web documents. This leads to an improvement of efficiency, see Table 3 and Table 4.

Table 2. Minimum and maximum boundary values for each category using Naïve Bayes (1,048 web documents)

Category	Minimum Boundary Value (Cmin)	Maximum Boundary Value (Cmax)
Attraction	0.08	20.06
Accommodation	0.08	4.35
Dining	0.13	14.97
Souvenir	0.02	0.47
OTOP	0.03	1.17
Events	0.05	4.21

2.2 Thai Tourism Ontology

Touristic spots have many different characteristics and attributes that have to be taken into account for the process of tourism recommendation system. For the tourist there are such criteria as cost, accommodation, attractions, restaurant, souvenir, traveling time or season and facilities that might be important for the purchasing recommendation. The data on tourism ontology comprise 10 classes including Province, Amphoe, Tambon, Event, Accommodation, Attraction, Restaurant, Souvenir, OTOP, and Transportation (Figure 1).

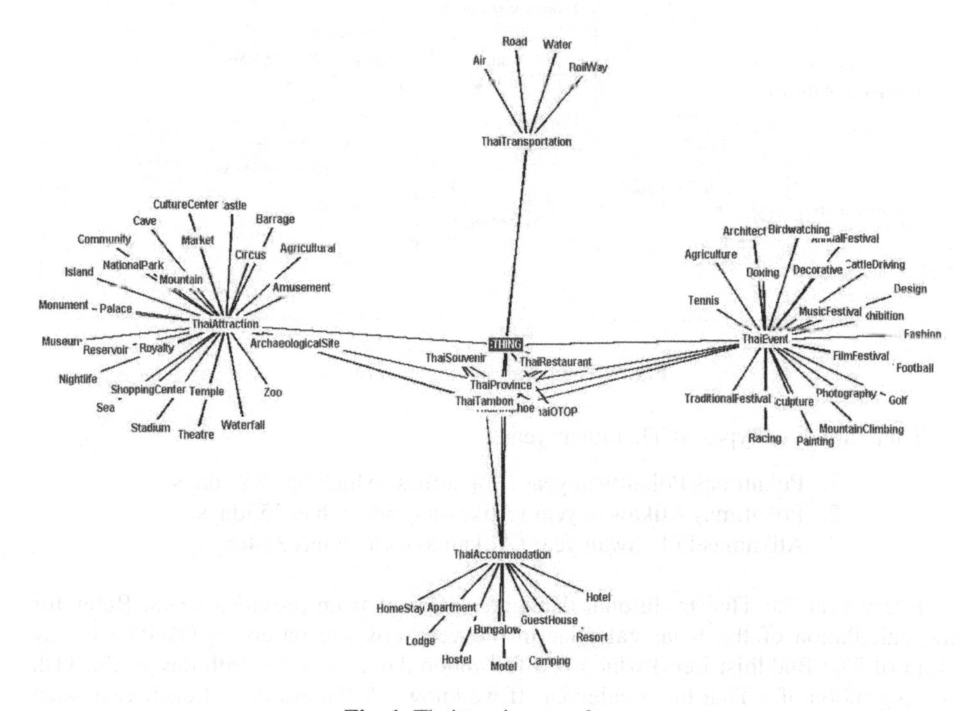

Fig. 1. Thai tourism ontology

The accommodation subclasses for Thailand that tourists can search for include Motel, Lodge, Resort, Guesthouse, Hotel, Hostel, Apartment, Camping, Home stay, and Bungalow. The event classes cover architecture, music festival, mountain climbing, sculpture, bird watching, decorative, racing, fashion, exhibition, photography, football, traditional festival, golf, design, annual festival, painting, film festival, tennis, boxing, agriculture and bull riding and trekking as sub classes. The attraction classes include mountain, waterfall, museum, cave, national park, circus, shopping center, barrage, theatre, reservoir, amusement, zoo, castle, nightlife, archaeological site, island, stadium, sea, community, royalty, temple, market, palace, culture center, monument and monastery (in Thailand mainly Buddhist temples).

2.3 Temporal Ontology

Information for seasons and time as well as important traditional Thai dates is something that tourists should consider, because they will not miss important events regarding their region of interest. Figure 2 shows the design of the Temporal Ontology that links the Attraction and Event (festivals) classes with the three main seasons in Thailand: summer, rainy season, and winter. When tourists input a travel time, the system checks traditional Thai events and attractions that tourists are recommended to visit during that time.

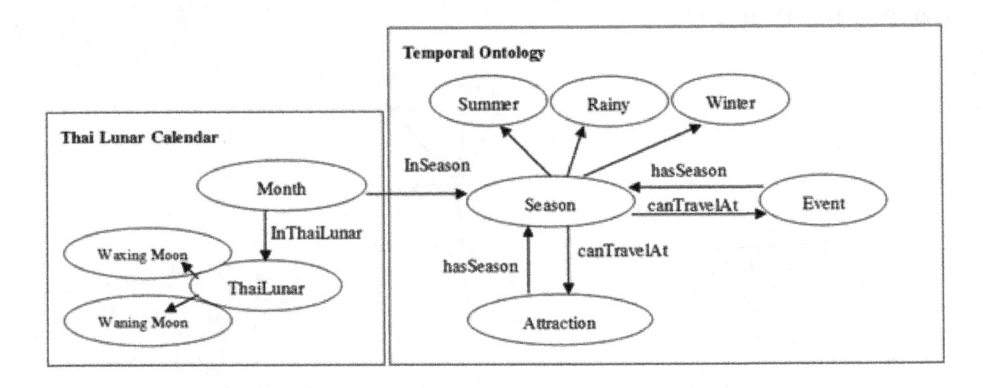

Fig. 2. Temporal ontology overview

There are three types of Thai lunar years:

1. Pokatimas Pokatiwan year (Pokatimas) which has 354 days
2. Pokatimas Atikawan year (Atikawan) which has 355days
3. Atikamas Pokatiwan year (Atikamas) which has 384days

Every year the Thai traditional dates are different from previous years. Rules for the calculation of the lunar calendar in the yearbook are based on Ok-Phansa day (End of The Buddhist Lent) which is a full moon day and is the 15th day of the 11th waxing moon of a Thai lunar calendar. If we know Ok-Phansa day of each year, then we can calculate other Buddhist days of each year as well. Therefore, we have implemented the Thai Lunar Calendar Algorithm to calculate Thai Lunar (Waxing moon and Waning moon) dates automatically to get exactly the Thai Buddhist Days described in the following.

The Thai Lunar Calendar algorithm:

1. Input: Tourist travel date e.g. 25/11/2015
2. take the year from 1 (2015 in this case)
3. identify types of the Thai lunar year above (Pokatimas, Atikawan and Atikamas) by calculating the number of days in the input year (from tourist) using following formula [14].

$$days = \sum_{i=1}^{12} dm_i + d_{in} - d_{pre} \tag{4}$$

where days is number of days in the year of input year, dm_i is number of days in each month of the input year, i is the number of month. (1-12),d_{in} is the number of the Ok Phansa (òk pan-săa) day in the input year, and d_{pre} is the number of the Ok Phansa (òk pan-săa) day in the previous year of the input year. For example, the Ok Phansa day (15th day of the 11th waxing moon) for input year and previous year are 27th October 2015 and 8th October 2014 respectively. Thus, the type of Thai lunar year is 365+27-8 = 384 which can be identified as Atikamas year using the following conditions

   3.1. If days equal to 354, then identify as the Pokatimas year
   3.2. If days equal to 355, then identify as the Atikawan year
   3.3. If days equal to 384, then identify as the Atikamas year

4. Find other Buddhist days from Ok-Phansa day using following conditions
   4.1. increase number of month of Ok-Phansa day (October, i=10) by 1 (i of Ok-Phansa day +1 which is)until i is equal to 12
   4.2. decrease number of month of Ok-Phansa day by 1 (i-1) until i equal to 1
   4.3. find the 15th day of the waxing moon and the 1st day of the waning moon
      4.3.1. increase i by 1, we find Loy Krathong day (15th day of the 12th waxing moon)which is 25th November 2015
      4.3.2. decrease i by 3 we find Asalha Puja day (15th day of the 8th waxing moon) which is 30th July 2015 and Buddhist lent day (1st day of the 8th waning moon), which is 31th July 2015.
      4.3.3. decrease i by 4 we find Visakha Puja day (15th day of the 7th waxing moon), which is 1st June 2015.
      4.3.4. decrease i by 6 we find Makha Puja Day (15th day of the 3th waxing moon) which is 4th March 2015
5. Display all Thai lunar dates according to Buddhist days.
6. Compare results from 5 and if the input day and month are exact day and month or even close to Buddhist days then recommend tourist for traveling
7. Display results and terminate.

3 Tourism Recommender System Architecture

Figure 3 shows the architecture of CAT-TOURS (Context-Aware Tourism Recommendation System), which can be described as follows:

The Temporal Ontology helps find the Thai festivals during the time of the year. NBB enables to find more information regarding six categories as accommodation, restaurant, attraction, etc. The system can help save time by finding various pieces of information and dismissing unwanted data.

1. Tourists choose travelling dates, and the system checks with temporal ontology whether or not the dates match with important Thai lunar dates
2. Users provide words or a sentence, and the system checks whether they are related to a province or other tourist spot, in which case the system crawls relevant Web documents from Google related to the tourist information of the province/spot and stores them in the database to prepare the Web document classification.
3. Classify those websites using NBB and show the results of classification of travel sites in six categories: Attractions, Accommodation, Dining, Souvenir, OTOP, and Events.
4. From 2, if it is not a province the system checks the input data and search for more detail of the region concerned for identifying a name of the province related to input data and then shows fundamental information of this province using tourism ontology pattern that are stored in OWL.
5. The system recommends tourist information based on the time and season applying the temporal ontology.

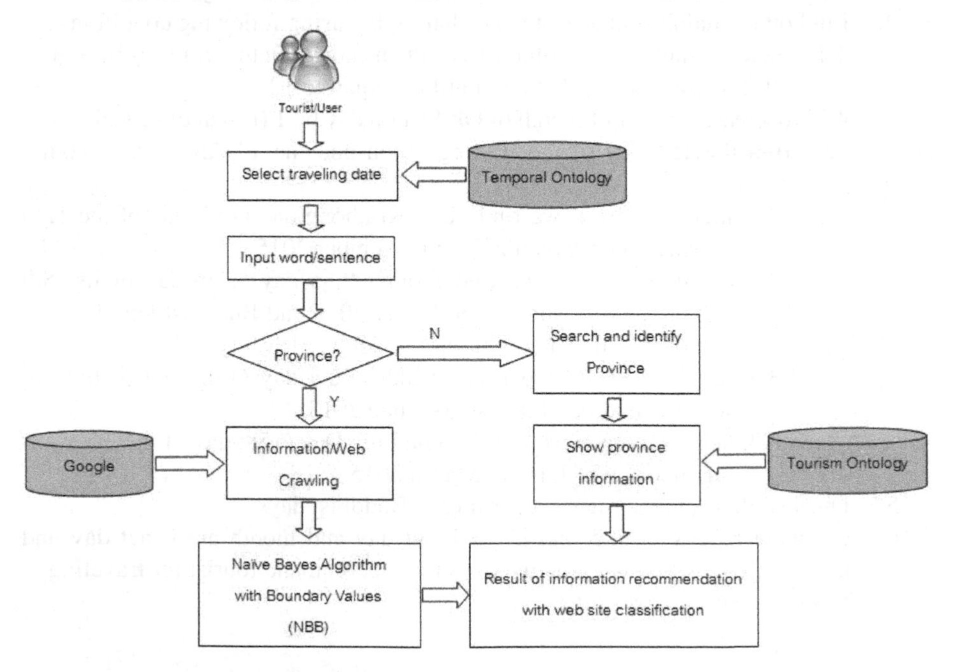

Fig. 3. CAT-TOURS system architecture

4 Testing and Results

4.1 Classification Results

The results relating CAT-TOURS are divided into two parts:

- Tourist websites classification results, and
- The interface of the system showing results suggesting travel sites based on the classification algorithm of NBB, result for travel information by time or season (Thai lunar calendar).

We used 500 tourism Web documents for classification and calculated the F-Measure for measuring the efficiency of NBB by Equation 5, for which results are shown in Table 3.

$$F = 2 * \left(\frac{P * R}{P + R} \right) \quad (5)$$

P is True Positive/(True Positive + False Positive)
R is True Positive/(True Positive + False Negative)
True Positive is web site in category and system was in that category
False Positive is web site not in category and system was in that category
False Negative is web site in category and system was not in that category

Another rather mature approach for automated document classification is the Latent Semantic Indexing (LSI), which uses a strictly mathematical method and is, therefore, independent of the natural languages involved.

Table 3 shows a comparison of tourism Web classification using LSI, Naïve Bayes (NB) and Naïve Bayes with boundary values (NBB). Precision P, recall R and F-Measure F have been calculated. LSI and NB algorithms result in 74.37% and 78.25% for precision, 62.88% and 70.04% for recall, 67.01% and 72.51% for F-Measure, respectively, because LSI and NB algorithms classify the Web documents into one category. The LSI algorithm produces some incorrect site classifications, such as those in the category Events. (41.83% accuracy), but classifies in the category of tourist or some other category, due to the content of those pages not being consistent with the frequency of the actual content. NB algorithm only uses P (NB) value and not so much the frequency of the search term in this category, which results the low efficiency of classification, especially in categories of Souvenir (Sou) and Event (Even) with accuracy 55.59% and 64.21%, respectively. This may be due to websites with rich content on the various categories such as a web site about the attraction but including other contents or categories as hotels and restaurants. However, the frequency of words appearing in the section of the restaurant is higher than attraction section. As a result, the classification does not correspond to the attraction website content rather it is classified in the restaurant category. NBB algorithm leads to 100% for precision,

97% approximately for F-Measure, and Web documents can correctly be in more than one category (Table 3). Moreover, the F-Measure data show that the NBB algorithm is more efficient in classification Web documents for Thailand tourism information.

Table 3. Performance of the classification algorithms

Category	LSI			NB			NBB		
	P	R	F	P	R	F	P	R	F
Att	81.82	57.91	67.82	70.42	88.57	78.46	100	95.71	97.81
Acc	87.76	54.34	67.12	94.76	63.21	75.83	100	100	100
Res	71.87	96.18	82.27	83.27	91.25	87.08	100	100	100
Souv	82.58	66.72	73.81	55.62	55.57	55.59	100	82.01	90.12
OTOP	80.34	63.23	70.77	100	58.61	73.90	100	95.34	97.61
Even	41.83	38.88	40.30	65.45	63.01	64.21	100	93.54	96.66
Average	74.37	62.88	67.01	78.25	70.04	72.51	100	94.43	97.03

4.2 System Testing

Figure 4 shows the interface of Tourism Recommendation System for Thailand in English. A user types in "travel in Phitsanulok" in English (or in Thai for the Thai Interface) and enters a travel date 03-11-2557 (3th November, 2014).The system shows basic information about Phitsanulok Province on the left screen. On the right screen, six tourist information categories ranked by popularity with URL websites are displayed. On the bottom of the screen, traveling recommendation based on Thai lunar calendar is shown, which mentions correctly the Pakthong Chai tradition, the Loy Krathong Festival and Buddha Day on 5th, 6th and 14th November 2014, respectively.

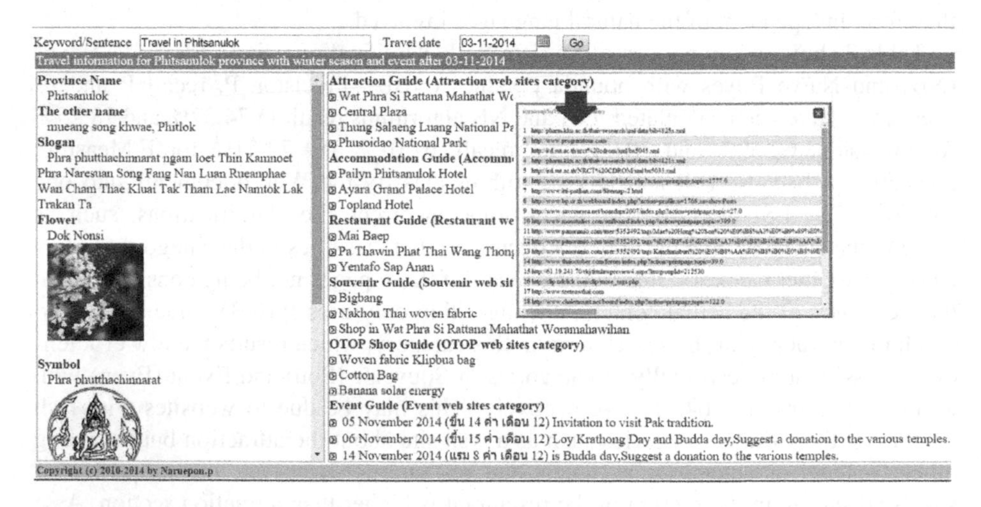

Fig. 4. Result example of Tourism Recommendation System in Thailand English version

5 Conclusion and Further Work

This paper introduces the Context Aware Thai Tourism Recommendation System (CAT-TOURS) with a Temporal Ontology and a novel Naïve-Bayes Algorithm with Boundary values (NBB). The NBB exhibits increased efficiency, which has been tested for Thailand tourism Website classification. We used tourism Web pages as a learning data set and defined thresholds based on minimum and maximum boundary values for probabilities for each of the six categories used in the classification testing with LSI, NB, and NBB algorithms. The result shows that NBB algorithm is the most efficient resulting in 100% for precision, 94.43% for recall and 97.03% for F-Measure. We have implemented the Temporal Ontology which gave additional details for attractions, events and festivals that tourists should visit during the time of their stay. Tourists specify only the province and the travelling dates, which results in the system suggesting where and when special and important traditional events are located near the destination.

This research also has some limitations. Due to the location-unaware setting via conventional Internet access, we did not consider the user's current location as a parameter for recommendations. This is planned for one of the next rollouts of CAT-TOURS, which will be based on a similar road segment ontology as described in [14]. Moreover, we will allow users to choose types of tourism, accommodation and foods by taking into account gender, age, and the number and duration of travels. Then, the system will be able to use the data and compare them with the rules of the tourism ontology based on SWRL, which is also stored in the data model. Finally, the system calculates the travel recommendation to meet the needs of users with the help of data mining techniques.

Another limitation is the rudimentary usability test of CAT-TOURS. However, we plan to set up a similar test suite as has been described by Zins et al. [15].

References

1. Ceci, M.: Classifying Web documents in a hierarchy of categories: a comprehensive study. Journal of Intelligent Information Systems 28, 37–78 (2007)
2. Rocchio, J.: Relevance feedback in information retrieval. In: The SMART Retrieval System: Experiments in Automatic Document Processing, pp. 313–323. Prentice Hall, Englewood Cliffs (1971)
3. Groza, A., Dragoste, I., Sincai, I., Jimborean, I., Moraru, V.: An ontology selec-tion and ranking system based on the Analytical Hierarchy Process. In: The 16th International Symposium on Symbolic and Numerical Algorithms for Scientific Computing, Timisoara, Romania (2014)
4. McCallum, A., Nigam, K.: A Comparison of Event Models for Naïve Bayes Text Classification. Journal of Machine Learning Research 3, 1265–1287 (2003)
5. Beineke, P., Hastie, T., Vaithyanathan, T.: The sentimental factor: improving review classification via human-provided information. In: Proceedings of the 42nd Annual Meeting of Association for Computational Linguistics (ACL 2004), Barcelona, Spain. Association for Computational Linguistics (July 2004) (2014), doi:10.3115/1218955.1218989

6. Hemalatha, M., Uma, V., Aghila, G.: Time ontology with Reference Event based Temporal Relations (RETR). International Journal of Web and Semantic Technology 3(1), 23–31 (2012)
7. Perventis, A.P., Polyxeni Marki, E., Petrakis, G.M., Sotirios, B.: CHRONOS: A Tool for Handling Temporal Ontologies in Protégé. In: IEEE 24th International Conference on Tools with Artificial Intelligence, pp. 460–467 (2012)
8. Eade, J.C.: The Calendrical Systems of South-East Asia. Leiden, Brill (1995) ISBN 9789004104372
9. Brückner, M.: Multilingual semantic modeling for cultural heritage organizations in ASEAN. ACM Journal of Computing and Cultural Heritage (in press, 2015)
10. Khan, A., Baharudin, B., Lee, L.H., Khan, K.: A review of machine learning algorithms for text-documents classification. Journal of Advances in Information Technology 1(1), 4–20 (2010)
11. Panawong, N., Snae Namahoot, C.: Performance Analysis of an Ontology-Based Tourism Information System with ISG Algorithm and Name Variation Matching. Naresuan University Science Journal 9(2), 47–64 (2013) (in Thai)
12. Panawong, N., Snae, C.: Search System for Attractions in Thailand with Ontolo-gy and Name Matching. Journal of Information Science and Technology 1(2), 60–69 (2010) (in Thai)
13. Siri, C.: Auspicious Thai calendar. Bangkok, Kurusapa Printing Ladphrao (2006)
14. Sadeghi Niaraki, A., Kim, K.: Ontology based personalized route planning system using a multi-criteria decision making approach. Expert Systems with Applications 36, 2250–2259 (2007)
15. Zins, A.H., Bauernfeind, U., Messier, F.D., Venturini, A., Rumertshofer, H.: An experimental usability test for different destination recommender systems. In: Information and Communication Technologies in Tourism (ENTER), Cairo, Egypt, pp. 228–238. Springer (2004)

The Research on Webpage Ranking Algorithm Based on Topic-Expert Documents*

Peng Lu[1] and Xiao Cong[2]

[1] Department of Media Technology and Communication, Northeast Dianli University, China
peng.lu2008@gmail.com
[2] College of Science, Northeast Dianli University, China
lup595@nenu.edu.cn

Abstract. A search engine returns a ranked list of documents for a query. If the query is broad then the returned list is usually too long to view fully. Studies show that users usually read only the top 10 to 20 results. Also, web search results can be much improved by using the information contained in the link structure between pages. The two best-known algorithms are HITS and PageRank. In this paper, based on the analysis of links-based and content-based sorting algorithm, a new scheme-Topic_ExpertRank for improving the accuracy and the efficiency of the search engine is contributed.

Keywords: Web search, Ranking algorithm, Link analysis, Web graph, Topic_ExpertRank.

1 Introduction

With the rapid development of the Internet, the vast online information, ordinary Internet users want to find the information they need is difficult needle in a haystack. There is an urgent need for an excellent search service which take the online content of complicated becomes easy access to information, and search engine is in such a background. At present, search engine is in a period of rapid development, and according to the latest survey report shows CNNIC [1]: the services in the user frequently used, the use ratio of Search engine has reached 79.3%. In the US, the search engine usage has reached 93%. As can be seen, the search engine has become a hot spot of the Internet, more and more people's attention. However, the quality of documents to the retrieval results is good or bad, and high-quality documents in the result set can have a better ranking has very important implications for users, these are also an important indicator to measure the pros and cons of search engine technology. Therefore, to assess the importance of the documents and sorting is one of the key technologies to solve of the search engines.

The importance index of documents include the number of backlinks, the correlation between the query and the distance between documents, etc.[2]. Here, the number

* This work was supported by the Northeast Dianli University Dr. Scientific Research Foundation Project (No. BSJXM-201219).

H. Unger et al. (eds.), *Recent Advances in Information and Communication Technology 2015*,
Advances in Intelligent Systems and Computing 361, DOI: 10.1007/978-3-319-19024-2_20

of backlinks of documents refers to the index of popularity; the correlation refers to the association of the documents with the query topic which to determine whether the document that satisfy the query requirements; and the distance of documents are used to evaluate the impact of the link. How to use these indicators to measure the importance of a document and excluding the impact of human factors on the search results, and find the most important and most useful web pages for Internet users is the problem of ranking algorithm to solved [3].

2 Related Works

In the Modern Information Retrieval (IR) [4], the system based on the user's query to retrieve relevant documents, and IR retrieval algorithms are usually based on the matching of words in the document. Web search is one of the important research of IR. However, due to the quality of the page is not balanced and constantly changing. Therefore, using traditional information retrieval techniques, such as TFIDF, produced very satisfactory results on the web. On the other hand, the web is composed by a large number of unstructured pages and links between them, and Broder showed that the web with the "bow-tie" shaped structure [5]. A large number of studies suggest that the information between the link structure of pages can be improved the search results. Therefore various link-based ranking strategies have been proposed in recent years, and two of the most famous algorithms are PageRank algorithm [6][7][8] and HITS algorithms[9][10]. Henzinger's study showed that the use of the links between the pages will produce very good results for sort [11].

The PageRank algorithm computes a grade vector in advance for each page to estimate the importance of each page on the web. This value is calculated in the offline and does not rely on the query. At query time, the sort results depend on the importance of scoring and detailed information retrieval query score, such as TFIDF. The advantage of this algorithm is that the generated of PageRank value is calculated on the entire web and not a small subset, so that it can improve the quality of search results. Moreover, the combined between PageRank values and traditions score takes little time during the query. However, since the query-independent, so PageRank cannot rely on their own to distinguish the authoritative pages of general sense and related to the query, as a result, ignore the relevance between links page and query topic and some pages with high PageRank value in itself, thus affecting the relevance and accuracy of search results.

Kleinberg proposed HITS algorithm. When a user enters a query, the HITS algorithm first to get a collection of related web pages which called the root set, then use the forward link and backward link of pages in the root set to expand and form a basic set, then try to find two types of pages in the basic set, which called hubs and authorities. The former is a set which point to the pages of the authority of the high value, and the latter is a set which are authority pages of the high value. Iterative computation hubs and authorities for each page until convergence, finally, the basic set of queries relevant to sort, and returned to the user according to the sorted results. However, since this algorithm is performed in real time in the query process, so it is not feasible to the search engine which would be processed thousands of queries current day. Bharat and Henzinger had discussed the HITS algorithm combines content analysis can compensate for some of the shortcomings of the algorithm and improve the accuracy of the query [10].

The studies of Chakrabarti[12] have shown that, the links to the page's subject is sensitive, that is, the link to the page usually points to other pages with the same subject[13]. This property not only explains why PageRank score was not rely on query so useful for sorting, but also pointed out the subjects can be used to improve the results which based on the calculation of the link. Matthew Richardson and Pedro Domingos proposed model of smart surfing [14], which called QD-PageRank. It will assigned greater weight of the subject to meeting relevant links in the model for the query. Take the content of the pages and analysis of the links for improve the quality of search. However, the disadvantage of this method is that for each query needs to be calculated in real time and the query response time is too long of the intelligent surfing. Taher H. Haveliwala proposed Topic-Sensitive PageRank [13] algorithm. On the basis of PageRank algorithm combine topic pages and query context to improve the accuracy of the query results [15][16]. This is a relatively more innovative ideas, it addresses a number of key issues of the quality of the search results. However, this method also has some problems, such as how to effectively define the categories, and how to determine a query condition may correspond to the topics. In order to improve the retrieval quality, Bharat and Mihaila proposed HillTop algorithm [17], namely to determine the sort weights of search results by quantity and quality of backlinks. The algorithm considers that the contribution from the relevant page links with the same topics of higher value than links from unrelated topics for weight calculation, and only the links from these expert pages to the target documents decided the main part of weight score linked pages. HillTop algorithm consists of two distinct phases which are calculated the score of expert pages and calculated the score the target pages. Therefore, it is combined with the PageRank and able to get more accurate search results [18].

In this paper, we propose a novel sorting algorithm which called Topic_ExpertRank, on the basis of link analysis, take advantage of the Topic-sensitive PageRank and HillTop and to make up for the shortcomings of both algorithms, but also take the distance of pages [19][20] as an index of importance. So, make the rank of important pages more forward and also further improve the overall accuracy of the results.

3 Program Improvement

In order to improve the accuracy of search results, the rankings of important documents are more front, we propose a novel method Topic_ExpertRank to sort the documents. The algorithm is divided into the following four processes.

- *Step One:* calculate the average distance of each page in the offline $Distance\_Score$ as a measure factor of the importance of pages.
- *Step Two*: calculate the $Target\_Score$ of target document which associated with the query use I-Hilltop algorithm.
- *Step Three*: calculate the content similarity score $Match\_Score$ of target document and query topics.
- *Step Four*: combined with the results of $Distance\_Score$, $Target\_Score$ and $Match\_Score$ get the final score of the pages, and take the results $Fina\_Score$ in descending order and returned to the users.

$$Fina\_Score = \left(0.2Match\_Score + 0.5Target\_Score + 0.3Distance\_Score\right) \times P\left(c_{ji} \mid q\right) \quad (1)$$

3.1 The Distance of Documents

About web analysis studies assume that the length between two pages is defined the number of clicks. However, with intervening clicks to measure the distance between two pages and do not reflect the intuitive distance of users [20]. For example, some pages having a large number of links, on the contrary, some pages only a few. For users, the cost of select a link from page with a lot of links and with fewer links are not the same. Its principle is that if a page has many backward links or it linked by many pages which have low values of distance, then the value of distance for this page is low. The following definitions illustrate this idea.

Definition 1. If the page *i* point to the page *j*, then the weight of link between the two pages defined as $\log_n O(i)$, and $O(i)$ is output degree of page *i*, that is the number forward links of page *i*. Set n is 7, because usually an average of 7 per page links to other pages.

Definition 2. The shortest distance between pages *i* and *j* is the minimum weight of path. The logarithmic distance denoted by the $D_{ij}(\min)$.

Where D_j represents the average value of shortest distance of all pages to the page *j*, and N represents the number of all pages. Instead of the traditional definition of distance, we put this definition is called Average-Clicks[20]. In this method, the weight for each link is $\log_7 O(i)$; when there is no path between the two pages, Set $D_{ij}(\min)$ to maximum $\log_7 O(N)$, and N is the number of pages.

Thus, if the distance between the page i and page j less than page i and page k, then the impact of the level of page i to page j must be greater than the impact to the page k. Do the issues of overhanging links in the PageRank does not affect the calculation of the distance values. Each page of the corresponding distance value stored in the distance database, as shown in Figure 1:

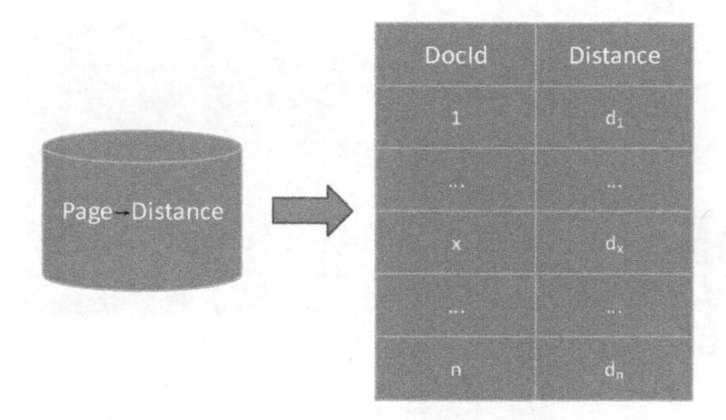

Fig. 1. The stored of distance values

3.2 ODP-Biasing

Open Directory Project(ODP) is the largest and most comprehensive human-edited directory in the Internet. It is a largest global directory of community which common maintenance and construction by volunteers from around the world, and the data is available free of charge. Up to now, it has 89,987 editors and produced over 1,023,693 categories. Based on ODP classification, according to the similarity with various directories and take the crawl pages categorized into directories in offline.

3.3 Improved Algorithm I-Hilltop of Hilltop

I-Hilltop was improved algorithm of Hilltop, it ensure quality, easy to make up for the noise and the recall rate and other shortcomings. The algorithm can guarantee the quality, and make up shortcomings such as prone to noise and the recall rate and others to the original algorithm at the same tiem.

In the select and index expert pages, the I-Hilltop handle pages database of search engine and pick out pages set which linked by good source of a particular topic, and examine all the pages which degrees above a certain threshold k ($k = 5$), detect these URLs whether or not point to m ($m <= k$) non-affiliated independent host, each qualifying pages are called expert document. After find the expert documents, I-Hilltop creates an inverted index to storage key phrases appear in the expert documents, such as title, heading and anchor etc.The URL was found by key phrases, and inverted index organized the match position of the expert document and key phrases in accordance with a list, and each of match position corresponding to the keywords appear in the expert document. In each match location is also stored the unique identifier of the phrase in the document, the code which represent the phrase category, such as title, heading or anchor, and offset of words within the phrase. In addition, the algorithm also provides a list of internal URL for each expert page, and provides identifier of keywords limited role for each URL, as shown in Fig.2.

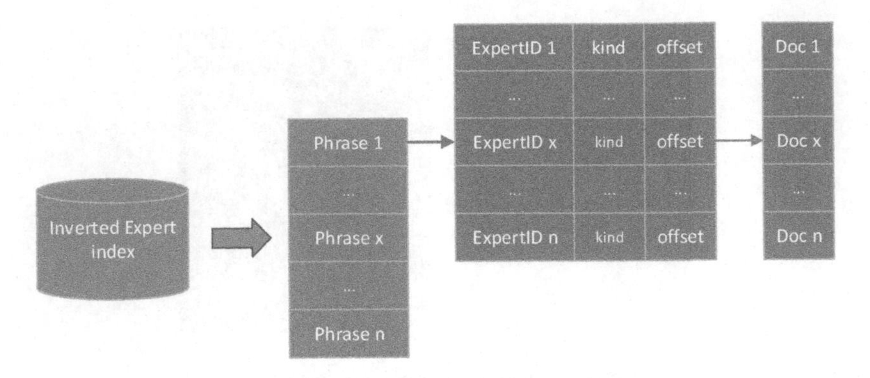

Fig. 2. Expert Documentation Index

4 Query Process

4.1 Select Topics

When a user enters a query q, the system uses native-Bayes algorithm and computing the similarity between top classification topic of ODP and query.

$$P(c_j \mid q) = \frac{P(c_j) \bullet P(q \mid c_j)}{P(q)} \infty P(c_j) \bullet \prod_i P(q_i \mid c_j) \tag{2}$$

Where, c_j is the j-th classification topics, q is the query entered by the user, q_i is the i-th keyword of the query entered in.

Select three topics which closest query, take all the pages under this topics as a candidate set of search results. According to the similarity with the query, and given a similarity rating scores for each candidate topic. Based on experience, set $P(c_{j1} \mid q) = 1$, $P(c_{j2} \mid q) = 0.8$, and $P(c_{j3} \mid q) = 0.6$.

4.2 I-Hilltop Algorithm Execution

Selected the pages which marked as an expert documents from the candidate set,and if there is no expert document, then according to the traditional information retrieval method to combined the distance score, and output the results in descending order. Otherwise, found all the expert documents and then find all the target pages by linking, merge all the expert documents and remove duplicate documents as the candidate set for the output results.

4.2.1 Calculate the Scores of Expert Documents

With the same HillTop, in order to reflect the quantity and quality of the key phrases of query key contains in the expert documents, and the degree of matching the query phrase,I-Hilltop calculate the score of expert documents.

Set k is the number of words in the input query q, and S_i is calculate the score of $k-i$ words contained in the key phrase accurately.

$$S_i = \sum_{i=0}^{2} LevelScore(p) * FullnessFactor(p,q) \tag{3}$$

Where $LevelScore(p)$ is the type score of key phrases p (Key phrases p with $k-i$ query terms), the *LevelScore* of title is 6 in the I-HillTop, heading is 6 and anchor text is 1. This score is set based on the topics of the expert pages, the assumption is that title more useful than heading , and heading more useful than anchor text. $FullnessFactor(p,q)$ is the score of terms in the p which appears in the q, set *plen* is the length of p, and m is the number of terms which in the p but not appearing in the q, $FullnessFactor(p,q)$ is calculated as follows:

If $m \le 2$, $FullnessFactor(p,q) = 1$; else $m > 2$, $FullnessFactor(p,q) = 1-(m-2)/plen$.

The goal of the algorithm is that the expert pages can able to match all query keywords, and the total expert score is composed of these three parts:

$$Expert\_Score = 2^{32} S_0 + 2^{16} S_1 + S_2 \tag{4}$$

4.2.2 Calculate the Scores of Target Pages

According to scores of expert pages to rank, then check these pages which they point to called the target pages. Calculation the scores of target pages not only reflect both the quantity and relevance of expert pages which linked them, but also to reflect the relevance of key phrases which they point of. The score of the target page is calculated by the following three steps:

- Step One: For every expert pages E, if there is a link from E to the target page T, then draw a straight edge $Edge(E,T)$. For every keywords w, set $occ(w,T)$ is the number of key phrases which contain w but restriction it appear $Edge(E,T)$ in the expert page, $Edge\_Score(E,T)$ is calculated as follows:

$$Edge\_Score(E,T) = Expert\_Score(E) \sum \{w\} occ(w,T) \tag{5}$$

For any query keywords, when $occ(w,T)$ is 0, $Edge\_Score(E,T)$ equal to 0.

- Step Two: check attached properties of expert documents which point to the same target, if there are more than two expert documents with subsidiary, then only retain the highest value of $Edge\_Score$.
- Step Three: Finally, calculation the $Target\_Score$ from sum all $Target\_Score$ which point to the target page.

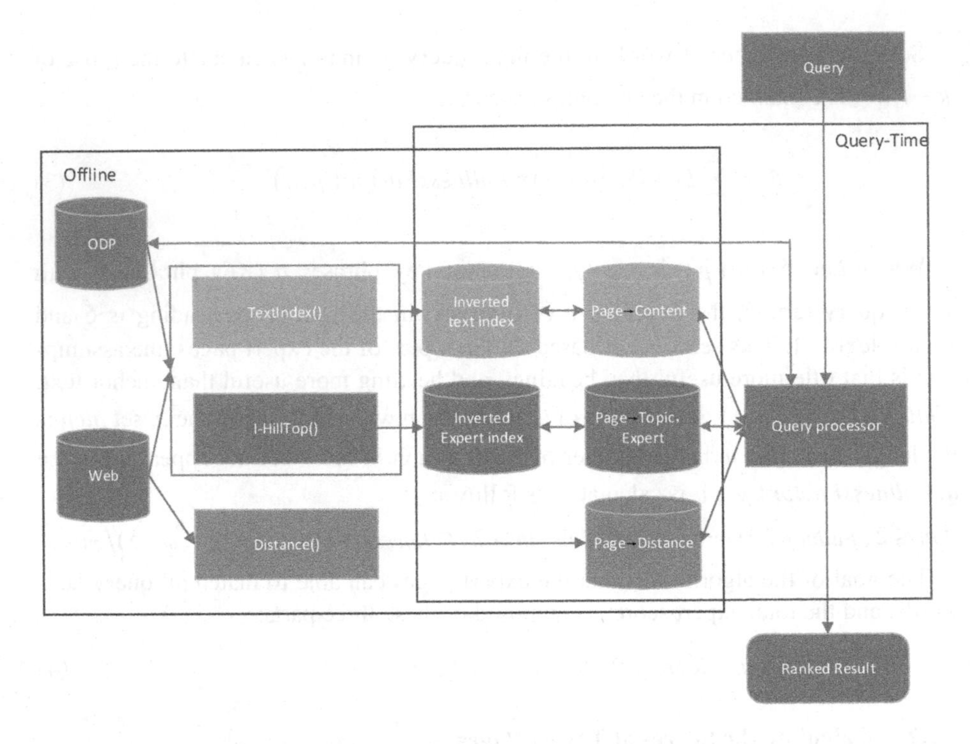

Fig. 3. Query process based on Topic-ExpertRank algorithms

4.3 Sort Results

1) Step One: If the expert documents are not in the target pages, then take these expert documents as part of the result set and the $Expert\_Score$ as *Target_Score* of this page. This would solve the problems of loss the query results and query results insufficiency.

2) Step Two: The use of content analysis method of traditional information retrieval for calculate the $Match\_Score$ based on query word and selected out of some experts and target pages.

3) Step Three: Combine distance score of offline computed, the final scores for result set as shown in Equation.

$$Fina\_Score = \left(0.2Match\_Score + 0.5Target\_Score + 0.3Distance\_Score\right) \times P\left(c_{ji} \mid q\right) \quad (6)$$

Where, $P\left(c_{ji} \mid q\right)$ is a topic score that matches the query, and select the most relevant three topics to the query $1 \le i \le 3$.

According the $Fina\_Score$ for each pages to order the result set in descending. And when did not find enough the expert pages, then

$$Fina\_Score = \left(0.2Match\_Score + 0.8Distance\_Score\right) \times P\left(c_{ji} \mid q\right) \quad (7)$$

Eventually, according to the $Fina\_Score$ of pages of the candidate set in descending order and returns to the user.

5 Conclusion

The search engine technology is the hot topic in the field of computer, and level or importance of pages has an important impact on the final ranking of search results. Although various sorting algorithms have been very rich, but so far, without any kind of ranking algorithm is perfect, so the search engines have stopped using only a valuable ranking algorithm to determine the results, but to absorb the essence of various ranking algorithms, and the proposed method in this paper also based on the idea.

In order to the results of rank more objective and reasonable, and the ranking algorithm still need to continue to do in-depth research in many aspects, the future we will conduct in-depth research on some smart algorithms and the depth of learning. It believe that there will be more and better research results appear in the near future, can help users find the information more quickly and accurately that they really need.

References

1. Report of China Internet Development Statistics, http://www.cnnic.net.cn/hlwfzyj/hlwxzbg/hlwtjbg/201407/P020140721507223212132.pdf
2. Pierre, B., Paolo, F., Padhraic, S.: Modeling the Internet and the Web: Probabilistic Methods and Algorithms. Wiley Press, Hoboken (2003)
3. Bharat, K., Broder, A.: A Technique for Measuring the Relative Size and Overlap of Public Web Search Engines. Computer Networks and ISDN Systems archive 30(1-7), 379–388 (1998)
4. Baeza, R.Y., Ribeiro, B.N.: Moderm Information Retrieval. Addison Wesley, New York (1999)
5. Broder, A.Z., Kumar, S.R., Maghoul, F., Raghavan, P., Rajagopalan, S., Stata, R., Tomkins, A., Wiener, J.L.: Graph structure in the web. The International Journal of Computer and Telecommunications Networking archive 33(1-6), 309–320 (2000)
6. The PageRank Citation Ranking: Bringing Order to the Web, http://ilpubs.stanford.edu:8090/422/1/1999-66.pdf
7. Brin, S., Page, L.: The anatomy of a large scale hypertextual Web search engine. In: 7th International World-Wide Web Conference. Elsevier Press, Brisbane (1998)
8. Wu, X.D., Vipin, K., Ross, J.Q., Joydeep, G., Qiang, Y., Hiroshi, M., Geoffrey, J.M., Angus, N., Bing, L., Philip, S.Y., Zhi, H.Z., Michael, S., David, J.H., Dan, S.: Top 10 algorithms in data mining. Knowledge and Information Systems archive 14(1), 1–37 (2007)
9. Kleinberg, J.M.: Authoritative sources in a hyperlinked environment. Journal of the ACM 46(5), 604–632 (1999)

10. Bharat, K., Henzinger, M.R.: Improved algorithms for topic distillation in a hyperlinked environment. In: Proceeding of the 21st Annual International ACM SIGIR Conference on Research and Development in Information Retrieval, pp. 104–111. ACM Press, Melbourne (1998)
11. Henzinger, M.: Hyperlink analysis for the web. IEEE Internet Computing 5(1), 45–50 (2001)
12. Soumen, C., Mukul, M.J., Kunal, P., David, M.P.: The structure of broad topics on the web. In: Proceeding of the 11th International World Wide Web Conference, pp. 251–262. ACM Press, Honolulu (2002)
13. Taher, H.H.: Topic-Sensitive PageRank. In: Proceeding of the 11th International World Wide Web Conference, pp. 517–526. ACM Press, Honolulu (2002)
14. The Intelligent Surfer: Probabilistic Combination of Link and Content Information in PageRank, http://research.microsoft.com/pubs/66874/qd-pagerank.pdf
15. Lawrence, S.: Context in Web Search. Data Engineering. IEEE Computer Society 23(3), 25–32 (2000)
16. Lev, F., Evgeniy, G., Yossi, M., Ehud, R., Zach, S., Gadi, W., Eytan, R.: Placing search in context: the concept revisited. ACM Transactions on Information Systems 20(1), 116–131 (2002)
17. Krishna, B., George, A.M.: When experts agree: using non-affiliated experts to rank popular topics. ACM Transactions on Information Systems 20(1), 47–58 (2002)
18. PageRank: meet Hilltop, http://isedb.com/20040127-658/pagerank-meet-hilltop
19. Ali, M.Z.B., Nasser, Y.: DistanceRank: An intelligent ranking algorithm for web pages. Information Processing & Management 44(2), 877–892 (2008)
20. Matsuo, Y., Ohsawa, Y., Ishizuka, M.: Average-clicks: A new measure of distance on the World Wide Web. Journal of Intelligent Information Systems 20(1), 51–62 (2003)

Measuring Opinion Credibility in Twitter

Mya Thandar and Sasiporn Usanavasin

School of Information Computer and Communication Technology(ICT),
Sirindhorn International Institute of Technology (SIIT),
Thammasat University, Thailand
{myathandar.ucsy,sasiporn.us}@gmail.com

Abstract. Today thousands of people in Hong Kong are protesting an election reform that would essentially mandate Beijing approval of candidates for Hong Kong's chief executive. Many people such as Hong Kong citizen, government staff, journalists and news channels express information and their opinions about Hong Kong Revolution over social media. In that event, we don't know whose opinions are strong or credential. Therefore when we take the information from opinion content, the identifier of an author may help to determine credibility. The opinions of specialists and recognized experts are more likely to be credential and to reflect a significant viewpoint. For that reason, we propose a new method to define the credibility of sentiment polarity based on their expertise or background knowledge and apply on Twitter: social media. Hence we identify the credibility of tweets polarity for a particular topic, we add weight of authors according to their expert knowledge. We classify tweets sentiment polarity using machine learning technique: Support Vector machine (SVM) and we combine it with weight of authors' background knowledge based on author's profile, twitter List feature and their tweets behavior for a given topic and then show the result as the percentage of credibility on their positive or negative views.

Keywords: Expert Knowledge, Machine Learning, Opinion Credibility, Sentiment Analysis, Social Media, Twitter.

1 Introduction

The world is changing with the explosion of social media nowadays. It is increasing connecting with each other, fast developing with effects on individual behavior. Using social media for information is a way to monitor and identify people opinions. Sentiment analysis is the main point of the social media research including Twitter, Facebook, blogs, and user forums. For example, 2014 Hong Kong Protest, known as the Umbrella Movement or Umbrella Revolution, began in September 2014. This revolution happened for the China's authorities that would limit who Hong Kongers can elect in 2017 elections. In this event social media played as important part. Hong Kong citizens post continuous information through the internet using social media. There are differences opinions in Hong Kongers who are supporters and skeptics. An opinion poll conducted by Chinese University said that 46% did not support this

H. Unger et al. (eds.), *Recent Advances in Information and Communication Technology 2015*,
Advances in Intelligent Systems and Computing 361, DOI: 10.1007/978-3-319-19024-2_21

revolution, while 31% backed the civil disobedience movement[1]. In that event, we didn't know whose opinions are strong or credential. Therefore when we take the information from opinion content, the identifier of an author may help to determine credibility. The opinions of specialists and recognized experts are more likely to be credential and to reflect a significant viewpoint. Therefore we proposed the method to determine credibility of tweets polarity based on authors' expert or background knowledge using one of the famous social media in Twitter. For example: Hong Kong Revolution tweets:

- **Dana Ryan:** More tourists are flocking to Hong Kong to watch the pro–democracy protesters, so business is good here. Merchants are happy.

- **RealHKNews:** #UmbrellaRevolution people's latest slogan "I want to go shopping". Picture showing disappointed shoppers sitting... http://t.co/VajNuZThnh

In this case, tweets intent to negatively affect Hong Kong business view for revolution issues. There are many issues arise. For example, whose opinion is more credential than the other, who represent about this tweet opinion is more credential than the other? Who represent about this tweet? Who is the author of these tweets? How much does this author know about this topic? What is their background knowledge? How many percentage of credibility on their positive or negative view? To achieve this, in this paper we calculate the tweet content polarity for a particular topic, first we use Support Vector Machine (SVM) to compute tweet polarity and we add the weight of authors according to their expert knowledge using the bio, List features and their tweets and then we calculate credibility of tweet sentiment polarity with authors' expert weight. We discuss our related work in section II. Section III presents our proposed method in details. Section IV shows our preliminary experiment result and Section V is our paper conclusion.

2 Related Work

Our area of research mainly focuses on sentiment analysis and we combine it with authors' expert knowledge to show the credibility of sentiment polarity. The previous researches only assumed the number of positive or negative points and they summarized the sentiment of tweets for the overall result [1, 3]. Go A. et al. [2] presented sentiment classification using machine learning method. They classified the tweets sentiment using emoticon for distant supervised learning. Pak A. and Paroubek P. [10] showed how to automatically collect a corpus to determine positive, negative and neutral sentiment for a document and rely on the multinomial Naïve Bayes classifier that used N-gram and POS-tags are features. But when they labeled for training dataset based on the emoticon for sentiment polarity. Liang P.W. and Dai B.R. [5] extracted the set of messages that

---

[1] http://www.scmp.com/news/hong-kong/article/1597646/one-five-hongkongers-may-emigrate-over-political-reform-ruling

contain opinions and filtered out non-opinion messages and determined their sentiment directions. They used unigram Naïve Bayes for extract tweets and determined opinion or not and used Mutual Information and X^2 feature selection for short text classification to discard some useless features. Neethu M. S. and Rajasree R. [7] analyzed sentiment from tweets in specific domain (electronic products) using two Machine learning approaches (Nave Bayes Classifier and SVM classifier). Pang B. et al. [9] classified documents (movie review) is whether positive or negative. They used standard machine learning (Naïve Bayes, maximum entropy classification, and support vector machines) to apply sentiment classification problem.

To explore expertise, it has little research area in social media and also it is ambiguous to decide characterize of expert. What kinds of background knowledge is impact their authority? Weng J. et al. [12] created TwitterRank, to identify influential users of Twitter. They utilized Latent Dirichlet Allocation (LDA) to calculate topical distribution of a user and made weighted user graph that point out topical similarity of users. They used PageRank algorithm to identify authorities on each topic. Pal A. and Counts S. [8] also proposed to find the most interesting and authoritative authors for any given topic in Twitter. They analyzed features such as user's content and combined it their friends and followers information to be affective to find topical experts. They ran Gaussian Ranking Algorithm to cluster users into two clusters over their features space for finding the most authoritative users. Sharma N.K. et al. [11] designed who-is-who service to infer characterize of individual Twitter users using Twitter List Features which allow a user to make groups of other users who related on a topic. List meta-data (names and description) provided to get semantic cues about who the users in it. Further it could also infer topical expertise users analyzing the meta-data of crowdsourced Lists that contains the user. Liang C. et al [4] designed a framework to help identify misinformation with the assessments of experts. They proposed a tag-based method extracting extracted the expertise of users from their microblog contents and matched the experts with given suspected misinformation. With the judgment of experts they defined the credibility of information and confuting of misinformation. Namihira Y. et al. [6] proposed to access the credibility of information based on topic and opinion classification depending on user's knowledge (expertise). They believed if they considered tweet of user knowledge, they handled as a more reliable opinion even if it is a minor opinion. Likewise our work is also focus on a user expert knowledge but different approach. We desire to assess the credibility of tweets opinions.

3 Proposed Method

Our system consists of two main components: sentiment analyzer and credibility calculator. Sentiment analyzer is to classify tweets polarity and opinion credibility calculator is to compute credibility of opinion based on an author's expert knowledge for a given topic. We combine with the results from sentiment analyzing with credibility calculation and show credibility of tweet's polarity result. Figure.1 show overview of our proposed system to find the credibility result of tweet polarity.

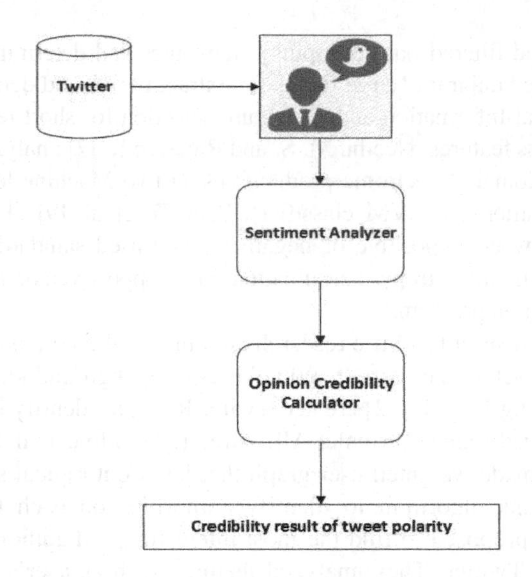

Fig. 1. Overview system of credibility of tweet polarity

3.1 Sentiment Analyzer

The task of sentiment analyzer is to achieve classification of tweets' polarity result. We perform sentiment classification to analyze which tweet is positive or negative using one of the machine learning approaches: support vector machine (SVM). SVM is a supervised learning method used for classification. SVM finds the hyperplane and separates two classes' points with the maximum margin. We use Linear SVM with Rapid Miner[2] software to classify sentiment polarity of our tweet dataset. w represents to find maximum hyperplane and separates document from one class or another. c_i corresponds to positive and negative with class of document d_i. α_i is to solve dual optimization problem. Our analyzer define positive for "1" and negative for "-1".

$$w = \sum_i \alpha_i c_i d_i \ , \alpha_i \geq 0 \quad (1)$$

3.2 Opinion Credibility Calculator

In this section, we calculate credibility of opinion based on an author's expert knowledge for a given topic. In Twitter, we consider the Twitter features: profile's bio, List feature and author tweets behavior to identify the weight of authors' expert knowledge. In List feature (allow a user to make groups of other users who related on a topic): name and description indicate valuable semantic cues about who has users included in the Lists are including their topics of expertise [11]. Based on Sharma N.K. et al. [11] approach, we use the author's profile and List name and description to

[2] www.rapidminer.org

infer who they are for a given topic. First we apply common language processing approaches: such as remove non-English word (using the WordNet to determine whether the word is English or not), remove url, # and @ (prefix, suffix), remove stop words, stemming, case folding. We apply N-gram approach (using unigram and bi-gram) to segment them and extract Noun and Adjective that are useful for characterizing users using NLP toolkit[3]. We used the ontology concept to filter a set of given topic to extract conceptual keyword that are related with given topic and calculate the ratio of number of related keywords to total number of raw keywords. This strategy produces first score w_{BL} (author expert score using bio and List) to define the author's expert knowledge.

$$w_{BL} = \frac{number\ of\ related\ keywords\ for\ given\ topic}{total\ number\ of\ raw\ keywords} \tag{2}$$

Next step is to discover the expert score based on an author tweets behavior and his/her social network activities. We focus on the following features that reflect the impact of user background knowledge (expert knowledge score) for a given topic when we calculate the tweet's credibility.

- author's topic related ratio (TR)
- ratio of author's tweet retweet by other users (RT) for a given topic
- ratio of author's friends and followers who are related with given topic (F_1, F_2)
- author's opinions ratio for given topic (OP)

We combine these features to make the expert score for a given topic. For author topic related ratio, we make the ratio of number of author's tweet related for specific topic to the number of his/her all tweet.

$$w_{TR} = \frac{number\ of author's\ tweet\ related\ from\ specific\ topic}{number\ of\ author's\ all\ tweet} \tag{3}$$

In Twitter, retweet is a re-posting someone else's Tweet. Using retweet features we compute how much times author's tweet has been retweeted by others.

$$w_{RT} = \frac{number\ of\ author's\ tweet\ retweet\ by\ others\ for\ given\ topic}{all\ author\ tweeets\ for\ givne\ topic} \tag{4}$$

w_{F1} and w_{F2} indicate the ratio of author's friends and followers who are related with given topic.

[3] www.nltk.org

$$w_{F1} = \frac{number\ of\ friends\ for\ given\ topic}{total\ number\ of\ author's\ friends} \tag{5}$$

$$w_{F2} = \frac{number\ of\ followers\ for\ given\ topic}{total\ number\ of\ author's followers} \tag{6}$$

For author's opinions ratio; we assume how many times author expresses this opinions based on his/her past all opinions for given topic. e.g., given tweet opinion is negative, we calculate number of author's negative tweet based on his/her all past tweet opinion.

$$w_{OP} = \frac{number\ of\ author's\ negative\ or\ positive\ tweets}{all\ past\ opinions\ of\ author's\ tweets} \tag{7}$$

To produce expert score, we combine all of these features by adding author's bio, List feature and his/her tweet behavior. Finally we compute expert score for given topic by adding author's bio, List features and author's tweet behavior.

$$Expert\ score = \frac{w_{BL}+w_{TR}+w_{RT}+w_{F1}+w_{F2}+w_{OP}}{6} \tag{8}$$

As the final result, we combine that expert score with the result of polarity from sentiment analyzer and calculate the credibility tweet polarity result $C(polarity)$.

$$C(polarity) = tweet's\ polarity \times Expert\ score \tag{9}$$

4 Preliminary Experiment

Dataset

We use Twitter API to collect data. We crawl HK revolution data in the period between Sep 30, 2014 and Nov 30, 2014 as our training set and testing set. We use 1000 sample data for training dataset to learn sentiment classification using RapidMiner. We run this dataset with sentiment analyzer and show the output result, which labeled with tweet's polarity (positive: 1, negative: -1). We use many set of SVM parameters for finding the best result and we apply 10 fold cross validation and the best results are shown in Table 2.

Table 1. Training Dataset

Dataset	positive	negative
Training Data	500	500
Test Data	100	100

In order to evaluate the accuracy of these polarity results, we use Linear SVM and calculate precision and recall. The following table shows the calculation results. We get precision 84.40% for positive and 96.93% for negative. We think the rest of the errors is emoticons. According to the SVM classifier, it can occur more weight and reduce accuracy.

Table 2. Result of Training Classification

	true positive	true negative	class precision
positive	487	90	84.40%
negative	13	410	96.93%
class recall	97.40%	82.00%	

The second part of experiment is credibility calculation. We take one sample tweet from training set. First, we compute this tweet polarity by using sentiment analyzer. Then we calculate weights and corresponding expert score for that tweet according to our equation 2 to 8. Finally, we get tweet polarity from sentiment analyzer and expert score, then we calculate credibility of tweet polarity. To evaluate the credibility of tweet polarity, we calculate accuracy of opinion credibility. To do so we define author's expertise level as maximum, medium and minimum trusted levels. We manually categorizes the author of HK revolution dataset as that three level. Accordingly we set the range for the credibility value as highest (h), middle (m) and lowest (l) range. The range setting depends on domain and classifier. In our evaluation we assume that (h >=70%), (l <=30%), (h<=m >=l) respectively.

Then we determine the range of our classification result and verify with the author's expertise categories. From our proposed method, the opinion credibility of author [csn216] and [dw11138375] are 29% and 17%. Likewise, as we can see in Table 3, the author [csn126] is in the minimum trusted author group. In addition author [Dana_Ryan]: 36%, [lailaoshi]: 33% and [natashkakhanhk]: 68% are in medium trusted group and author [krisic]: 72% is maximum trusted group. From this comparison, we can conclude that our proposed method can provide reliable credibility.

Author's tweet: [csn126] Gotta love this guy. #occupyhk http://t.co/xg28uAjhYf

$$C(polarity) = tweets\ polarity \times expert\ score$$

tweet polarity : Positive: 1

w_{bi}= 0.55; w_{TR}=0.02; w_{RT}=0.0001; w_{F1}=0.25; w_{F2}=0.08; w_{OP}=0.81

Expert score = 0.55+0.02+0.0001+0.25+0.08+0.81/6 = 29%
C(polarity) = 1 (positive) x 0.29 = 29%

Fig. 2. Example of how to measure the opinion credibility

Table 3. Example of trusted Author group

Maximum	Medium	Minimum
krisic	lailaoshi	csn126
Ramyinocencio	Dana_Ryan	HKAYPGOLD
JeromeTaylor	natashkakhanhk	RealHKnews
fion_li	jen1113	dw11138375

Table 4. Sample Results of opinion credibility

No	Authors	Tweets	Opinion	Opinion Credibility
1.	Dana_Ryan	It's so bad that 82% of Hong Kongers wanting to kill CY Leung for him playing the "Waiting Game". The police may not enforce the deadline.	negative	36%
2.	dw11138375	Heavy police presence and a lot of angry shoppers #OccupyCentral #UmbrellaRevolution #UMHK https://t.co/efPi74cvWF	negative	17%
3.	lailaoshi	More tourists are flocking to Hong Kong to watch the pro-democracy protesters, so business is good here. Merchant is happy.	negative	33%
4.	krisic	"I am proud to be a HKer" @rosetangy #OccupyHK	positive	72%
5.	natashkakhanhk	"The most powerful weapon in winning democracy for HongKong is the people of the Umbrella Generation" -Benny Tai OpEd http://t.co/QN5bctyJ2y	positive	68%
6.	csn126	Gotta love this guy. #occupyhk http://t.co/xg28uAjhYf	positive	29%

5 Conclusion

We proposed a system to measure credibility sentiment polarity based on authors' expert knowledge. To find credibility opinion, our system performed 2 steps: sentiment classification and opinion credibility calculation. We did preliminary experimentation using Hong Kong revolution dataset. Based on our preliminary experiment results; our sentiment classification got precision: 84.40% and recall: 97.40% for true positive, and precision: 96.93% and recall: 82.0% for true negative. In our future work, we will evaluate the accuracy of our credibility opinion results. We will do further evaluation with large tweets dataset and compare with other topics.

Acknowledgements. I would like to gratefully and sincerely thank my adviser, Dr. Sasiporn Usanvasin for her guidance, understanding, patience, and most importantly, her invaluable constructive criticism and friendly advice during the period of study. I would also like to thank everyone who supported me throughout this project. I am sincerely grateful to them for sharing their truthful and illuminating views on a number of issues related to the project. This research was partially supported by the Center of Excellence in Intelligent Informatics, Speech and Language Technology and Service Innovation (CILS) and by NRU grant at SIIT, Thammasat University.

References

1. Bafna, K., Toshniwal, D.: Feature Based Summarization of Customers' Reviews of Online Products. In: 17th International Conference in Knowledge Based and Intelligent Information and Engineering Systems, vol. 22, pp. 142–151 (2013)
2. Go, A., Bhayani, R., Huang, L.: Twitter sentiment classification using distant supervision. CS224N Project Report, Stanford (2009)
3. Hu, M., Liu, B.: Mining and summarizing customer reviews. In: Proceedings of the Tenth ACM SIGKDD International Conference on Knowledge Discovery and Data Mining, pp. 168–177. ACM Press, New York (2004)
4. Liang, C., Liu, Z., Sun, M.: Expert Finding for Microblog Misinformation Identification. In: Proceedings of COLING, Mumbai (2012)
5. Liang, P.W., Dai, B.R.: Opinion Mining on Social Media Data. In: IEEE 14th International Conference on Mobile Data Management (MDM), pp. 91–96. IEEE Press, Milan (2013)
6. Namihira, Y., Segawa, N., Ikegami, Y., Kawai, K., Kawabe, T., Tsuruta, S.: High Precision Credibility Analysis of Information on Twitter. In: Proceeding of the International Conference on Signal-Image Technology & Internet-Based Systems, Kyoto, pp. 909–915 (2013)
7. Neethu, M.S., Rajasree, R.: Sentiment analysis in twitter using machine learning techniques. In: Proceeding of the Fourth International Conference on Computing, Communications and Networking Technologies, pp. 1–5 (2013)
8. Pal, A., Counts, S.: Identifying topical authorities in microblogs. In: Proceedings of the Fourth ACM International Conference on Web Search and Data Mining, pp. 45–54. ACM Press, New York (2011)

9. Pang, B., Lee, L., Vaithyanathan, S.: Thumbs up? Sentiment Classification using Machine Learning Technique. In: Proceedings of the ACL 2002 Conference on Empirical Methods in Natural Language Processing (2002)
10. Pak, A., Paroubek, P.: Twitter as a Corpus for Sentiment Analysis and Opinion Mining. In: Proceedings of the Seventh Conference on International Language Resources and Evaluation (2010)
11. Sharma, N.K., Ghosh, S., Benevenuto, F., Ganguly, N., Gummadi, K.: Inferring Who-is-Who in the Twitter Social Network. In: Proceeding of the 2012 ACM Workshop on Online Social Networks, pp. 55–60. ACM Press, New York (2012)
12. Weng, J., Lim, E.P., Jiang, J., He, Q.: TwitterRank: finding topic-sensitive influential twitterers. In: Proceedings of the Third ACM International Conference on Web Search and Data Mining, pp. 261–270. ACM Press, New York (2010)

Detect the Daily Activities and In-house Locations Using Smartphone

Sittichai Sukreep, Pornchai Mongkolnam, and Chakarida Nukoolkit

School of Information Technology
King Mongkut's University of Technology Thonburi
Bangkok 10140, Thailand
{sittichai.s,pornchai,chakarida}@sit.kmutt.ac.th

Abstract. Falls are a key cause of significant health problems, especially for elderly people who live alone. Falls are a leading cause of accidental injury and death. To help assist the elderly, we propose a system to detect daily activities and in-house location of a user by means of a smartphone's sensor and Wi-Fi access points. We applied data mining techniques to classify activity detection (e.g., sitting, standing, lying down, walking, running, walking up/downstairs, and falling) and in-house location detection. Health risk level configurations (threshold model) are applied for unhealthy activity detection with an alarm sounding and also short messages sent to those who have responsibility such as a caregiver or a doctor. Moreover, we provide various forms of easy to understand visualization for monitoring and include health risk level summary, daily activity summary, and in-house location summary.

Keywords: Activity of Daily Living (ADL), Smartphone, Accelerometer, Access Point, Wireless Signal, In-house Location, Data Mining, Classification, Visualization.

1 Introduction

Fall accidents can cause severe health problems and can happen to anybody. The World Health Organization (WHO) reported in 2012 that falls were the second leading cause of accidental injuries and deaths worldwide; 424,000 individuals died from falls (1,160 persons/day), and over 80% of such deaths occurred in low and middle income countries [1-3].

Currently, the world is changing toward ageing societies. More elderly people live alone rather than with their families. When those people fall during their daily activities, they cannot help themselves. Without anybody taking notice and providing help, their injuries could be fatal. Therefore, timely assistance and care may reduce the severity of the injuries.

We aimed to improve the detection of Activities of Daily Living (ADL) and in-house location in real-time, using a smartphone's accelerometer sensor, three in-home's Wi-Fi access points (APs), and data mining classification methodology. We provided monitoring and warning of health risk levels. Moreover, we created various

H. Unger et al. (eds.), *Recent Advances in Information and Communication Technology 2015*,
Advances in Intelligent Systems and Computing 361, DOI: 10.1007/978-3-319-19024-2_22

data visualization tools for generating frequently used in-house locations and providing summarized health risk level reports.

2 Related Work

There are several systems available which detect daily activities and indoor locations. However, those systems or methods require rather sophisticated hardware or infrastructure. This has motivated our work to overcome their shortcomings.

In 2013, Jian et al. [4] developed an automatic fall warning system. This system used an accelerometer and gyroscope sensors attached to a vest or other garment and collected the activities and fall data from elderly people. Daniel et al. [5] proposed a methodical algorithm which classified 11 activities and posture transitions (i.e., stand, sit, sit to stand, stand to sit, bend down, bend up, walking, lying, lying to sit, sit to lying, and bent) using an inertial tri-axial accelerometer located on the waist. The support vector machine (SVM) algorithm was used for classifications. Several researchers established elderly people received injuries or in some cases death from falling [6-16]. In addition, more recent research has focused on activity or movement detection by using smartphones or sensors such as accelerometer or gyroscope.

In 2014, Stephen et al. [16] developed a system for rehabilitation and diagnoses to understand the patients' activities (e.g., walk or sit) by carrying a phone in different positions, including belt, pocket, hand, and bag. The authors used a smartphone's accelerometer and SVM classifier to classify the activities. Guiry et al. [17] proposed a method to accurately detect human activities, including sitting, standing, lying, walking, running, and cycling using two accelerometers and to compare activity recognition classifiers using C4.5, CART, SVM, Multi-Layer Perceptrons and Naïve Bayes with accuracies as high as 98%. Quoc et al. [18] developed a wireless sensor system and algorithm to identify falls such as forward fall, backward fall, and sideway fall (left and right) by using ADXL345 (3-axis digital accelerometer sensor) and ITG3200 (3-axis digital gyroscope sensor), MCU LPC17680 (ARM 32-bit cortex M3), and Wi-Fi module RN13. Paliyawan et al. [19] developed a prolonged sitting detection system for office worker syndrome by using Kinect.

Liu et al. [20] developed a technique and system for surveying wireless indoor positioning. Premchaisawatt [21] proposed machine learning techniques for enhancing indoor position in an experiment area of 30x10 meter^2 with 77.32% accuracy. Zhongtang et al. [22] developed an in-house location detection system with an experiment area of 16x29 meter^2 and with 5 marked spots.

3 Methodology

3.1 System's Overview

We propose a system which classifies falls, basic daily movement activities, and in-house location detection, and provides health risk feedback with several easy to understand visualizations. The system obtains input data from a smartphone's accelerometer sensor and the Wi-Fi signal strength from 3 APs in real-time. Detection of the daily activities and in-house locations are done using data mining classification as shown in Fig. 1(a). Knowing the activities and locations, an alarm sounds and SMS is sent when a high risk situation or unhealthy condition is detected. The system can also provide summary reports on the safety level of the user's activities and in-house locations.

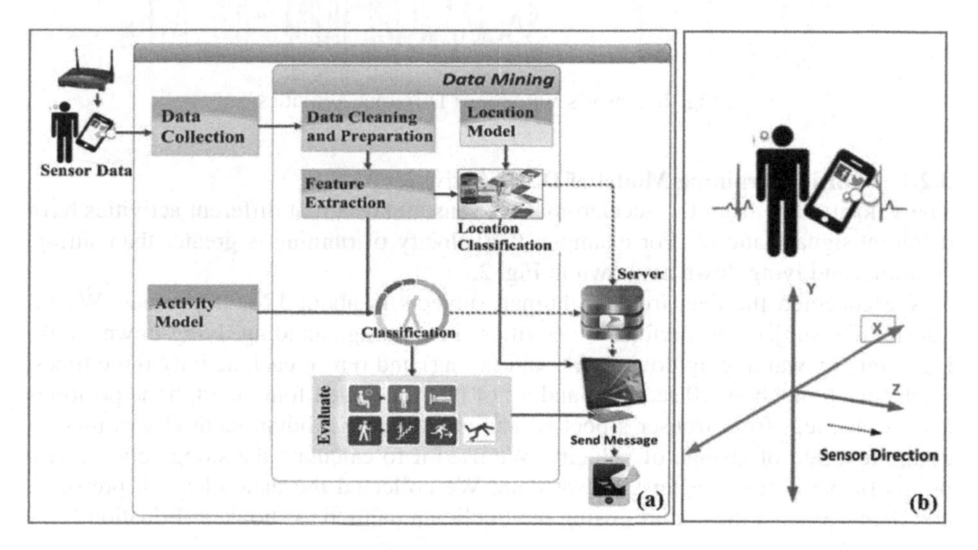

Fig. 1. (a) System's Architecture and (b) Accelerometer's Sensor Coordinates

3.2 Daily Activity Detection Algorithm

We develop an Android application for collecting acceleration data from a smartphone sensor and use a Windows application for processing the data. The smartphone sends the acceleration data of occurred movements to a server. The data are obtained by a tri-axial accelerometer as X, Y, and Z coordinates as shown in Fig. 1(b). The system supports 7 activities (sitting, standing, lying down, walking, running, walking up/downstairs, and falling) and 4 selectable positions for the smartphone (i.e., waist, leg, front trouser's pocket, and arm). It is necessary that the smartphone be attached to one of those positions.

The method we use is based on machine learning. We divide the activity detections into two parts. First, in an offline phase, we build a learning system by letting the users perform activities and collecting the data when they are performed. After that we clean and prepare the data, extract their features, compare the classifiers provided by WEKA [23], including decision tree (J48), naïve Bayes, support vector machine (SVM) and k-nearest neighbor (KNN), and select the optimal classifier. Second, in a real-time phase, we use the program developed for the Android smartphone to classify the 7 daily activities.

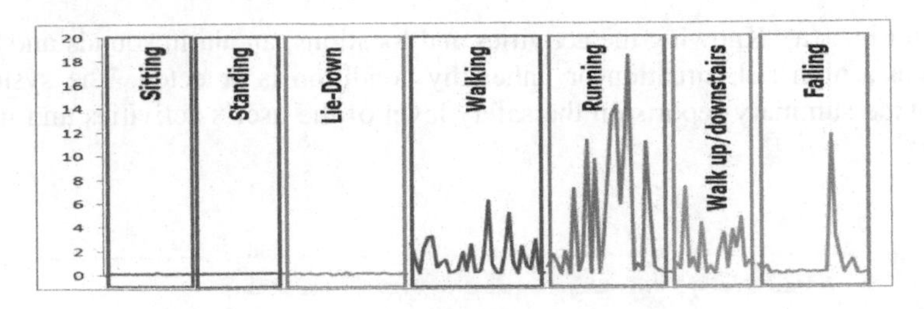

Fig. 2. Sensor's Velocity of Different Activities

3.2.1 Offline Training Model of Daily Activities

The velocity data from the accelerometer's sensor shows that different activities have different signal patterns. For example, the velocity of running is greater than sitting, standing, and lying down as shown in Fig. 2.

We collected the data from 25 human subjects in about 120,000 frames. We requested the subjects to perform 7 activities (i.e., sitting, standing, lying down, walking, running, walking up/downstairs, and falling) and repeat each activity three times. Each time took about 10 seconds, and all of them included four smartphone positions (i.e., waist, leg, front trouser's pocket, and arm). The Euclidian method was used to compute a rate of change of velocity. We used it to calculate the slope between two points of the accelerometer's sensor data. We collected the data, cleaned, prepared, and then extracted the features using the Euclidian method as shown in Equation (1).

$$Acc(i,j) = \sqrt{(Xi - \mathrm{Xj})^2 + (Yi - Yj)^2 + (\mathrm{Zi} - \mathrm{Zj})^2} \quad , \tag{1}$$

where Acc(i, j) represents the accelerometer from the ith and j^{th} records, and X, Y, and Z represent the coordinates.

We trained the data using WEKA with various classifiers such as decision tree (J48), naïve Bayes, support vector machine (SVM), and k-nearest neighbor (KNN). Then we compared the results and chose the optimal classifier.

3.2.2 Real-Time Activity Classification

We applied our optimal classifier, the KNN model (using 3 nearest neighbors for classification), on our system for detecting various kinds of activities. The system collected the acceleration data (the X, Y and Z coordinates) from the smartphone in real-time every 0.2 second. The data were processed as the input of the KNN model, and then the system predicted the activities.

3.2.3 Noise Filtering

Acceleration of movement data from the accelerometer's sensor can be adversely affected by the abrupt transitions in activity detection and interference. We used the following method to prevent this problem. For example, when the user changed an activity immediately from one activity to another, such as from sitting to standing, the system would accept the state change from sitting to standing only when the activity transition state had been continuously changed for more than 5 frames (~1 second). Otherwise, the state change would be rejected and it remained as the previous activity.

3.3 In-house Location Detection Algorithm

Our in-house location detection algorithm used the Wi-Fi signal strength to position the user's whereabouts. We developed an Android's application for collecting the Wi-Fi signal strength levels from 3 APs. APs were set up on the same side of the house because we could reduce interferences and consider only one side of the Wi-Fi signal coverage as shown in Fig. 3(a).

The algorithm used for classifying the in-house locations was based on machine learning. We divided our in-house location detection into two phases. First, in offline phase, we collected the data in the two-story house (with 24 marked locations). We collected the signal strength 50 frames for each marked location and repeated 10 times around those spots. A data mining process was used to clean and prepare the data, and extract the features. We then compared four classifiers, including decision tree (J48), naive Bays, support vector machine (SVM) and k-nearest neighbor (KNN) and chose the optimal one. Second, in a real-time phase, we developed the system to detect in-house locations and evaluated the accuracy.

3.3.1 Offline Training Model for Location Detection

The Wi-Fi signal strength data from APs showed that different locations had different values. We used 3 APs for the in-house location detections, and we experimented with them in the two-story house. The house had an area of 3.5x10.5 meter^2 on both floors as shown in Fig. 3(b).

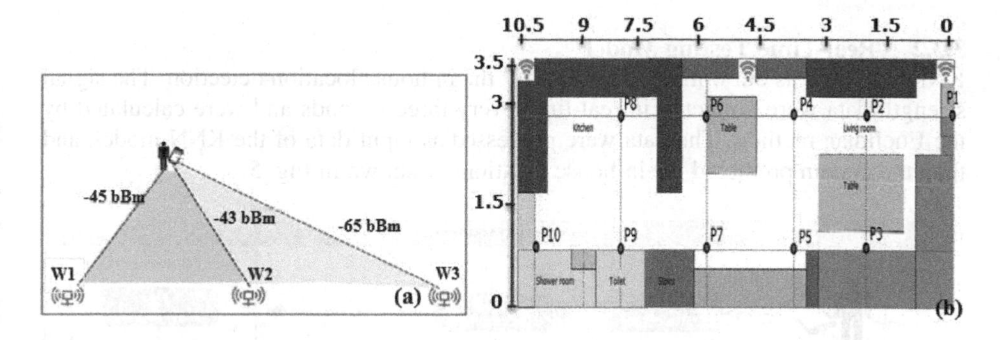

Fig. 3. (a) Signal Strength and Positions of Access Points and (b) Marked In-house Locations

We collected the data from 24 marked spots (11 spots on the first floor and 13 spots on the second floor) as shown in Fig. 3(b). The distance between each spot was about 2 meters, both vertically and horizontally. We used the smartphone to collect the Wi-Fi signal strength data also known as Received Signal Strength Indication (RSSI) from 3 APs. For each marked spot we collected the data 10 times, and each time we collected 50 frames/spot. So we had data equal to 50x10x24 = 12,000 frames in total. The data underwent a data mining process. They were cleaned and prepared, and the features were extracted by the Euclidian method. The Euclidian calculation involved two parts. First, we calculated signal strength from the first AP and the second AP as shown in Fig. 4(a) using Equation (2). Second, we calculated it from the second AP and the third AP as shown in Fig. 4(b) using Equation (3).

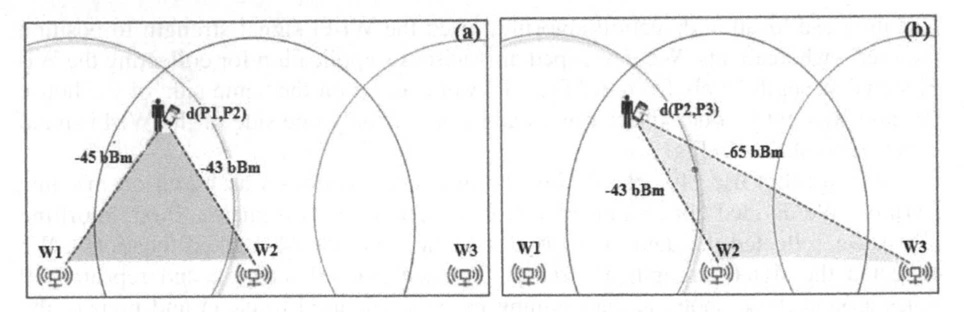

Fig. 4. Signal Strength from (a) First AP and Second AP and (b) second AP and third AP

$$Distance(W1, W2) = \sqrt{(W1 - \mathrm{P1})^2 + (W2 - P2)^2} \quad , \tag{2}$$

$$Distance(W2, W3) = \sqrt{(W2 - \mathrm{P2})^2 + (W3 - P3)^2} \quad , \tag{3}$$

where W1, W2, and W3 represent the mean value of signal strength from the 1st, 6th and 11th marked spot, and P1, P2, and P3 represent the signal strength from APs.

We trained the data using WEKA with four classifiers, including decision tree (J48), naive Bays, support vector machine (SVM) and k-nearest neighbor (KNN). Then we compared the results of those classifiers and chose the optimal one.

3.3.2 Real-Time Testing Model

KNN was used as our optimal classifier for the in-house location detection. The signal strength data were collected in real-time every three seconds and were calculated by the Euclidian method. The data were processed as input data of the KNN model, and then the system predicted the in-house locations as shown in Fig. 5.

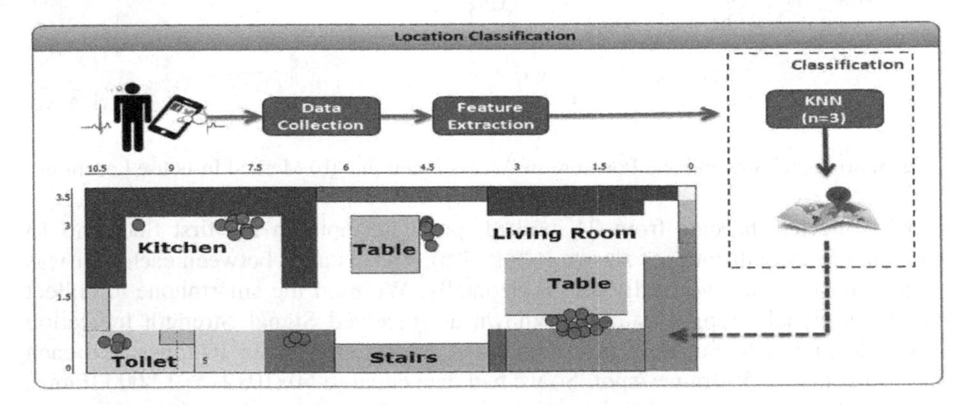

Fig. 5. In-house Location Detection

4 Experiment and Results

4.1 Experiment Setup

In our test, we set up the 3 APs in a two-story house. The house floor area was approximately 3.5x10.5 meter^2. There were 24 marked spots, with 11 marked points on the first floor and 13 marked points on the second floor. All 3 APs were on the first floor and on the same wall. All activities and in-house locations were covered. Ten volunteers were asked to perform various activities, using a Samsung Galaxy S5 phone equipped with the accelerometer and Wi-Fi signal receiver.

4.2 Data Collection

4.2.1 Activity Data Collection

We collected the accelerometer data for the 7 activities (sitting, standing, lying down, walking, running, walking up/downstairs, and falling). The activities were performed by 10 volunteers. We let the volunteers with the attached smartphones move freely from one position to another position. The smartphone was attached to four different positions, including the front trouser pocket, arm, leg, and waist.

The volunteers performed activities 10 seconds (~5 frames/second) for normal activities without fall and 3 times for falling. So there was a total of 12,580 frames for testing. All data were saved to a database server in real-time.

4.2.2 In-house Location Data Collection

We collected location data from the Wi-Fi signal strength of the smartphone. We requested the volunteers to stand at the marked spots for collecting the data. At each marked spot, 50 frames were collected and saved to the database. There were 24x50 = 1,200 frames per volunteer. So there was a total of 1,200x10 = 12,000 frames altogether.

4.2.3 Evaluation

4.2.3.1 Activity Detection Evaluation. The accelerometer data that we collected from the volunteers were feature-extracted and evaluated using the KNN (K=3) model and recorded an average accuracy of 97.48% of all activities combined. The accuracy of each activity is listed in Table 1.

Table 1. Accuracy of Activity Detection

No.	Gender	Age	Weight (kg.)	Height (cm.)	Sit	Stand	Lie-down	Walk	Run	Walk Up/Downstairs	Fall
1	Male	43	60	167	100.00	100.00	100.00	92.69	98.02	88.94	100.00
2	Female	24	54	152	100.00	100.00	100.00	95.79	99.50	89.24	100.00
3	Female	48	80	158	100.00	99.03	99.04	94.62	100.00	91.23	100.00
4	Male	63	45	165	100.00	99.01	100.00	93.84	100.00	88.33	100.00
5	Female	36	73	165	100.00	100.00	100.00	92.73	98.48	92.89	100.00
6	Female	60	69	155	98.57	100.00	99.50	92.56	98.50	88.09	100.00
7	Female	38	58	157	100.00	100.00	98.58	93.12	99.04	92.69	100.00
8	Female	45	65	167	98.04	99.02	98.05	95.85	100.00	93.75	100.00
9	Male	42	68	170	99.51	100.00	99.04	95.81	95.00	94.69	100.00
10	Male	50	60	172	100.00	100.00	99.03	91.96	98.53	91.11	100.00
Average (%)					**99.61**	**99.71**	**99.32**	**93.90**	**98.71**	**91.10**	**100.00**

4.2.3.2 Location Detection Evaluation. The in-house location data which we collected from the volunteers were feature-extracted and evaluated using the KNN (k = 3) model. We evaluated each marked spot (F01-F24) and recorded an average accuracy of about 94.11%. The accuracy table is shown in Table 2.

Table 2. Confusion Matrix Showing Accuracy of In-house Location Detections

Marked Spot	F01	F02	F03	F04	F05	F06	F07	F08	F09	F10	F11	F12	Accuracy
F01	98.80	0.00	0.80	0.00	0.40	0.00	0.00	0.00	0.00	0.00	0.00	0.00	**98.80**
F02	0.00	96.02	0.20	2.19	1.59	0.00	0.00	0.00	0.00	0.00	0.00	0.00	**96.02**
F03	2.65	1.43	92.87	0.20	2.85	0.00	0.00	0.00	0.00	0.00	0.00	0.00	**92.87**
F04	0.00	1.06	0.00	92.39	0.42	5.50	0.63	0.00	0.00	0.00	0.00	0.00	**92.39**
F05	0.80	3.99	2.00	0.20	92.42	0.20	0.20	0.00	0.20	0.00	0.00	0.00	**92.42**
F06	0.18	0.73	0.91	2.73	0.73	94.36	0.36	0.00	0.00	0.00	0.00	0.00	**94.36**
F07	0.00	1.67	0.42	1.88	3.55	2.92	86.85	1.04	1.67	0.00	0.00	0.00	**86.85**
F08	0.00	0.00	0.00	0.00	0.00	0.00	2.88	88.89	3.29	0.62	4.12	0.00	**88.89**
F09	0.19	0.00	0.00	0.00	1.33	0.19	3.05	3.05	91.24	0.00	0.95	0.00	**91.24**
F10	0.00	0.00	0.00	0.00	0.00	0.00	0.00	1.21	0.00	98.79	0.00	0.00	**98.79**
F11	0.00	0.00	0.00	0.00	0.00	0.00	0.00	3.25	2.43	1.42	92.90	0.00	**92.90**
F12	0.00	0.00	0.00	0.00	0.00	0.00	0.00	0.00	0.00	0.00	0.00	95.61	**95.61**
Marked Spot	**F13**	**F14**	**F15**	**F16**	**F17**	**F18**	**F19**	**F20**	**F21**	**F22**	**F23**	**F24**	**Accuracy**
F12	0.60	0.60	1.80	0.20	0.40	0.00	0.80	0.00	0.00	0.00	0.00	0.00	**95.61**
F13	96.79	0.20	1.80	0.00	0.20	0.00	0.20	0.00	0.00	0.00	0.00	0.00	**96.79**
F14	1.41	93.33	0.40	0.20	0.61	0.00	0.00	0.00	0.00	0.00	0.00	0.00	**93.33**
F15	3.59	0.80	91.43	0.00	1.59	0.00	1.59	0.00	0.00	0.00	0.00	0.00	**91.43**
F16	0.00	0.00	0.00	96.89	0.97	0.00	1.95	0.00	0.00	0.00	0.00	0.00	**96.89**
F17	2.43	1.77	4.86	2.65	86.31	0.00	1.10	0.00	0.00	0.00	0.00	0.00	**86.31**
F18	0.00	0.00	0.00	0.00	0.00	99.56	0.00	0.44	0.00	0.00	0.00	0.00	**99.56**
F19	0.89	0.89	3.12	0.67	4.45	0.00	89.31	0.00	0.00	0.00	0.00	0.00	**89.31**
F20	0.00	0.00	0.00	0.00	0.00	1.22	0.00	98.09	0.70	0.00	0.00	0.00	**98.09**
F21	0.00	0.00	0.00	0.00	0.00	0.00	0.00	1.05	97.26	1.05	0.42	0.21	**97.26**
F22	0.00	0.00	0.00	0.00	0.00	0.00	0.00	0.80	3.21	95.99	0.00	0.00	**95.99**
F23	0.00	0.00	0.00	0.00	0.00	0.49	0.00	0.00	1.23	0.62	96.54	0.86	**96.54**
F24	0.00	0.00	0.00	0.00	0.00	0.00	0.00	0.00	0.60	1.39	1.98	96.03	**96.03**
Average (%)													**94.11**

It could be seen that some locations were misclassified because the signal strength from APs were affected by the surrounding environment, such as door opening, door closing, or postures of the volunteers; therefore, data from the signal strength could change, resulting in misclassified locations.

5 Conclusions and Future Work

In this paper, we propose a practical and affordable system using a smartphone's accelerometer sensor and the Wi-Fi signal strength from APs to detect and visualize the in-house locations and daily activities such as sitting, standing, lying down, walking, running, walking upstairs and downstairs, and falling. We apply data mining techniques for those daily activities and in-house locations of the user. The changes of activities and in-house locations are detected by the threshold model. We provide an easy to use and understand user interface and visualization to monitor the activities and in-house location in real-time by displaying the related information on a separate monitoring computer screen. Moreover, our system can warn the user when his or her health risk level exceeds the preset level. The achieved accuracy of activity detections is 97.48%, and the accuracy of the in-house location detections is 94.11%. Moreover, our proposed system can detect the fall activity, the detection of which is crucial for one's well-being, with 100% accuracy. In addition, our proposed system is much easier to set up than previous systems.

We hope to use this system with elderly people in order to track their daily physical movement activities so that we can learn more about them, e.g., used at home, nursing homes, or hospitals; for example, some elderly may prefer sitting idly on the couch while watching television. Some might spend more time in the bedroom than in the kitchen. This information is very helpful to family members and healthcare providers, particularly in an ageing society.

Acknowledgments. We would like to acknowledge School of Information Technology at KMUTT and its staff for a supportive working environment, especially John Francis Lawry and Vichchuda Tedoloh for their help in English proofreading. This work passed the institutional review board (IRB) process of KMUTT in December 2014. We thank all the volunteers for helping with the data collection.

References

1. World Health Organization, WHO Global Report on Falls Prevention in Older Age, http://www.who.int/ageing/publications/Falls_prevention7March.pdf?ua=1
2. Centers for Disease Control and Prevention, National Center for Injury Prevention and Control, Web–based Injury Statistics Query and Reporting System (WISQARS) (2013), http://www.cdc.gov/homeandrecreationalsafety/falls/adultfalls.html

3. ASTV-Manager Online, Eight ways to prevent the elderly falls (May 26, 2014), http://www.manager.co.th/QOL/ViewNews.aspx?NewsID=9570000058682
4. Jian, H., Chen, H., Yang, L.: An autonomous fall detection and alerting system based on mobile and ubiquitous. In: IEEE 10th International Conference on Ubiquitous Intelligence and Computing and 10th International Conference on Autonomic and Trusted Computing (UIC/ATC), pp. 539–543 (2013)
5. Daniel, R., Albert, S., Carlos, P., Andreu, C., Joan, C., Alejandro, R.: SVM-based posture identification with a single waist-located tri-axial accelerometer. Expert Systems with Applications 40(18), 7203–7211 (2013)
6. Jovanov, E., Milosevic, M., Milenkovi, A.: A mobile system for assessment of physiological response to posture transitions. In: Proceeding of the 2013 35th Annual International Conference of the IEEE Engineering in Medicine and Biology Society (2013)
7. Zigel, Y.: A method for automatic fall detection of elderly people using floor vibration and sound. IEEE Transactions on Biomedical Engineering 56(12), 2858–2867 (2009)
8. Jonghun, B., Byoung-Ju, Y.: Posture monitoring system for context awareness in mobile computing. IEEE Transactions on Instrumentation and Measurement 59(6), 1589–1599 (2010)
9. Jennifer, R., Gary, M., Samuel, A.: Activity recognition using cell-phone accelerometers. ACM SIGKDD Explorations Newsletter 12, 74–82 (2010)
10. Yue, S., Yuanchun, S., Jie, L.: A Rotation based method for detecting on-body positions of mobile devices. In: Proceeding of the 13th International Conference on Ubiquitous Computing, pp. 559–560 (2011)
11. Petar, M., Roman, M., Marko, J., Hrvoje, H., Aimé, L., Patrizia, V.: System for monitoring and fall detection of patients using mobile 3-axis accelerometers sensors. In: IEEE International Workshop on Medical Measurements and Applications Proceedings (MeMeA), pp. 456–459 (2011)
12. Stefano, A., Marco, A., Francesco, B., Guglielmo, C., Paolo, C., Alessio, V.: A smartphone-based fall detection system. Pervasive and Mobile Computing 8, 883–899 (2012)
13. Ying-Wen, B., Siao-Cian, W., Cheng-Lung, T.: Design and implementation of a fall monitor system by using a 3-axis accelerometer in a smart phone. In: Proceeding of the IEEE 16th International Symposium on Consumer Electronics (ISCE) (2012)
14. Lee, H., Lee, S., Choi, Y., Youngwan, S.: A new posture monitoring system for preventing physical illness of smartphone users. In: Proceeding of the 2013 IEEE Consumer Communications and Networking Conference (CCNC), pp. 657–661 (2013)
15. Lina, T., Quanjun, S., Yunjian, G., Ming, L.: HMM-based human fall detection and prediction method using tri-axial accelerometer. Sensors Journal 3, 1849–1856 (2013)
16. Stephen, A., Mark, V., Konrad, P.: Hand, belt, pocket or bag: practical activity tracking with mobile phones. Journal of Neuroscience Methods 231, 22–30 (2014)
17. Guiry, J.J., van de Ven, P., Nelson, J., Warmerdam, L., Riper, H.: Activity recognition with smartphone support. Medical Engineering & Physics 36, 670–675 (2014)
18. Quoc, T.H., Uyen, D.N., Su, V.T., Afshin, N., Binh, Q.T.: Fall detection system using combination accelerometer and gyroscope. In: International Conference on Advances in Electronic Devices and Circuits – EDC, Kuala Lumpur, Malaysia (2013)
19. Paliyawan, P., Nukoolkit, C., Mongkolnam, P.: Prolonged sitting detection for office workers syndrome prevention using Kinect. In: Proceeding of the 11th International Conference on Electrical Engineering/Electronics, Computer, Telecommunications and Information Technology (2014)

20. Liu, H., Darabi, H., Banerjee, P., Liu, J.: Survey of wireless indoor positioning techniques and systems. IEEE Transactions on Systems, Man and Cybernetics, Part C: Applications and Reviews 37(6), 1067–1080 (2007)
21. Premchaisawatt, S.: Enhancing indoor positioning based on partitioning cascade machine learning models. In: Proceeding of the 11th International Conference on Electrical Engineering/Electronics, Computer, Telecommunications and Information Technology (2014)
22. Zhongtang, Z., Yiqiang, C., Shuangquan, W., Zhenyu, C.: FallAlarm Smart Phone Based Fall Detecting and Positioning System. Procedia Computer Science 01, 617–624 (2012)
23. The University of Waikato, Weka 3 - Data Mining with Open Source Machine Learning Software in Java, http://www.cs.waikato.ac.nz/ml/wek/

Factors Influencing Users' Willingness to Use Cloud Computing Services: An Empirical Study

Ahmad Asadullah, Ishaq Oyebisi Oyefolahan, Mohammed A. Bawazir, and Seyed Ebrahim Hosseini

Information System Department, Faculty of Information and communication Technology, International Islamic University Malaysia
Assad1989@hotmail.com,
ishaq@iium.edu.my,
{bawazir333,Hosseini.qalati}@gmail.com

Abstract. Cloud computing technology is one of the newest technologies widely used by consumers globally due to its advantages. One of the common advantages is that users can get access to applications and their own files and data on demand. However, there are privacy concerns which discourage users from being active in cloud computing. This paper attempts to investigate the factors that influence consumers' willingness to use cloud computing services by using a quantitative research approach. A total of 340 cases were gathered from a sampled population of students. Based on SPSS analysis, the study found that perceived privacy control, perceived effectiveness of privacy policy, and information privacy concerns in a cloud computing environment have significant influence on users' willingness to use cloud computing services.

Keywords: Willingness to Use Cloud Services, Privacy Concern, Cloud Computing, Privacy Policy, Privacy Control, Security, Cloud, Privacy Risk in Cloud.

1 Introduction

Cloud computing is a recent trend in the field of information and communication technology. It is defined by the US National Institute of Standards and Technology (NIST) as *"a model for enabling ubiquitous, convenient, on-demand network access to a shared pool of configurable computing resources (e.g. networks, servers, storage, applications, and services) that can be rapidly provisioned and released with minimal management effort or service provider interaction"*. According to the Criminal Justice Information Services department (CJIS), this is the most accepted definition. Cloud computing deals with computation, software, data access and storage services that may not require end-user knowledge of the physical location and the configuration of the system that is delivering the services [1].

1.1 Background of Study

Cloud computing technology offers various types of services and applications for the public and organizations. One of the most commonly used cloud computing services

H. Unger et al. (eds.), *Recent Advances in Information and Communication Technology 2015*,
Advances in Intelligent Systems and Computing 361, DOI: 10.1007/978-3-319-19024-2_23

is social networking such as Facebook, LinkedIn, Twitter, Instagram and many others, although these were not initially considered cloud computing services. Another example of cloud computing services is E-mail services. Other major example of cloud computing services is Google Drive. All of Google's services could be considered as cloud computing such as Gmail, Google Calendar, Google Reader, Google Voice, Google glass and so on. In addition, Apple's cloud service is primarily used for online storage and synchronization of user mail, contacts, calendar, and more. All the data you need is available to you on your iOS, Mac OS, or Windows device. Prior studies found that information privacy concerns is an essential factors that discourage and influence users from transiting through online channel [2]. Researchers found that the growth and success of a new technology and e-commerce is linked to users' willingness to disclose personal information to service providers [3] [4].

2 Research Model

2.1 Willingness to Use Cloud Computing Services

In this study, willingness to use cloud computing services is a dependent variable which has been addressed by allocating five items. Overall, almost all previous studies investigated willingness to disclose personal information in online banking and e-commerce whereas this study assesses it in the context of a cloud computing environment. According to Wu, Huang, Yen, & Popova (2012), the success or growth of e-commerce is linked inextricably to consumers' willingness to provide personal information to service providers [5]. Most researches have focused on privacy and risk generally as factors that influence users' intention to provide personal information to an online service provider [6]. Providing personal information to an online service provider has been studied in the domain of online shopping globally [7].

2.2 Information Privacy Concern

According to the Oxford Dictionary, privacy means *"The state or condition of being free from being observed or disturbed by other people"*. Mishra, Ranjita; Dash, Sanjit K (2011) indicated that privacy is the protection of transmitted data from passive attacks [8]. However, as cloud computing systems usually offer services (e.g. DaaS, SaaS, IPMaaS, PaaS, and so on) on the Internet, the secret information of individual users' is stored and managed by the service providers in the cloud, and consequently results in privacy concerns [9]. In this study, the researcher investigates the impact of users' information privacy concern on willingness to use cloud computing services. Previous studies found that there is negative relation between the privacy concern and willingness to disclose personal information. Therefore, hypotheses 1: Willingness to use cloud computing services is negatively affected by consumers' information privacy concerns [4] [10].

2.3 Perceived Privacy Control

Cloud computing technology involves distributed computation on multiple large-scale data sets across a large number of computer nodes. Every Internet user is able to share his or her personal data to the Cloud Computer systems which are located on the other side of the Internet. For instance, a user's click stream across a set of webs (e.g., Amazon book store, Google search web pages, etc.) can be used to provide targeted advertising [11]. However, few studies tried to clarify the nature of control in the privacy context. For instance, in privacy literature, control has been utilized to refer to several objectives such as social power and procedural fairness of an organization's privacy [12]. Studies have specified perceived privacy control based on three dimensions: 1) knowledge: users should be aware of a service provider's information practices. It is assumed that without this knowledge, a consumer is unable to make a decision as to either disclose personal information or not. 2) Choice/access: users should be provided with choices as to how their personal information is utilized beyond the use for which the information was provided. 3) Use of privacy tools: when privacy tools like protocol for privacy preferences (P3P) or privacy seals are used by some service providers then the user begins to believe that his control over his personal information is growing. Previous studies assumed that there is positive relationship between perceived privacy control and willingness to disclose personal information. Therefore, hypotheses 2: Perceived privacy control will positively influence consumers' willingness to use cloud computing services [4] [10].

2.4 Perceived Effectiveness of Privacy Policy

Privacy policies are notices that are displayed on an online service provider's website, accessible to the public, and describe an organization's information practices. Previous studies found that privacy policies and strategies that online service providers adopt and implement can have an impact on users' perception of online providers' fairness and trustworthiness and on users' willingness to engage in online transactions [13]. Meanwhile, other studies confirmed that the presence of a privacy policy had a significant influence on information disclosure in online shopping [13]. Generally, privacy policies should be designed to illustrate the steps that would employ to ensure obedience with the fair information principles. This is because the ultimate objective of any privacy policy is to provide consumers with comfort and security. Therefore, hypotheses 3: Perceived effectiveness of privacy policy will positively influence consumers' willingness to use cloud computing services.

2.5 Perceived Privacy Risk

Risk has been defined as the uncertainty resulting from the potential for a negative outcome and the possibility of another party's opportunistic behaviour that can result in losses for oneself. Therefore, usually negative perceptions are related to risk and it may influence an individual emotionally and physically [14]. Previous studies agreed that since online service providers required users' personal information as a part of online transactions, users express their concern about such information being misused, sold, disclosed, or exchanged with other parties without authorization from the owners [15]. It has been proved that such risk perceptions of online users constitute an obstacle to the widespread acceptance of online transactions mainly when sensitive information is required [16]. There is thus a consensus among researchers that risk is vital factors in making any kind of online decisions. Thus, hypotheses 4: Perceived privacy risk will negatively influence consumers' willingness to use cloud computing services [15] [17].

2.6 Individuals' Factors

In this study, the researcher investigated whether the individuals' factors (age, gender, educational level) influence users' willingness to use cloud computing services. Nevertheless, prior studies assumed that the demographic factors influencing users' willingness to disclose personal information [4]. Therefore, hypotheses 5: There will be a significant relationship between the demographic information (age, gender, educational level) and willingness to use cloud computing services.

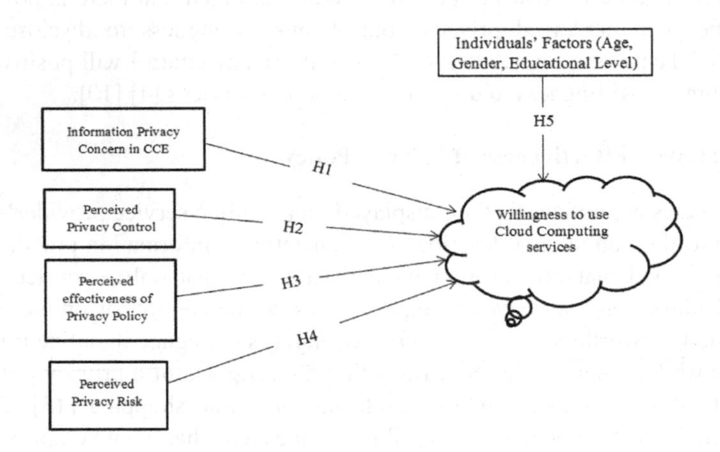

Fig. 1. Research Model

3 Methodology

A quantitative survey method has been adopted to collect data from targeted respondents. For that, we conducted a self-administered and online questionnaire to be answered by undergraduate and postgraduate students from the International Islamic University Malaysia as our sample with 340 cases as sample size. The survey items were adopted from prior studies where a five point-Likert scale was used for the items

4 Result

4.1 The Result of Demographic Information

The demographic profile of the respondents illustrates that the largest number of cases was from female students' 51.2.1% and 35.6% of the total respondents were aged in the group of below 20. Moreover, the majority of the respondents were undergraduate students (70.9%).

4.2 Reliability Test

Cronbach's alpha test was used out to measure the reliability of the items. It was agreed upon the lower limit which is 0.7 across all constructs. Table 1 illustrates that.

Table 1. Reliability result

Construct	Number of Items	Reliability
Privacy concern	4	.841
Perceived Privacy Control	4	.845
Perceived Effectiveness of Privacy Policy	3	.842
Willingness to Use Cloud Computing Services	5	.708

4.3 Cloud Computing Services Usage Rate

Based on previous studies, cloud computing services were categorized into five categories: Social Networking (e.g., Facebook, Twitter and others), Email Services (e.g., Hotmail, Gmail and others), Google Services (e.g., Google Drive, Google Apps, and others), Apple Services (e.g., iCloud, iTunes and others) and Data Storage Services (e.g., Box, Dropbox, and others). However, it can be seen that Social Networking services has the highest percentage of respondents with 29.93%, whereas apple services has the lowest percentage of respondents with 7%. Between the highest and the lowest percentage of respondents are categories that include Email services 26.70%, Google services 21.64% and Data storage services 14.73% respectively.

4.4 Testing Research Hypotheses

A multiple linear regression technique was used for testing the hypotheses. The result provides two tables ANOVA and Coefficients. ANOVA table indicates the model is significant ($p = .000 < 0.05$). Nevertheless, the coefficients table shows that that there is significant relationship between perceived privacy control, perceived effectiveness of privacy policy, and willingness to use cloud computing services (perceived privacy control $p = .000$, perceived effectiveness of privacy policy $p = .000, < 0.05$, information privacy concern $p = .017$). On the other hand, the result illustrates that there is no significant relationship between perceived privacy risk and willingness to use cloud computing services (p = .681, > 0.05). Therefore, hypotheses 1, 2, and 3 are supported except hypotheses 4. Table 2 provides the result of multiple linear regression.

Table 2. Multiple linear regression result

ANOVA[a]

Model		Sum of Squares	df	Mean Square	F	Sig.
1	Regression	33.371	4	8.343	21.907	.000[b]
	Residual	127.575	335	.381		
	Total	160.946	339			

a. Dependent Variable: Willingness_To_Use
b. Predictors: (Constant), Privacy_Control, Privacy_Concern, Privacy_Policy, Priavcy_Risk

Coefficients[a]

Model		Unstandardized Coefficients		Standardized Coefficients	t	Sig.
		B	Std. Error	Beta		
1	(Constant)	2.410	.252		9.571	.000
	Privacy_Concern	-.123	.051	-.155	-2.404	.017
	Priavcy_Risk	.021	.052	.027	.412	.681
	Privacy_Policy	.233	.043	.296	5.444	.000
	Privacy_Control	.152	.043	.191	3.545	.000

a. Dependent Variable: Willingness_To_Use

4.5 Individuals' Factors

In this study, the relationship between the demographic information and willingness to use cloud computing services is examined. It is assumed that *"There will be a significant relationship between the demographic information (age, gender, educational level) and willingness to use cloud computing services"*. Nevertheless, in this research the age is categorized into four groups (Below 21, 22 to 25, 26 to 30 and 30 and above). Therefore, the ANOVA technique was conducted. The result shows that there is no significant differences among the means' of age groups (p = .061 > 0.05), as illustrated by Table 3.

Table 3. ANOVA result for age group

ANOVA

Willingness

	Sum of Squares	Df	Mean Square	F	Sig.
Between Groups	3.482	3	1.161	2.476	.061
Within Groups	157.464	336	.469		
Total	160.946	339			

Multiple Comparisons

Dependent Variable: Willingness

(I) Age	(J) Age	Mean Difference (I-J)	Std. Error	Sig.
	22 to 25	.06300	.08470	.879
Below 21	26 to 30	-.13182	.12052	.694
	30 and above	-.26061	.13444	.214
	Below 21	-.06300	.08470	.879
22 to 25	26 to 30	-.19481	.11812	.352
	30 and above	-.32360	.13229	.071
	Below 21	.13182	.12052	.694
26 to 30	22 to 25	.19481	.11812	.352
	30 and above	-.12879	.15765	.846
	Below 21	.26061	.13444	.214
30 and above	22 to 25	.32360	.13229	.071
	26 to 30	.12879	.15765	.846

The t- test technique was carried out to test the significant differences between the means across educational levels. The result illustrates that there is significant different in the means across the two groups ($p = .034 < 0.05$), as seen in Table 4. The result shows that the mean for the postgraduate group is higher than the undergraduate group (Postgraduate Mean = 3.3939, Undergraduate Mean = 3.2199).

Lastly, in this study gender has two groups (Male and Female). The t- test technique was utilized to determine the significant differences between the means across the two groups. Table 4 shows the result of the t-test, which illustrates that there is no significant different in the means across the two groups ($p = .265 > 0.05$).

Table 4. *T-test* result for educational level

Group Statistics

	Educational_Level	N	Mean	Std. Deviation	Std. Error Mean
Willingness_To_Use	Undergraduate	241	3.2199	.68015	.04381
	Postgraduate	99	3.3939	.69837	.07019

Table 4. *(continued)*

Independent Samples Test

		Levene's Test for Equality of Variances		t-test for Equality of Means				
		F	Sig.	t	df	Sig. (2-tailed)	Mean Difference	Std. Error Difference
Willing-ness_To_Use	Equal variances assumed	.029	.865	-2.127	338	.034	-.17402	.08183
	Equal variances not as-sumed			-2.103	178.199	.037	-.17402	.08274

Lastly, the gender in this study the gender has two groups (Male and Female). Thus, t- test technique was utilized to determine the significant differences between the means across the two groups. Table 5 shows the result of t- test, the result illustrates that there is no significant different in the means across the two groups ($p = .265 > 0.05$).

Table 5. T-test result for gender

Group Statistics

	Gender	N	Mean	Std. Deviation	Std. Error Mean
Willingness	Male	166	3.3133	.70061	.05438
	Female	174	3.2299	.67731	.05135

Independent Samples Test

		Levene's Test for Equality of Variances		t-test for Equality of Means				
		F	Sig.	T	df	Sig. (2-tailed)	Mean Difference	Std. Error Difference
Willingness	Equal variances assumed	.355	.552	1.116	338	.265	.08337	.07473
	Equal variances not assumed			1.115	335.798	.266	.08337	.07479

5 Discussion

This study revealed that Social Networking services have the highest percentage of usage rate among the participants. The outcome supports prior studies [18]. This study identified that there is a statistically significant relationship among information privacy concerns in cloud computing environment, perceived effectiveness of privacy policy, perceived privacy control, and willingness to use cloud computing services. Meanwhile, perceived privacy risk is not statistically significant upon users' willingness to use cloud computing services. This is similar to prior studies [19]. Our finding indicates that users' willingness to use cloud computing services is positively and significantly influenced by perceived effectiveness of privacy policy. This result is similar with Mollick, Joseph Sudeep (2005) who found that customers' willingness to transact online will be higher when venders have policies that allow customers the ability to authorize or give informed consent to data collection than when venders do not have effective and clear policies [20]. As for the demographic information such as age and gender, it has no significant influence on users' willingness to use cloud computing services, whereas, educational level has a statistically significant influence.

6 Conclusion

In this study, the primary objective was to identify the factors that influence users' willingness to use cloud computing services. Prior studies focused on the factors that influence users' willingness in e-commerce and online banking environment, whereas, this study investigated it in a cloud computing environment. Our findings proved that information privacy concerns, perceived privacy control, and perceived effectiveness of privacy policy have a statistically significant relationship on users' willingness to use cloud computing services. Meanwhile, perceived privacy risk has no significant influence on willingness to use cloud computing services. It is recommended that the quality of cloud services should be improved. One of the most interesting findings of our study is that users are uninformed of the process of data collection processes that take place while using cloud services as well as the possible reuse of the data by the cloud service provider. Hence, cloud computing service providers should provide clear and effective privacy policies to their clients; otherwise, users are not willing to use cloud services.

References

1. Jadeja, Y., Modi, K.: Cloud Computing - Concepts, Architecture and Challenges, pp. 877–880 (2012)
2. Morton, A.: Measuring Inherent Privacy Concern and Desire for Privacy - A Pilot Survey Study of an Instrument to Measure Dispositional Privacy Concern. In: Proceeding of the 2013 International Conference on Social Computing (SocialCom), pp. 468–477 (2013)
3. Barto, T.P.: Distress and Other Situational Factors that Influence Consumer Willingness to Provide Personal Information in an Online Buyer-Seller Exchange: an Equityty Theory Perspective (2011)

4. Odeyinde, O.: Information Privacy Concerns of Undergraduate Students in a Nigerian University and their Willingness to Provide Personal Information to Transact on the Internet (November 2013)
5. Wu, K.W., Huang, S.Y., Yen, D.C., Popova, I.: The effect of online privacy policy on consumer privacy concern and trust. Computers in Human Behavior 28(3), 889–897 (2012)
6. Zimmer, J.C.R., Arsal, E., Al-Marzouq, M., Grover, V.: Investigating online information disclosure: Effects of information relevance, trust and risk. Information & Management 47(2), 115–123 (2010)
7. Mancha, R., Beebe, N., Yoon, H.S.: Customers' Personalit, Their Perceptions, and Green Concern on Internet Banking USE, pp. 1–13 (2012)
8. Mishra, R., Mishra, D.P., Tripathy, A., Dash, S.K.: A privacy preserving repository for securing data across the cloud. In: Proceeding of the 2011 3rd International Conference on Electronics Computer Technology (ICECT), pp. 6–10 (2011)
9. Chen, D., Zhao, H.: Data Security and Privacy Protection Issues in Cloud Computing. In: Proceeding of the 2012 International Conference on Computer Science and Electronics Engineering, vol. (973), pp. 647–651 (2012)
10. Zorotheos, A., Kafeza, E.: Users' perceptions on privacy and their intention to transact online: a study on Greek internet users. Direct Marketing: An International Journal 3(2), 139–153 (2009)
11. Zhou, M., Zhang, R., Xie, W., Qian, W., Zhou, A.: Security and Privacy in Cloud Computing: A Survey. In: Proceeding of the 2010 Sixth International Conference on Semantics Knowledge and Grid, pp. 105–112 (2010)
12. Xu, H., Dinev, T., Smith, J., Hart, P.: Information Privacy Concerns: Linking Individual Perceptions with Institutional Privacy Assurances. Journal of the Association for Information Systems 12(12), 798–824 (2011)
13. Hui, K.L.H., Teo, H., Lee, S.Y.T.: The value of privacy assurance: An exploratory field experiment. MIS Quarterly 31(1), 19–33 (2007)
14. Moon, Y.: Intimate Exchanges: Using Computers to Elicit Self-Disclosure From Consumers. Journal of Consumer Research 26, 323–339 (2000)
15. Dinev, T., Hart, P.: An extended privacy calculus model for e-commerce transactions. Information Systems Research 17, 61–80 (2006)
16. Eastlick, M.A., Lotz, S.L., Warrington, P.: Understanding online B-to-C relationships: An integrated model of privacy concerns, trust, and commitment. Journal of Business Research 59, 877–886 (2006)
17. Kim, D.J.D., Ferrin, L., Rao, H.R.: A trust-based consumer decision-making model in electronic commerce: The role of trust, perceived risk, and their antecedents. Decision Support Systems 44, 544–564 (2008)
18. Alhamed, M., Amiri, K., Omari, M., Le, W.: Comparing privacy control methods for smartphone platforms. In: Proceeding of the 2013 35th International Conference on Software Engineering, pp. 36–41 (May 2013)
19. Salleh, N., Hussein, R., Mohamed, N., Aditiawarman, U.: An Empirical Study of the Factors Influencing Information Disclosure Behaviour in Social Networking Sites. In: Proceeding of the 2013 International Conference on Advanced Computer Science Applications and Technologies, pp. 181–185 (2013)
20. Mollick, J.S.: Privacy Policies, Fairness, Trustworthiness and Willingness to Transact With Firms Online (2005)

On the Distributions of User Behaviors in Complex Online Social Networks

Suwimon Vongsingthong[1], Sirapat Boonkrong[1], Mario Kubek[2], and Herwig Unger[2]

[1] King Mongkut's University of Technology North Bangkok, Bangkok, Thailand
suwimonv@yahoo.com, sirapatb@kmutnb.ac.th
[2] FernUniversität in Hagen, Hagen, Germany
{mario.kubek,herwig.unger}@fernuni-hagen.de

Abstract. Understanding user behavior is an important issue to make any prediction of the resource utilization and the distribution of information in social networks as well as to determine approaches to optimize the networks. Based on the results of surveys, a typical group of users, their behaviors and activities in the online social network Facebook have been analyzed and classified. The results confirm the validity of power laws and small-world properties in various areas of social network systems and will later allow the establishment of useful models for further simulations and investigations in the area of social network mining.

Keywords: Online Social Networks, User Behavior, User Activities, Power-law Distribution, Small-world Networks, Network Optimization.

1 Introduction

The opportunities presented by online social networking systems (SNSs) have led billions of users to congregate around these sites. This is mainly due to the set of available operations for registering and setting up a personal profile, searching, adding or removing friends, joining communities, communicating with other users, and disseminating of content.

This study quantifies and analyzes the user behavior in a very popular online social network, Facebook. Facebook is an integral part of most college students' lives which constitute our focus group. It is a cloud-based web application where its services and storage are distributed on different servers located in different geographical locations. As the number of users and activities increases over time, the network becomes dynamically evolving, complex, and denser with irregular structure inheriting the properties of complex network. Regarding the exceptional growth in size and complexity, this network shares many common characteristics such as small diameter, high clustering coefficient and a power-law degree distribution of node degrees [1], also known as *small-world* properties. As a consequence, the existing of communities and the average distance between nodes tend to increase logarithmically over time as a function of the number of nodes. This network is also considered as scale-free because it exhibits a power-law degree distribution, a useful model to explore the utilization of the network resources.

H. Unger et al. (eds.), *Recent Advances in Information and Communication Technology 2015,*
Advances in Intelligent Systems and Computing 361, DOI: 10.1007/978-3-319-19024-2_24

The existence of these properties has drawn massive attention from many researchers that examined the diffusion of information and rumors as well as the duplication of content [2] in those complex networks. Besides, the analysis of the evolution and structural properties of online social networks and the identification of user groups and their behaviors are topics of increasing interest.

However, none of the studies reviewed have investigated whether the power law was applicable to SNSs or have considered users and their behaviors as a unity to harness their structure. Moreover, due to technical restrictions, it is very difficult to obtain relevant data from SNSs like Facebook directly. Therefore, the idea to be conducted in this research is to select a sample of Facebook users and collect information on their behaviors in this online social network through a developed questionnaire. The two underlying aims are to explore the complexity of college student networks in SNSs and to test the validity of the power-law distribution of their activities. Resemble different scenarios in the fields of communication networks, biology and economics which the distribution of data can be explained by the power law, the characterization of the dynamic growth of traffic originating from activities users perform will be explored by using the power law distribution.

The organization of this paper is as follows: Section 2 focuses on the presentation of the complexity of online social networks. Some of the common properties of the topology and structure of complex networks are given and explained how they are measured based on literature review. Section 3 presents the research methodology deployed to analyze the behaviors of college students in Facebook, demographic analysis and results of the experiment. Section 4 is the discussion and implication of the analysis results. The last section, Section 5 concludes the article and suggests future research direction.

2 Complexity of Social Networks

Many different phenomena such as the topology of the Internet, the human brain, the bacterial metabolism and social relationships [3] can be described using network terminology. The complexity of these networks is mainly determined by the number of nodes and evolving patterns of bilateral connections among individual entities. The state of each entity or node can vary over time in many ways, and each node can represent anything ranging from individuals, web pages and activities related to economics. To create an understanding of how these networks are complex, the basic notions of network complexity are introduced in the following sections.

2.1 Structure of Complex Networks

Even if these complex networks may be formed from dissimilar origins, they share a surprising number of common structural properties of "small worlds" and a power-law distribution of node degrees:

2.1.1 Small-World Property

The small-world property has been found in a variety of real networks, including biological and technological ones with two structural properties of interest, the small-world effect and the presence of clusters. The small-world effect is the concept where any two nodes in a network are connected to each other through a short path based on the metric called average path length. Given a network as a graph $G = (V, E)$ where V is a set of nodes and E is a set of edges. The average path length is

$$L_G = \frac{2}{n(n-1)} \sum_{i \neq j} d(v_i, v_j) \tag{1}$$

where n is the number of vertices in G, and $d(v_i, v_j)$ is the distance between node i and j. The second property is a high clustering coefficient. The clustering coefficient is a measure of the presence and density of triangles which is a closed loop of a length of three. The clustering coefficient is

$$C = \frac{1}{n} \sum_{i=1}^{n} C_n \qquad \text{where } C_n = \frac{|\{e_{jk}: v_j, v_k \in \text{neighborhood of vertex } n\}|}{k_n(k_n-1)} \tag{2}$$

where k_i is the number of vertices linked to by vertex n (neighbors) and e_{jk} indicates an edge between vertices j and k that are in n's neighborhood. A high value implies that if two nodes are having a common neighbor, they also have a high tendency to be connected to each other. This value tends to decrease with the increase of scale of the network, however the larger the scale of the network, the slower the speed of decrease. A value close to 0 means that there is hardly any connection in the neighborhood and a value of 1 indicates a fully connected neighborhood.

2.1.2 Power-Law Distribution and Scale-Free Networks

A first step towards a deep understanding of SNSs is the exploration of the node degree distribution. Insight studies of SNSs show that the distribution of their node degree possesses the characteristic of a heavy tail. Also, a small proportion of nodes has a high degree while the vast majority of nodes has a low degree. Hence, the standard linear plot of all nodes is not able to display the coexistence of widely different degrees of nodes properly. The usage of double logarithmic axis (log-log plot) accounts for power-law distributions of node degrees is more appropriate.

A network with a power-law distribution of node degrees is also known as a scale-free network. Its two main characteristics are: nodes add themselves to the network on a continuous basis and nodes add themselves preferentially to other nodes following a power-law distribution. Here, the simplest mathematical form of $P_k \sim k^{-\gamma}$ will be used. P_k is the probability that a randomly selected node has exactly k edges and γ is what is called the "*scale-free exponent*", a constant parameter of the distribution. This implies that scale-free networks have few nodes with a large number of first-order connections k, and a large number of nodes with few first-order connections. Some of real-world examples that exhibit this characteristic are the number of e-mails received by a person, metabolic reaction networks, the telephone call graph and the World Wide Web.

2.2 State of the Art in User Behaviors Analysis

Previous studies had assumed that user behavior, especially in SNSs, were too complex to understand and tried to prove that they held the property of random networks with the probability p of link distribution following a Poisson distribution. Recently, many researchers have emphasized on the analysis of the structure of SNSs as well as the behavior of users and obtained a variety of interesting results.

Falck-Ytter and Øverby [4] compared the number of followers who followed popular contents in Twitter and Youtube. The result depicted that Zipf's law was not suitable to describe followers in Twitter. In contrast, it was suitable to describe Youtube viewers. Gyarmati and Trinh [5] compared the degree distribution of nodes for users' online sessions in Bebo, MySpace, Netlog, and Tagged. The authors claimed that the distribution of session times of users and the number of sessions follow power-law distributions. Zhong et al. [6] explored user behavior and interests in different SNSs and found that they influenced one another. Besides, only a small percentage of users actually performed many activities in SNSs leading to distribution of users with a long-tail property.

Yan et al. [7] studied the behavior of users in posting microblogs and found that the interval time distribution of people posting did not fit a normal distribution, instead it was a power-law distribution. Ding et al. [8] studied behaviors of users in the BBS social network by analyzing the read and reply data and realized that the distributions of discussion sizes and user participation levels followed a power-law distribution. Morales et al. [9] analyzed the effects of user behavior on social structure emergence using the information flow on Twitter and concluded that community structure was formed inside the network. In addition, the distribution of the number of posts by users was found to fit an exponentially truncated power law. Liu et al. [10] discovered that the number of daily active users using applications in Facebook followed power-law distribution. Feng et al. [11] explored pin distribution (visual bookmarks of images) and board distributions as well as number of comments, likes, repins in Pinterest and detected that those characteristics followed a power law.

Much work has been put into characterizing user behavior in SNSs due to their merits to socioeconomics and the optimization of involved application. Those studies generally acquired large volumes of data through application programming interfaces (APIs) or crawling. The first has restriction on obtainable functionalities and the latter is limited by specific mechanisms implemented in the services to prevent automatic scraping of data to a large extent. To avoid such problems, the experiment described and explained in this article gathered data from a survey.

Most of the existing analyses focused on coarse-grained user activities such as duration and frequency of login, and number of friends but none of those have explored into real-world activities of users such as browsing, playing games, and shopping. Therefore, this study will explore a real-world network of college students to prove that a power law is not only applicable to describe the topology of SNSs but also applicable to the behaviors of user in SNSs by using Facebook as a precursor. The implications derived from our findings are expected to provide contributions in understanding the evolution of SNSs in general, but also to point out ways to technically simulate SNSs in a proper manner.

3 Tracking User Behaviors in Facebook

Evidently, the realization of user behavior is essential to predict how social networks change and evolve. Variety of approaches have been adopted technique to derive basic user behavior which nonetheless is universal enough to capture most of the dynamics of the complexity of social interactions. The steps towards the results are described in the following sections.

3.1 Methodology

How users behave in Facebook was fundamentally tracked through a questionnaire which was a modified version of a previous survey of teenagers' behaviors in SNSs [16]. The questionnaire was put forward because the accessibility to insights on how users behave in Facebook was not permitted. To analyze the effect of user characteristics to SNS structure, the target group selected is college students who clearly occupy approximately 90% of SNS population. The understanding of how users behave when they are online will not only give an indication of how to define probability model. It also technically assists to define behavioral rules that particularly influence the evolution of SNSs. Leading to the simulation on groups of agents/artificial users which can properly predict user behavior in SNSs while relying on a sound scientific basis.

The developed questionnaire is split into three sections. Section I focuses on demographic data, Section II is related to user preferences and invariable behaviors such as conditions for adding friends, sharing information, number of friends, etc. Because a user can perform a wide range of activities on a typical online social network, we further tried to semantically group similar activities into a category by utilizing the structure of Facebook. Hereby, Section III contains groups of primal activities such as playing games, posting/tagging of content, profile updating, visiting other pages, selling/buying merchandises, chatting, and posting LIKE. For each interaction question, user was requested to evaluate their activities and connections with friends, acquaintances and unknowns.

Errors in questionnaire were diminished in two stages; pre-testing and formal measurements. The idea of pre-testing was to judge whether the content of questionnaire needed modification. After the pre-test, its reliability was assured by Cronbach's ($\alpha > 0.96$). The designated participants were college students in universities in Bangkok who agreed to take part in the survey. The process of collecting data were actualized during October 2013. A total of 1,200 questionnaires were issued, and 1,173 valid questionnaires were obtained.

3.2 Analysis of College Student Behavior in Facebook

The analysis of user behavior in SNSs comprises three subtasks. The first task is a statistical analysis for a preliminary understanding of the distribution of user groups and their basic activities. Second, the importance of user social relationships is discussed and evaluated based on the attributes of average path length and network clustering coefficient derived from the obtained social network graph. Third, the analysis results regarding the distribution of user activities in SNSs are given. All the discussion regarding to the results will be proffered in Section 4.

3.2.1 Data Description

Among effective samples, female users account for 59%, and male users account for 41%. In respect of age, people aged 21 to 25 occupy the highest proportion (56.6%), follow by those aged 15-20 account for 33.4%. In case of occupation, the group of students are the most dominant group (82.8%). As for educational level, undergraduate students take the first place with 71.3%. Therefore, this dataset represents a real-world population of college students. Main purposes of using SNSs found in this study are to contact friends and search for entertainment. On average, college students check SNSs 6-10 times a day and for each login consumes 2-3 hours, mostly using smart phone as accessing tool. 98% of the participants designate Facebook as primary SNS and almost all (99.2%) check up their timeline as soon as they have new alerts or anyone responses to their posts.

There is an obvious selection bias in this experiment because this selected group is naturally inclined towards technology. Even though these young college students cannot be inferred as the whole population of Facebook in Thailand, they are of interest to be studied. Their habits of checking Facebook as first thing when they wake up and last thing before they go to bed can circulate diverse activities and behaviors.

3.2.2 Identification of Small-World Characteristics

The interconnection within a specified area of the college student network is shown in Fig.1. How well the nodes (representing the students) are connected is illustrated. The graph contains 1,173 nodes with 12,731 edges. Activities in SNS are classified into five components. Approximately 75% of all occupy by browsing/posting and peeking to friend's page. The graph is also highly connected with average path length of 3.234 as derived from (1). The clustering coefficient, 0.317 descended from (2) indicates that the probability that college students will introduce friends to any other friends is approximately 32%. These values also indicate that the college student network is a closely connected group which possesses small-world characteristics.

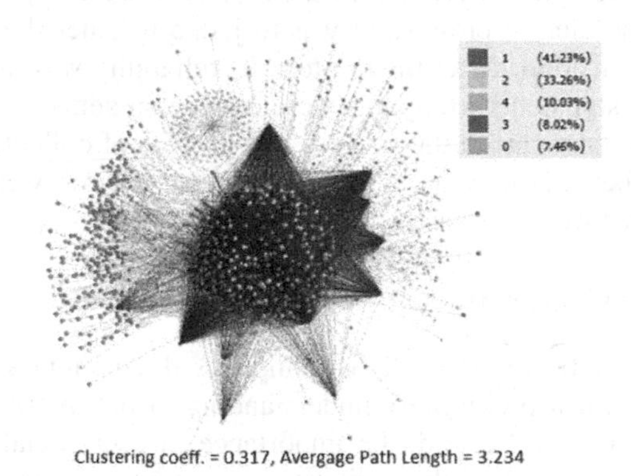

Fig. 1. Clusters of college students based on activities

Explore into clusters of the characteristics in Figure 1, the first cluster (1) is "the post/browse sweetheart", this group loves to express its feelings, be the center of attention from friends and in flavor of looking around for something of interest. The second cluster (2) is considered "the inquisitive crowd", very curious to learn new things by peeking into other friend pages or interesting links. The third set (4), "the game lover", this group is fond of playing online games and making friends with those who share similar game interest. The forth group (3) is "the profile update clan" who likes to evince personal events and new experience to friends as appeared in forms of photos, text and videos. The smallest group (0) in student community is "the shopping mania", refers to those who prefer online shopping or posting in personal pages what they think friends will be interested to buy.

3.2.3 Distributions of User Activities

The log-log plots in Fig. 2(a-e) show the distribution of user activities in Facebook gained from the questionnaire. Activities displayed comprise playing online games, updating profiles, browsing/posting/tagging, visiting other pages and selling/buying merchandises. All exhibit power-law characteristics which can be interpreted that the mention activities fit a power-law distribution with the exponent γ in Table 1. It is worth to mention that the average power exponent is 1.65 and complies well with Barabasi's theory of heavy tails in human behaviors [12]. This means that each activity is generally performed by few college students and only a few is actualized by many. The majority (70%) has interaction within Facebook less than 3 times a day with an average of 2-3 times a day. Very few interacts more than 10 times a day. Considering purposes of those who login, 10% posted messages, 6% played online games, 6% updated profiles, 6% visited other pages, and 5% shopped online.

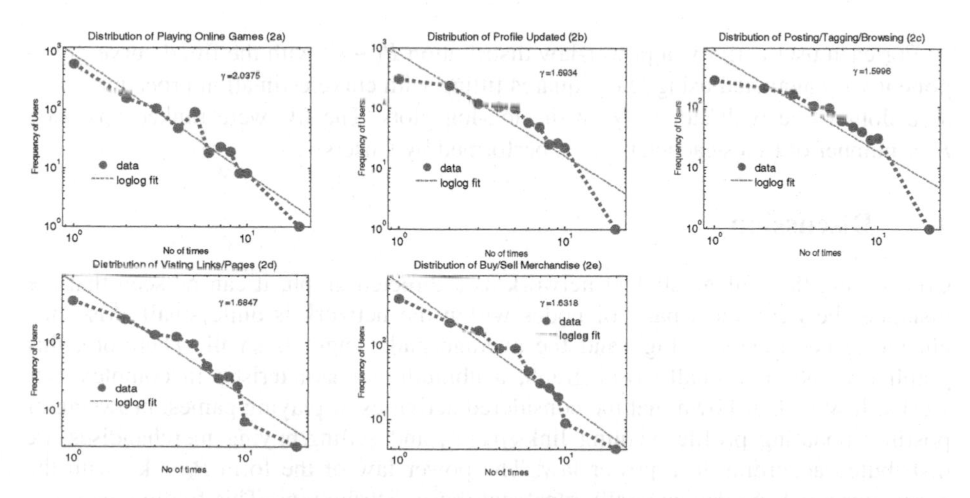

Fig. 2(a-e). Distributions of daily college students' activities

The mentioned scaling exponent (γ) is shown with more detail in Table 1 along with average value and standard deviation of each activity and group respectively.

Table 1. Summary of activities with power-law distributions

Gr	Activity Group	Activities	γ	Avg. γ	SD.
1	Games	playing online game alone	2.04	1.81	0.33
		playing online game with friends	1.95		
		playing online games with unknown	1.43		
2	Browse/Post /Tag	posting photos	1.69	1.56	0.21
		posting videos	1.66		
		browsing friends tag	1.31		
3	Profile Update	changing profile photo	1.60	1.59	0.02
		update cover page photo	1.60		
		update own status	1.57		
4	Visit Other Pages	browsing of advertisement pages	1.68	1.51	0.11
		visiting popular pages	1.48		
		visiting friend of friend page	1.44		
		visiting link posted	1.46		
5	Sell/Buy Merchandises	searching for merchandises	1.63	1.63	0.13
		post for selling products	1.80		
		questioning on products	1.72		
		actual buy	1.51		
		response to merchandise sold by friends	1.66		
		response to merchandise sold by unknown	1.46		

For each user activity, a power-law distribution $P_k \sim k^{-\gamma}$ with the fitted curve of exponent γ is calculated using least squares fitting with curve estimation procedure, seen as a dotted line with slope - γ on the log-log plots. The fits were ranked based on P_k = number of times each activity is performed by k users.

4 Discussion

Considering the college student network as a directed graph, it can be seen that the distances between most pairs of nodes within the network is quite small. Also, the clustering coefficient is high and the average path length is small. Therefore, this graph becomes the small-world graph, a ubiquitous characteristic in complex networks. It was also shown that the considered activities of playing games, browsing or posting, updating profile, visiting links/pages, and selling/buying merchandises are distributed according to a power law. The power law of the form $P_k \sim k^{-\gamma}$ with the parameter γ = 1. 65 can generally represent those distributions. This finding also suggests that activities users perform in Facebook occur with different frequencies. The activities of posting/tagging/browsing make up the largest share (>50%) of all activities considered.

Knowing that a quantity does or does not follow a power law provides important theoretical clues about the underlying generative mechanisms to be considered. When it does, the power law's exponent γ is the only parameter in question. In our experiments, the number of times k users perform each activity grows as a power function with the exponent γ. The experiments suggest that power-law distributions can be generally used to describe and elucidate human behavior.

Boccaletti [13] mentioned that for most real scale-free networks, the exponent's value is $2 < \gamma < 3$. In this study, the specific exponents lie between 1.31 and 2.04 or ($1 < \gamma \leq 2$) whereas $\gamma < 2$ is observed only in networks that are relatively small (as it is the case here) or in which the power-law behavior has some cutoff.

In interpreting the results of this study, one must pay attention to a number of limitations. The data used was self-selected from 1,173 college students at the age of 20-25 that were active on Facebook. Therefore, this data contains some bias regarding the sample group. However, this may be an advantage considering that this group is technology-oriented and often active in SNSs in various ways. Even though this data may not be used to draw conclusions about different groups of users in other SNSs, it is an insightful dataset suitable for modeling user behavior in simulations of SNSs. A first simulation for SNSs [14] derived from this dataset comprising four models for social convergence, contact and communication relationships and information distribution. This simulation uses a modified Barabási-Albert-model (BA-model) to simultaneously take into account in each simulation step several parameters from the questionnaire such as personal relationships, friendships, interests, sex, age and residence at once. By this means, the evolution and growth of social networks can be examined in a realistic way because many parameters influencing personal relationships are elegantly factored in. The main results of the simulation show that the artificial social network exhibits small-world characteristics after 20 simulated days and is scale-free. Also, the so-called "Matthew effect" has been observed that implies that "the rich get richer and the poor get poorer". After 5 simulated days, the number of nodes with a high degree increased drastically, which confirms the existence of this effect. Additionally, triadic closures of nodes occurred throughout the simulation. Therefore, the simulation confirmed all results from section 3. In a future article, these simulation results will be discussed in detail.

With this dataset, relatively few nodes are needed to control the entire network because of the exponent's values $\gamma < 2$, whereas many nodes would be required if $\gamma > 2$. In contrast, networks with large exponent values ($\gamma > 2$) have faster exponential decay, where hubs are weakly connected or almost absent, and therefore require more nodes to be fully regulated. Lastly, the data for this study has not been acquired over a long period of time, instead the data collection was carried out during one month. Albeit, we expect that the users' behavior in SNSs does not change much over time.

5 Conclusion and Future Work

The analysis of the dataset on social network usage of college students acquired through the survey mentioned uncovered a number of interesting findings. Based on the metrics described, this network of students explicitly exposes small-world properties whereas the user activities stated in a questionnaire occurred according to power-law distributions. This indicates that power laws can be used to describe the frequency of activities of users in SNSs. These findings allow us to divide users into different groups and represent a proper basis for easily and effectively model users in large-scale simulations using artificial agents. A deep knowledge of the underlying topology of SNSs in conjunction with user behavior, methods of community formation, content distributions and network parameters will obviously be contributed to an understanding of information flows and emergent patterns affecting the structure in those networks.

In the future, personality will be treated as a key factor for the individual behavior in SNSs regarding to content generation and propagation. The introduced dataset and the analyses are expected to serve as a proper foundation for the design and simulation of user behavior model. A more realistic analysis of large and complex networks with respect to their topologies and the dynamics of the processes that take place in them is as well anticipated. The first simulation mentioned will therefore be extended to account for fine-grained user behavior model for even more realistic one.

References

1. Guo, L., Tan, E., Chen, S., Zhang, X., Zhao, Y.E.: Analyzing Patterns of User Content Generation in Online Social Networks. In: Proceedings of the 15th ACM SIGKDD International Conference on Knowledge Discovery and Data Mining, pp. 369–377. ACM (2009)
2. Kawarabayashi, K.-i., Nazir, F., Prendinger, H.: Message Duplication Reduction in Dense Mobile Social Networks. In: Proceedings of 19th International Conference on Computer Communications and Networks (ICCCN), pp. 1–6. IEEE (2010)
3. Ochoa, X., Duval, E.: Quantitative Analysis of User-Generated Content on the Web. In: Proceedings of the First International Workshop on Understanding Web Evolution, Beijing, China, pp. 19–26 (2008), `http://citeseerx.ist.psu.edu`
4. Falck-Ytter, M., Øverby, H.: An Empirical Study of Valuation and User Behavior in Social Networking Services. In: World Telecommunications Congress, pp. 1–6. IEEE (2012)
5. Gyarmati, L., Trinh, T.A.: Measuring User Behavior in Online Social Networks. IEEE Network, 26–31 (2010)
6. Zhong, E., Fan, W., Wang, J., Xiao, L., Li, Y.: ComSoc: Adaptive Transfer of User Behaviors over Composite Social Network. In: Proceedings of the 18th ACM SIGKDD International Conference on Knowledge Discovery and Data Mining, pp. 696–704. ACM (2012)
7. Yan, Q., Wu, L., Zheng, L.: Social network based microblog user behavior analysis. J. Physica A 39, 1712–1723 (2013)
8. Ding, F., Liu, Y., Cheng, H., Xiong, F., Si, X.-M., Shen, B.: Read and Reply Behaviors in a BBS Social Network. In: 2nd International Conference on Advanced Computer Control, pp. 571–576. IEEE (2010)
9. Morales, A.J., Losada, J.C., Benito, R.M.: Users structure and behavior on an online social network during a political protest. J. Physica A 391, 5244–5253 (2012)
10. Liu, H., Nazir, A., Joung, J., Chuah, C.-N.: Modeling/Predicting the Evolution Trend of OSN-based Applications. In: The International World Wide Web Conference Committee (IW3C2), May 13-17, pp. 771–780. ACM (2013)
11. Feng, Z., Cong, F., Chen, K., Yu, Y.: An Empirical Study of User Behaviors on Pinterest Social Network. In: International Conferences on Web Intelligence and Intelligent Agent Technology, pp. 402–409. IEEE/WIC/ACM (2013)
12. Barabási, A.-L.: The origin of bursts and heavy tails in human dynamics. J. Nature 435, 207–211 (2005)
13. Boccalettia, S., Latorab, V., Morenod, Y., Chavezf, M., Hwanga, D.-U.: Complex networks: Structure and dynamics. Physics Reports, 175–308 (2006)
14. Beetz, J.: Einfluss von nutzerspezifischen Parametern auf die Evolution sozialer Netzwerke. Master's thesis. FernUniversität in Hagen (2014)

Decentralized Growing Self-Organizing Maps

Hauke Coltzau, Mesut Yildiz, and Herwig Unger

Fernuniversität in Hagen, Department of Communication Networks, Hagen, Germany
{hauke.coltzau,herwig.unger}@fernuni-hagen.de,
mesut@mesutyildiz.com

Abstract. We present an algorithm based on the Growing Self-Organizing Map (GHSOM) that is able to build and maintain hierarchical SOMs in a decentralized manner. Time- and message complexity of both finding the best matching unit during the training phase and of navigating to any part of the structure is in $O(\log N)$.

Keywords: Self-Organizing Map, Decentralized Systems.

1 Introduction

Navigation in decentralized systems is usually only provided by either exact queries or range queries. Especially in document-oriented systems, in which the dimensionality is neither fixed nor known, an intuitively understandable way to move and orientate in document space would be very helpful for users. In such a system, documents (or more general speaking: datasets) with similar contents could be found in each other's neighborhood without forcing the user to deal with the high dimensionality of the document space. In other words: An adaptive low dimensional projection of the high dimensional data, which is neighborhood preserving and through which the users can navigate as they can in a map, would help to get orientation even in large distributed systems with large amounts of documents.

From centralized systems, these kinds of mechanism for dimensionality reduction are well known as self-organizing maps (SOM) [1]. While static in size in their classical form, extensions for adaptive SOMs have been discussed, as e.g. in [2] and [3]. Since large document systems usually can hardly be projected on flat 2-dimensional space, the growing hierarchic self-organizing-map (GHSOM) [4][5] has been proposed, which allows to navigate through document space along a hierarchical structure, giving overview over the system first and letting users find more details for document clusters and subclusters.

The GHSOM approach does have the remarkable advantage over classic 2d SOMs that it scales with the number of documents and lets users find any document within $O(\log D)$ timesteps in a GHSOM with D documents. But it is still designed as a centralized system with the root of the GHSOM being both a bottleneck and a single point of failure.

In this article, we discuss an extension of the GHSOM approach to make it feasible for use in decentralized systems. Our approach has the same benefits as the centralized

H. Unger et al. (eds.), *Recent Advances in Information and Communication Technology 2015,*
Advances in Intelligent Systems and Computing 361, DOI: 10.1007/978-3-319-19024-2_25

GHSOM but adds redundancy and scalability and allows to build a GHSOM on top of a DHT as, for example, Chord. The remaining part of this article is organized as follows: In section 2, an overview of related work is given. The approach to provide a distributed GHSOM is described in section 3. In section 4, the time and message complexities of the approach as well as simulation results are discussed.

2 Related Work

To the best knowledge of the authors, no approach exists so far to use self-organizing maps for navigation in or structuring of decentralized systems. Neither do decentralized approaches for training SOMs exist. Only distributions of SOMs over parallel infrastructures are described (as, e.g., based on PVM in [10]). These approaches rely on a rather fixed infrastructure, and, more importantly, on a central coordinator and thus are not decentralized.

In this section, we focus on the *growing self-organizing map* (GHSOM), which is the base structure for our approach. Additionally, because the proposed decentralized structure develops in a data-driven and not in a peer driven manner, an approach for peer virtualization is discussed, which is implicitly assumed to be used as P2P fundament for the approach to provide additional fault tolerance.

2.1 Growing Hierarchical Self-Organizing Map

Self-organizing maps are used to represent high-dimensional data in a low-dimensional space while preserving the topological structure of the data itself. The map is built by a set of neurons, classically arranged in a lattice such that each neuron is connected with 4 direct neighbors. Each neuron holds a vector of the same dimensionality as the training data, the so-called reference-vector for this neuron. Each reference-vector represents the average of the fraction of the input data that a single neuron currently represents. After training, the distribution of the reference vectors reflects the distribution of the original data. Additionally, neighbored neurons will have similar reference-vectors after the training, i.e., the dimensionality reduction done by the SOM is neighborhood-preserving.

The core problem in training self-organizing maps is to find the best matching unit (BMU) for any given training data. When the BMU is found, the reference vector of it and its neighbors is adapted towards the training data. In classic lattice-structured SOM approaches, finding the BMU is done by comparing the training value with the reference-vector of every single neuron. Hence, with n neurons, the time complexity to find the BMU is in $O(n)$.

Tree-oriented approaches like the GHSOM use multiple small lattice-structured SOMs, which are arranged in a tree-like manner. The specific advantage of the GHSOM approach is that map sizes and map hierarchies can grow during training and therefore allow to adapt to the actual structure of the training data.

The SOMs in a GHSOM are arranged in layers, each SOM being assigned to a single neuron of a SOM on the previous layer of the hierarchy. Initially, a single neuron on layer 0 is used as root, which has a single child on layer 1 serving as root- or overview map. No further child maps exist at this time. In addition to the reference vector, each neuron stores the average value over the training data it has been trained with so far.

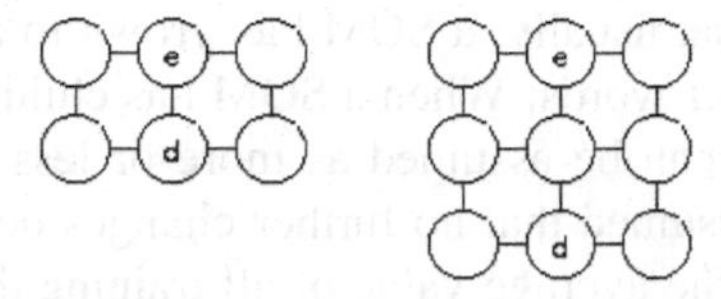

Fig. 1. Insertion of a line of neurons in GHSOM (source: [4])

The training starts with the layer 1 SOM and is done by comparing each training value with every neuron of the map – just the way as it is done in any classic SOM approach. After λ training cycles, the neuron with the largest deviation between its weight vector and the average of the input data it represents, is selected as error unit e. Then, among e's direct neighbors, the unit d with the most dissimilar average input vector in comparison to e's average input vector is selected. A new line of neurons between e and d is inserted into the SOM, with each neuron being initialized with their reference- and input data vectors as the average of their neighboring, already existing neurons.

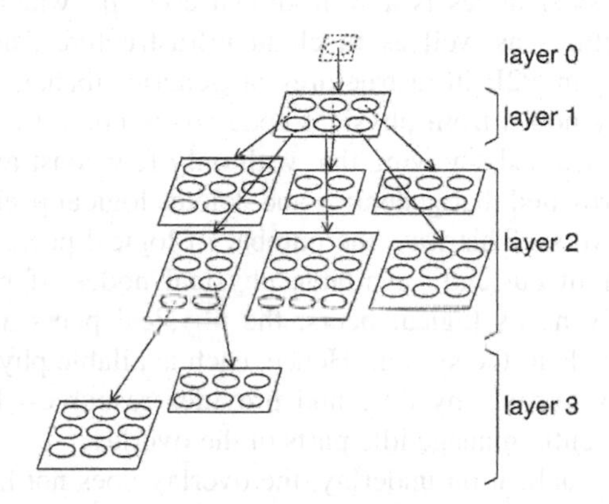

Fig. 2. GHSOM hierarchy of SOMs (source: [4])

The quantization error q_i of a neuron i is calculated as the average distance of the training data assigned to this neuron and the reference vector of i. The mean quantization error MQE of the whole SOM is consequently calculated as the average of all q_i in the SOM. If the MQE of the SOM is below a threshold τ_{mqe}, no further lines of neurons are inserted, the map is assumed to represent the training data in good enough detail.

Independently from the overall quantization error of the SOM, the quantization error $q_{i,\Delta}$ of a *single* neuron Δ can be above a threshold τ_{max}. In this case, it is assumed that Δ is not able to represent the respective training data in high enough details. Hence, a new SOM is created on the next layer and assigned to Δ. This new map is trained, whenever Δ is the BMU of the previous layer. The training is done according to the same rules as described above.

It is important to note that usually, a SOM has grown to a size of its needs, before children are created. In other words: When a SOM has children, changes to this SOM will only seldom occur, it can be assumed as more or less stable. In the centralized approach, it can even be assumed that no further changes occur at all, because before the actual training begins, the average value of all training data is stored in the single neuron on layer 0. Once the map on layer 1 has reduced its mean quantization error below τ_{mqe}, one can be sure that it represents all training data in an appropriate way, even the datasets that have not yet been trained.

Child maps can have child maps themselves, depending on the input data. Thus, a hierarchy of maps is created, each child map representing a single neuron of a parent map in more detail. This way, it is possible to navigate through the training data from an overview point of view into details intuitively. This is not only possible for numeric data, but also for documents represented by keyword vectors.

2.2 Peer Virtualization

Virtualization of physical nodes is a well-known concept, which is widely used in client-server architectures as well as in cloud infrastructures, and even on private computers. However, in P2P infrastructures, a generic, managed and system-wide abstraction of overlay nodes from physical nodes does not yet exist. In [6], such an abstraction layer is discussed, showing that with only few constraints, a decentralized architecture can be provided that generates and deletes logical peers on demand of the overlay and not vice versa. This way, the number of logical peers is totally independent from the number of currently available physical nodes. If more physical peers exist than the overlay needs logical peers, the physical peers add redundancy and communication channels to the system. Hence, each available physical peer is always integrated into the system at any time and provides resources. Idle physical nodes only exist, if they currently manage idle parts of the overlay.

By using such a virtualization underlay, the overlay does not have to take care for churn or other node-related changes on the structure. Instead, algorithms to build decentralized systems can concentrate on the required functionality and which structure fits their needs best. Since the structure of a decentralized GHSOM as discussed in this article develops in a data-driven way, using this concept as middleware would be a helpful enhancement and shall be investigated in future work.

3 Decentralized Growing Self-Organizing Map

As shown in e.g. [5], the GHSOM approach itself is suitable for navigation and orientation in high-dimensional datasets, including large numbers of documents. It groups similar data in neighbored areas of the GHSOM and provides different levels of detail due to the tree-like structure.

When looking at the GHSOM approach from the perspective of distributed systems, the SOM on layer 1 (which we will further reference to as *root map*) is a single point of failure and congestion, since every training-iteration as well as every matching starts on layer 1. To build a distributed GHSOM, it is therefore necessary to maintain multiple copies of the root map and keep them synchronized in a scalable way.

Additionally, without further proof, we assume that the map on layer 1 might change even after longer training periods, because in a decentralized system, the neuron on layer 0 cannot be initialized with the average of all training data beforehand. Instead, this average value is calculated dynamically over time, making it possible that a previously stable map on layer 1 needs further adaption after new training data comes in. We discuss our approach to handle these possible changes in section 4.

3.1 General Idea of the Distributed GHSOM-Approach

Peers in our approach manage maps, i.e. no peer interaction is necessary for the training of a single map. Each peer manages a single map with no children, i.e. a leaf map in the GHSOM tree. For this leaf-map, the respective peer also stores the related documents or knows the peers storing them. Since every map with children can be a potential bottleneck during training as well as navigation, parent maps are replicated on multiple peers. When the distributed GHSOM grows, a peer manages copies of multiple parent maps of its leaf map, going up to the root map. The number of replicas for a map depends on the total number of children (including grandchildren, grand-grandchildren, etc.) and therefore indirectly on its layer in the GHSOM tree. The closer a map is to the root, the more often it is replicated. The root map itself is replicated onto every peer.

Maps without children are only stored on a single peer. This is, because during training, these maps undergo changes more often than maps with children (see section 2) and would need high efforts to keep replication on multiple peers synchronized. One might also argue that the fewer children a map has in comparison to the overall number of maps, the smaller a portion of the input data it represents, i.e. the less important the map is. If additional fault tolerance and redundancy is needed for the leaf maps, we suggest to rely on a generic virtual peer concept as discussed in section 2.

3.2 Map Replication

The basic infrastructure is a Chord ring [7] with the Chord-typical shortcuts into far distant areas allowing a message complexity of $O(\log N)$ for routing in a network with N peers. In difference to the Chord-approach, the keyspace is not meant to represent hashvalues, but only addresses for maps. These addresses do have no further meaning than providing a 1-dimensional order for maps.

Each map in the hierarchy is assigned to a fixed address space in the Chord. Each of its s neurons is assigned to a fraction of $1/s$ of the address space. This assignment is only done, when the map is trained, i.e. its mean quantization error is below the system-wide threshold τ_{mqe} (see section 2.2). In an untrained map, the number of neurons may still change, so that the address space splitting is not yet sensible in such a case. This does not have negative influence, because untrained maps usually do not have child maps, i.e. the address assignment for child maps is not yet necessary (see also section 2.2).

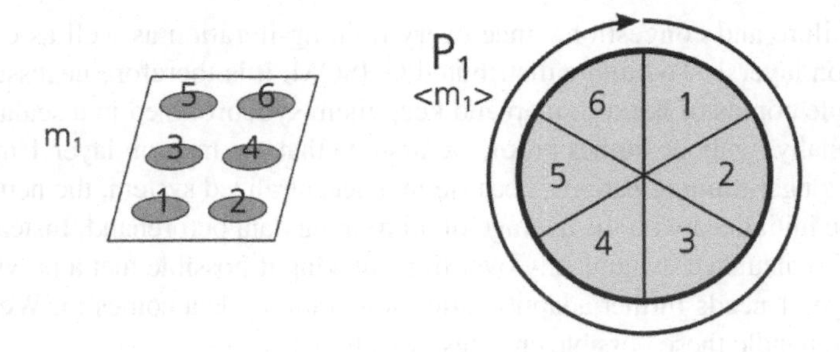

Fig. 3. Address space splitting and peer assignment with only a single map

When a child map is created during training as a result of the GHSOM protocol, also a new peer is created to manage it and is assigned to the address space fraction of the respective child map. The peer managing the parent map also manages the remaining address space. For practical reasons, the first child map created does not result in the creation of a new peer. Instead, the first child is managed by the parent-map's peer, also.

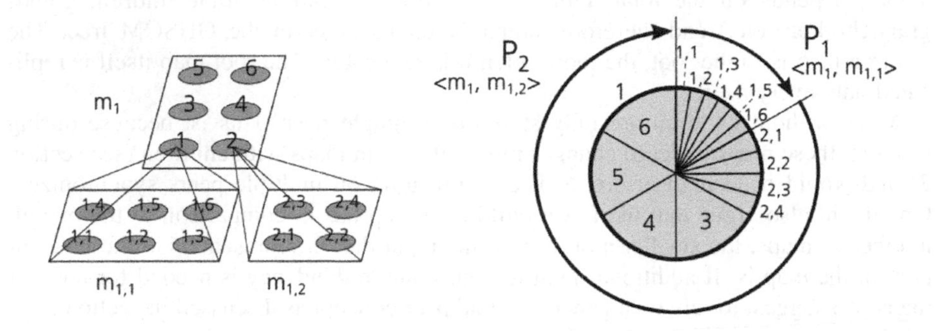

Fig. 4. Address space splitting and peer assignment with multiple maps

When a new peer is created, it receives a copy of all maps "above" it in the hierarchy, including its own parent map and including the address spaces of all these parent maps. It will be integrated into the Chord depending on the address space assigned to the map it manages and will also add far-distance links as requested by the Chord protocol.

Figure 4 shows an example network with a two-layer GHSOM, where layer 2 contains two maps. The first child map $m_{1,1}$ is managed by the same peer that manages the parent map m_1. Since $m_{1,1}$ is bound to 1/6th of the parent map's neurons, it is also assigned 1/6th of the overall address space. Each neuron of $m_{1,1}$ may have another child map, which would be assigned 1/6th of the address space of $m_{1,1}$. In comparison, $m_{1,2}$ only contains four neurons, i.e. each of its neurons is assigned 1/4th of the address space of $m_{1,2}$.

The result of this mechanism is that whenever a new child map is created, each of the parent maps of it is duplicated. The root map, being an ancestor of every other map in the system, is thus replicated onto every peer in the system. Layer 1 maps are copied to $1/s^{th}$ of the peers with s being the size of the root map.

4 Evaluation

The core problem during training and when doing an exact search is finding the BMU (see above). The centralized GHSOM approach provides a time complexity of $O(\log N)$ in a GHSOM with N maps to find the child-map, in which the best matching unit resides. Since the child maps are relatively small in comparison to the overall number of neurons, the necessary comparison of the training data with every neuron of the respective child map is almost of no consequence.

4.1 Complexities

In our decentralized scenario, each peer carries a copy of the root map. For simplicity reasons and without losing generality, we assume that every child map has the same number of s neurons. In the first step for finding the BMU, it is necessary to navigate to the peer that manages the layer 2 child map that fits the training/search data best. If no such map exists, the BMU is on the root map and can be found within a single step. If navigation to a layer 2 child map is necessary, it is helpful that each neuron of the root map is assigned to a well-known section of the $1/s^{th}$ of the Chord's address space. Since Chord itself needs only $\log N$ messages in average to find any peer from any starting point, the peer managing the respective child map can also be found with only $\log N$ hops in average. If navigation to one of the children of the currently found child map is necessary, it can be found within $(\log N)/s$ hops, because the peer managing this map is in the same $1/s$ fraction of the overall address space. The child's child can be found in another $(\log N)/s^2$ hops following the same principle, and so on. Overall, the maximum number of hops to reach any child map is

$$\lim_{l \to \infty} \sum_{k=0}^{l} \frac{1}{s} \cdot \log N = \frac{s}{s-1} \cdot \log N \in O(\log N) \tag{1}$$

independent from the starting point and the current training or search data. Hence, for both training and navigation, the message complexity is in $O(\log N)$.

However, it needs to be pointed out that in contrast to the centralized approach, training a child map may influence its parent map as well further ancestors up the GHSOM tree (see section 2). In previous works [8], the authors have discussed a similar problem and have shown that by introducing a very small threshold, changes from child maps have to overcome to be reflected in the s copies of their parent map, the average number of parent maps that need to be updated is constant and less than 2, depending on the threshold value used. Since the updates of the s copies of a direct parent map for any child can be done in parallel, the average time necessary to handle such an update is constant. If changes are not big enough to overcome the threshold, they are cumulated until they do so. In very few cases, updates of child maps may have to be propagated multiple layers up the GHSOM tree and therefore result in a broadcast through the Chord, because (almost) all peers may be effected by such an update. But since these broadcasts are rare exceptions and each update message is small, they can be tolerated and do not have a notable influence on the Chord's overall performance.

4.2 Simulation

The decentralized approach has been tested with well-established training sets from the UCI machine learning repository [9], as for example the *Corel Image Features Data Set* with 68040 instances and 89 attributes and the *Letter Image Recognition Data* with 20.000 instances and 17 attributes. The results for the latter are shown in table 1 and 2.

Table 1. Distribution of hops necessary for training the *letter image recognition* dataset

hops	6	5	4	3	2	1
occurrences	204	693	4039	9179	5760	125

The letter dataset contains of handwritten letters and numbers that SOM is trained with and then has to classify previously unknown images. Table 1 shows the physical hops necessary to find the BMU during training, resulting in a GHSOM with 223 maps and an average hopcount of only ~3.00.

Table 2. Classification quality comparison between centralized and distributed GHSOM approach for the *letter* dataset, all values are in percent

digit to recognize	0	1	2	3	4	5	6	7	8	9	**Overall**
GHSOM	94,8	98,1	88,0	78,0	63,9	75,2	94,0	82,0	72,5	79,1	**82,6**
distributed GHSOM	96,3	96,0	87,0	76,9	67,3	79,8	89,8	78,3	79,8	83,1	**83,4**

Using the same dataset, table 2 shows that the quality of classification after training is the same as in the centralized approach. For simplicity reasons, only the classification of the number 0 to 9 is shown, the results for letters are the same. The results for the simulations of other datasets from the repository are of the same quality as the above mentioned. They are omitted here in favor to compactness.

5 Conclusion and Outlook

A SOM-based approach for intuitive navigation and orientation in decentralized systems has been presented. It is based on the centralized GHSOM-approach and uses an underlying Chord DHT. By using map replication, bottlenecks and single points of failure can be avoided. Both during training and when used for search and navigation, the system is scalable with time and message complexities in O (log N), even though a high level of redundancy is maintained and the root of the GHSOM structure is even stored on every peer in the network. In future works, the approach shall be combined with a virtual peer architecture as described in section 2.2 to build a generic navigation layer for distributed systems independent from the actual kind of data stored in it.

References

1. Kohonen, T.: The Self-Organizing Map. Proceedings of the IEEE 1990(9), 1464–1480 (1990)
2. Ghaseminezhad, M.H., Karami, A.: A novel self-organizing map (SOM) neural network for discrete groups of data clustering. Applied Soft Computing 4, 3771–3778 (2011)
3. Montazeri, H., Sajjad, M., Reza, S.: Continuous state/action reinforcement learning: A growing self-organizing map approach. Neurocomputing 7, 1069–1082 (2011)
4. Merkl, D., Rauber, A.: Uncovering the Hierarchical Structure of Text Archives by Using an Unsupervised Neural Network with Adaptive Architecture. In: Terano, T., Liu, H., Chen, A.L.P. (eds.) PAKDD 2000. LNCS (LNAI), vol. 1805, pp. 384–395. Springer, Heidelberg (2000)
5. Dittenbach, M., Merkl, D., Rauber, A.: The Growing Hierarchical Self-Organizing Map. In: Proceedings of the International Joint Conference on Neural Networks (IJCNN 2000), Como, Italy (2000)
6. Berg, D.: A Generic Sublayer for Structured Peer-to-Peer-Networks. Autonomous Systems: Developments and Trends, pp. 201–211 (2011)
7. Stoica, I., Morris, R., Karger, D., Kaashoek, M.F., Balakrishnan, H.: Chord: A scalable peer-to-peer lookup service for internet applications. ACM SIGCOMM Computer Communication Review 31(4), 149–160 (2001)
8. Coltzau, H., Ulke, B.: Navigation in the P2Life Networked Virtual Marketplace Environment. In: Unger, H., Kyamaky, K., Kacprzyk, J. (eds.) Autonomous Systems: Developments and Trends. SCI, vol. 391, pp. 213–227. Springer, Heidelberg (2012)
9. University of California in Irvine (UCI) Machine Learning Repository, http://archive.ics.uci.edu/ml/
10. Lobo, V., Bandeira, N., Moura-Pires, F.: Ship recognition using distributed self organizing maps. In: Proceedings of the International Conference on Engineering Benefits from Neural Networks (1998)

An Algorithm for Min-Cut Density-Balanced Partitioning in P2P Web Ranking

Sumalee Sangamuang, Pruet Boonma, and Juggapong Natwichai

Data Engineering and Network Technology Laboratory
Department of Computer Engineering
Chiang Mai University, Chiang Mai, Thailand
sumalee.sa@cmu.ac.th, {pruet,juggapong}@eng.cmu.ac.th

Abstract. In P2P-based PageRank computing, each computational peer contains a partitioned local web-link graph and its PageRank is computed locally. Then, collaborative web ranking between any two peers will be proceeded iteratively to adjust the web ranking until converge. In this paper, the problem of partitioning web-link graph for web ranking in P2P is formulated as a minimal cut-set with density-balanced partitioning. Then, an efficient algorithm called DBP-dRanking is proposed to address such problem. The algorithm can solve the problem with computational complexity of a polynomial function to the web-link graph size. The results also confirm that the proposed algorithm can reduce the ranking error by partitioning web-link graph and perform faster than two other algorithms.

Keywords: graph partitioning, distribute PageRank, P2P-based PageRank.

1 Introduction

Ranking is an important operation for web searching; in particular, the search results are ranked according to their relevance to the search terms and also their importance. There are several approaches to compute ranking in web search result, e.g., [2,3,7]. One of the most notable approach for web ranking is Google's PageRank [9,8]. Generally speaking, web pages and their links are represented by web-link graph; a web page is represented by a node, and a link between two pages is represented by an edge. Then, PageRank determines the importance of a web page through the web page's incoming and outgoing links. Each page considers the summation of PageRank values from incoming links as its PageRank value while distributes it's PageRank value to the other pages through the outgoing links . Therefore, PageRank calculation has to be performed iteratively until the values are converged. Generally speaking, the convergence time of a PageRank calculation is affected by the web-link graph density, i.e., the ratio of the number of edges to the number of nodes, such that if a web-link graph has high density, it will take longer time to converge [9].

Because a web-link graph can be generally large, computing PageRank in a single computer is not efficient. Peer-to-Peer (P2P) is a viable choice to address

H. Unger et al. (eds.), *Recent Advances in Information and Communication Technology 2015*,
Advances in Intelligent Systems and Computing 361, DOI: 10.1007/978-3-319-19024-2_26

such limitation [10,13,14]. In P2P-based PageRank computing, each computational peer contains a local web-link graph, i.e., a sub-graph of the global web-link graph, and its PageRank is computed locally. To be able to compute the global ranking, a special node, so called world-node, is constructed to store the linkage information of the other peers. After the local web rankings of all peers are computed, a collaborative web ranking between any two peers will be proceeded to adjust the web ranking; this process is called peer-meeting. In a peer-meeting, the computation starts with merging of each two web-link graphs including the common world nodes. Then, the ranking is performed on such merged graphs. Finally, the merged graphs will be split into two local web-link graphs stored in the two participating peers as before. A peer has to perform the peer-meeting process with the other peers repeatedly until there is no changes in the ranking.

As suggested in [10], the cut-set, i.e., the set of edges between two local web-link graphs, needs to be minimized in order to reduce the number of iteration to perform peer-meeting. There are many approaches to reduce the cut-set between sub-graphs, for instance, in [12] Arora et al. proposed an approximation algorithm for separation graph into two sub-graphs with minimal cut-set.

In this paper, the problem of partitioning web-link graph for web ranking in P2P is formulated as a minimal cut-set with density-balanced partitioning. The problem is proved to be an NP-Hard, by reducing to the minimum bi-section problem[6]. Then, an efficient algorithm called DBP-dRanking is proposed to address such problem. The algorithm partitions a web-link graph into sub-graphs with minimal cut-set and balanced density. Notably, DBP-dRanking finds partitions with small cut-set; thus, the error of P2P web ranking will be very small such that the peer-meeting process can be eliminated. Also, because the sub-graphs are balanced, the differences of local PageRank iteration is minimal; the overall number of iteration to performs local PageRank will be minimal as well. In particularly, this algorithm uses tree-decomposition technique in order to reduce the search space. Thus, the algorithm can solve the problem with computational complexity of a polynomial function to the web-link graph size. The evaluation results show that the proposed algorithm allows web ranking in P2P perform significantly faster compared with two other algorithms. Also, the results also confirm that the proposed algorithm can reduce the ranking error by partitioning web-link graph into sub-graphs with minimal cut-set.

2 Minimal Cut-Set Density-Balanced Partition Problem

A web-link graph is represented as a directed graph which is an ordered pair $G = (V, E)$. The set of web pages of G is denoted as $V(G)$ while the set of web-links of G is denoted as $E(G)$. An element in $E(G)$ is an ordered pair (u, v) such that $u, v \in V(G)$. The density of web-link graph G, denoted $d(G)$, is the ratio of the cardinality of the web-links set and the web pages set of G_i, as shown in equation 1.

$$d(G) = \frac{|E(G)|}{|V(G)|}. \tag{1}$$

A subgraph $G' = (V', E')$ of G is a local web-link graph that satisfied following conditions: $V' \subseteq V$ and $E' \subseteq E$. Each local web-link graph is denoted as $G_i = (V_i, E_i)$; $i \in I^+$. The set of local web pages of G' is denoted as $V(G_i)$ where $V(G_i) \subseteq V(G)$. On the other hand, the set of web-links of G' is denoted as $E(G_i)$ where $E(G_i) \subseteq E(G)$.

Cut edges are web-links between local web-link graphs. The set of cut edges, called cut-set, is denoted as $C(G_i, G_j)$ where $C(G_i, G_j) \subseteq E(G)$ and $C(G_i, G_j) \notin E(G_i) \wedge C(G_i, G_j) \notin E(G_j)$ and G_i is not a subgraph of G_j and vice versa.

A partition p of G is a set of local web-link graphs which satisfies the following properties: $\bigcup_{i=1}^{k} V(G_i) = V(G)$ and $\bigcap_{i=1}^{k} V(G_i) = \phi$; where k is the set cardinality. And P is the set of all possible partitions in G.

The partition p is a size-balanced partition according to ε constraint if and only if the number of web pages, i.e., $|V(G_i)|$, of all local web-link graph G_i satisfies the condition;

$$\frac{|V(G)|}{k}(1-\varepsilon) \leq |V(G_i)| \leq \frac{|V(G)|}{k}(1+\varepsilon). \tag{2}$$

; where k is p's cardinality.

The density unbalanced value can be used to represents the divergence of the local web-link graphs' density in the partition. It is the different between highest density and lowest density of local web-link graphs in the partition p, as shown in equation 3.

$$MAX\,(d(G_i)) - MIN\,(d(G_i))\,. \tag{3}$$

As a consequence, p is a density-balanced partition if and only if the density unbalanced value of p is zero.

2.1 Problem Definitions

In P2P-based PageRank [10][11], each computational peer contains a local web-link graph and its PageRank is computed locally. The web pages of a local web-link graph are gathered by a local web crawler of the peer. So the local web-link graphs might have different density and may contain duplicated web pages. After the local web ranking of all peers are computed, the peer-meeting will be performed.

In the naive P2P-based web ranking[10], the error in web ranking, compared with centralized approach, is eliminated by the peer-meeting process. However, it takes a long time until the web-link graph converge. In [12], the problem of $(k, 1+\varepsilon)$ balanced partitioning is discussed. In this problem, a web-link graph is partitioned into k size-balanced subgraphs while the cut-set is minimal. It is proved that the problem is NP-Hard by reduction from [6]. However, it is possible to find a polylogarithmic time approximation algorithm [12].

Based on the aforementioned problem, this paper defines a problem of P2P-based web ranking using minimal cut-set and density-balanced partitioning , i.e., $(k, \varepsilon, \alpha, \gamma)$-balanced problem, as followed: Given a web-link graph $G = (V, E)$, find partition p where number of the web pages in any local web-link graph is

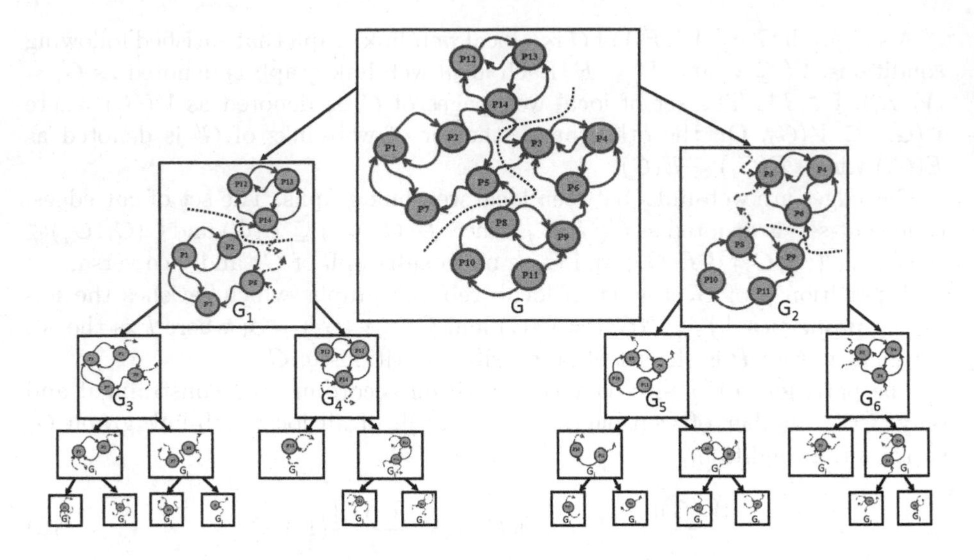

Fig. 1. Example of a binary tree decomposition from a web-link graph G

not greater than ε, density unbalanced value is not greater than α, cardinality of cut-set is not greater than γ and number of peer is k.

3 The DBP-dRanking Algorithm

Generally speaking, the cardinality of P can be very large because the number of possible partition from a web-link graph has higher asymptotic growth rate than the exponential function of graph size, according to Stanley-Wilf conjecture. Particularly, in $(k, \varepsilon, \alpha, \gamma)$-balanced problem, not all partitions are valid, i.e., according to k, ε, α and γ constraints. Thus, finding valid partitions is a tedious task. Therefore, this paper proposes an efficient algorithm called DBP-dRanking to solve the problem. The proposed algorithm uses graph decomposition in order to eliminate invalid partitions. In particular, DBP-dRanking will first decompose a web-link graph G into many binary tree decompositions that satisfy γ. A web-link graph G is decomposed to many binary tree decompositions with a recursive edges separation algorithm [4]. Such that each node of the tree contains a local web-link graph and cardinality a cut-set of a local web-link graph of the original web-link graph is not greater than γ.

Definition of a binary tree decomposition, let G = (V,E) be a graph, a binary tree decomposition of graph G is a tree that satisfies the following properties, root of tree contains graph G ,for each child nodes c of a parent node p contains the subset of $V(G_p)$ in the parent nodes such that $\bigcup_{i \in c} V(G_i) = V(G_p) \wedge \bigcap_{i \in c} V(G_i) = \emptyset$ and for each external node of a tree contains an individual node of G

An example of a binary tree decomposition, is showed in Figure 1. From the figure, the root nodes of the tree contain G. Then, the graph in a tree node is

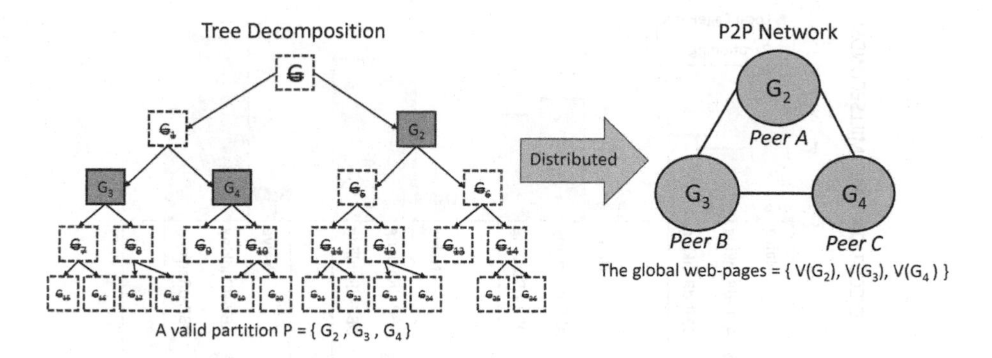

Fig. 2. Example of distributed the local web-link graphs in a partition p into the peers

partitioned into two subgraph and they are placed into the child nodes, . This separation of tree node $'$s content is performed by edge separation algorithm to guarantee that the partition satisfy γ. This node splitting is performed recursively until the leaf nodes contains subgraphs with only one node from G.

Second step is to find partitions that are size-balanced. Finding the size-balanced partitions and also minimal cut-set size concurrently is known to be NP-Hard [1]. Thus this algorithm will consider only the cut-set size in the first step, then consider the size-balanced, using the ε, in this step, separately. The ε is size-balanced constrain that only the local web-link graphs G_i that satisfy the following condition from equation 2. These local web-link graphs are valid. The value of ε is $[0, 1]$ and k is the partition size. All of the binary trees from first step are filtered in this step using ε. Then, only the binary trees that contain valid partition, e.g., partition which all sub-graph are valid according to ε, are considered for the last step.

The third step is to find partitions that are density-balanced. In this step, the density unbalanced constraint α is applied to all valid binary trees from the previous step. If the different of the highest and lowest density of local web-link graphs in the partition is higher than α, this makes the partition invalid.

Next, the algorithm random chooses a valid partition which has the same size as the number of peers in the target P2P network and distributes local web-link graphs in the partition into the peers, as shown in Figure 2 . After that, the algorithm computes PageRank locally without the peer-meeting process.

Given the values of ε and α used in Figure 2 are 0.3 and 2.0 , respectively and the tree in Figure 1 is considered. From the values of ε and α, DBP-dRanking chooses only partition with the number of nodes is in between 3 and 7 (Calculated from equation 2) and the density unbalanced value of the partition was not greater than 2. Therefore, there were 4 valid partitions; $P = \{p_1, p_2, p_3, p_4\}$ such that $p_1 = \{G_1, G_2\}, p_2 = \{G_1, G_5, G_6\}, p_3 = \{G_2, G_3, G_4\}, p_4 = \{G_3, G_4, G_5, G_6\}$ and the density unbalanced value of elements of P were 0.12, 0.42, 0.24 and 0.42 , respectively (Calculated from equation 3). Because the target P2P network has three peers, i.e., $k = 3$, the partition p_3 is selected.

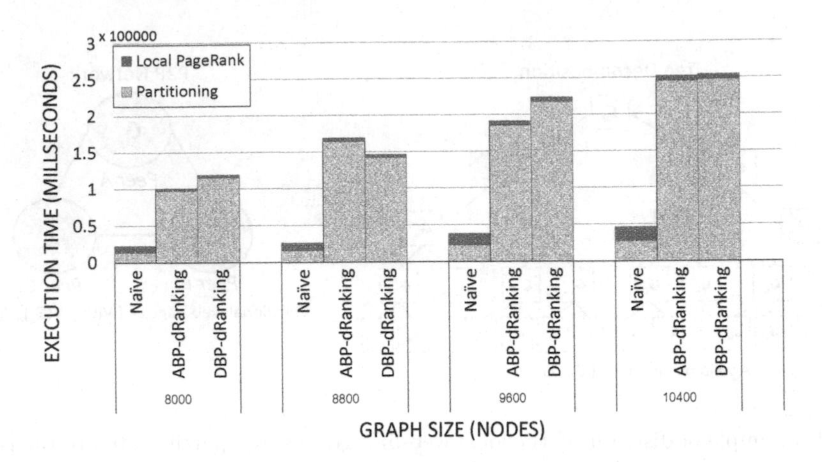

Fig. 3. Impact of web-link graph size on the execution time in the local web ranking (Partitioning web-link graph and local RageRank)

Finally, the local web-link graphs are distributed into peers in the target P2P networks. Then, to find a search result, each peer locally computes PageRank based on the search keyword. Finally, global search result is generated by combining and sorting all local search result.

4 Experiment Evaluation

This section evaluates DBP-dRanking against a naive P2P web ranking algorithm [10] and a P2P web ranking algorithm with size-balanced constraint [1]. They are label DBP-dRanking, Naive and ABP-dRanking, respectively.

4.1 Simulation Setup

This evaluation use synthetic data set where the web-link size is in range of 8000 - 32000 web pages (nodes) while the average number of web-links per web page is about 10. The value of k, ε, α and γ used in the experiments are 8, 0.2, 1.5 and 500, respectively. The results comes from 10 experiments. The P2P simulator used in this evaluation is PeerSim[1] that builds a static overlay topology. Chord protocol is used to maintain an overlay network among peers. A number of messages imply network traffic because a message size is fixed.

4.2 Experiment Results

Figure 3 shows the impact of web-link graph sizes on execution time of the local web ranking. The X-axis shows the web-link graph sizes, in the range of

[1] `http://peersim.sourceforge.net/`

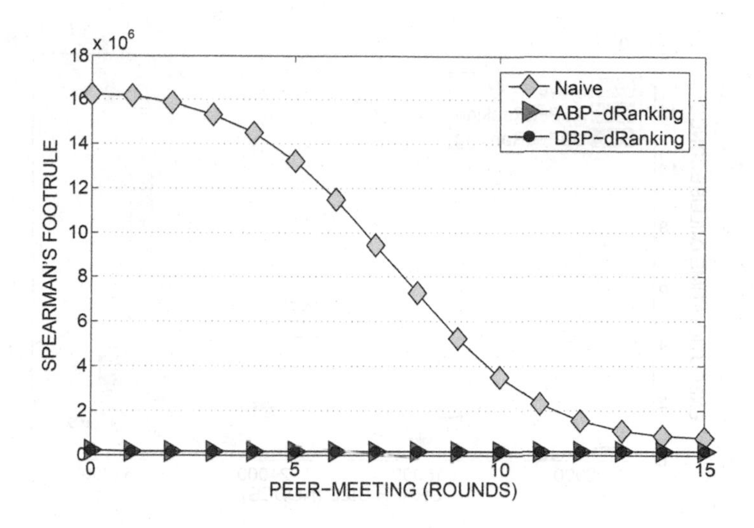

Fig. 4. Impact of the peer-meeting process on Spearman's Footrule distance of the P2P web ranking

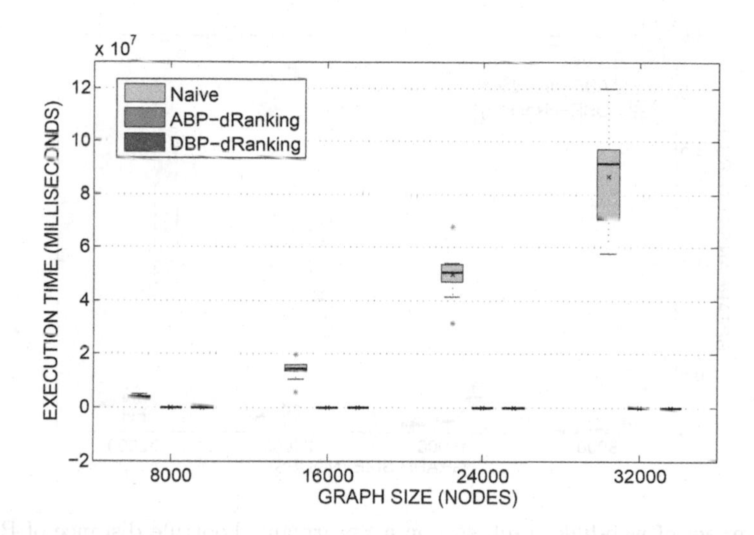

Fig. 5. Impact of web-link graph size on the execution time of the peer-meeting process

8000 - 10400 web pages. The Y-axis shows execution time of local web ranking which consists of the time to perform partitioning with binary tree decomposition and local PageRank execution. The result shows that ABP-dRanking is faster than DBP-dRanking because DBP-dRanking has more step to perform, i.e., to consider α, than ABP-dRanking. However, both are still slower than the naive P2P web ranking because the naive P2P web ranking partitions web-link graph by ordering nodes by their degree and assigning them to partitions in a round robin fashion. This method is faster than tree decomposition.

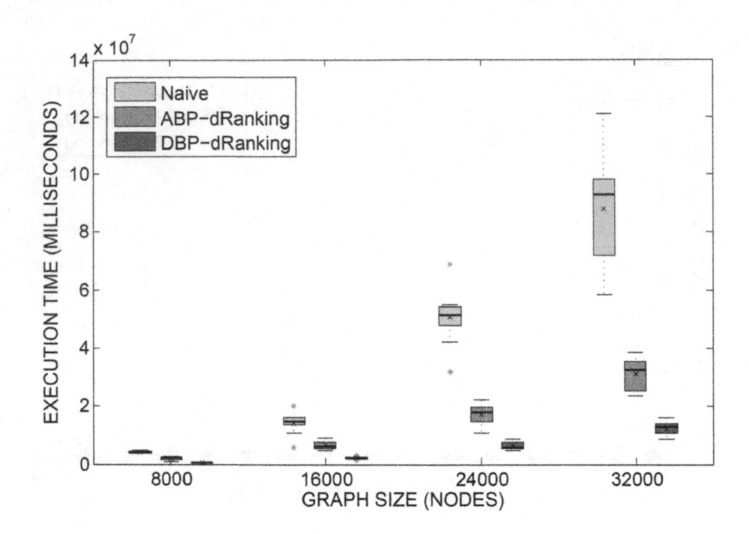

Fig. 6. Impact of web-link graph size on the execution time of the P2P web ranking

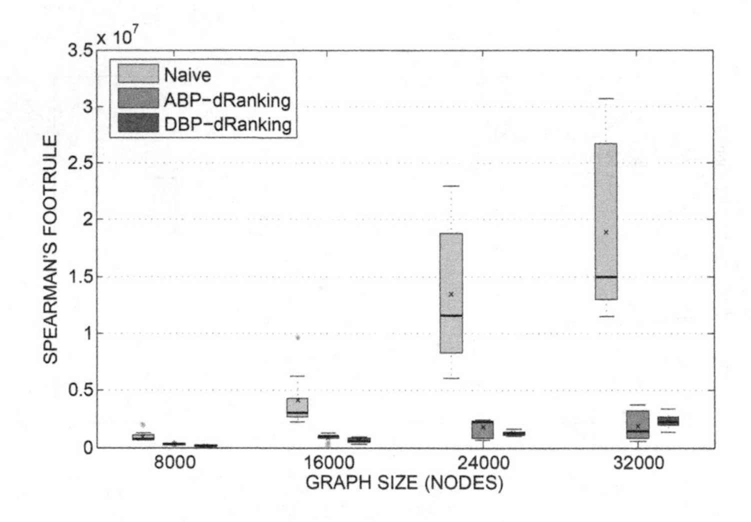

Fig. 7. Impact of web-link graph size on a Spearman's Footrule distance of P2P web ranking

Figure 4 shows the impact of the peer-meeting process on Spearman's Footrule distance of P2P web ranking. Spearmans footrule distance [5] is defined as $F(\sigma 1, \sigma 2) = \sum |\sigma 1(i) - \sigma 2(i)|$ where $\sigma 1(i)$ and $\sigma 2(i)$ are the rank of the web page i in the centralized web ranking and P2P-based web ranking, respectively. This value indicates the error from P2P-based web ranking, higher Spearman's footrule distance indicates higher error. In this figure, the X-axis shows the number of rounds to perform peer-meeting process and the Y-axis shows Spearman's Footrule distance. The result shows that ABP-dRanking and DBP-dRanking have very small Spearman's Footrule from the beginning. Naive approach, on the other hand, has very high Spearman's Footrule distance at the beginning, indicate very high ranking error, and the value becomes smaller after the peer-meeting process is performed. The result shows that ABP-dRanking and DBP-dRanking eliminate the peer-meeting process by considering cut-set in partitioning.

Figure 5 shows the impact of web-link graph size on the execution time of the peer-meeting process. The X-axis shows the web-link graph size and the Y-axis shows the execution time of the peer-meeting process. The execution time of DBP-dRanking and the ABP-dRanking are always zero because the algorithms abolish the peer-meeting process. On the other hand, the execution time of the naive approach is increasing with the graph size.

Although local web ranking execution time of DBP-dRanking and ABP-dRanking are higher than naive approach, as showed in Figure 4); however, when considered the whole process, both have lower execution time, as shown in Figure 6. Moreover, they also have lower Spearman's Footrule distance, as shown in Figure 7. When considered only DBP-dRanking and ABP-dRanking, DBP-dRanking is a much faster in the whole process. DBP-dRanking is faster because it considers α constraint which allow partitions to have slow density divergence; therefore, peers will have similar execution time for PageRank calculation. ABP-dRanking, on the other hand, does not consider α constraint, so some peers might take very long time to perform PageRank.

The results from the experiment indicates that DBP-dRanking and ABP-dRanking clearly outperform naive approach both in term of performance and error. DBP-dRanking allows peers to calculate local PageRank faster so the overall execution time is lower than that of ABP-dRanking.

5 Conclusion

In this paper, the problem of minimal cut-set with density-balanced partitioning is discussed. Then, an efficient algorithm called DBP-dRanking is proposed to address such problem. The algorithm partitions web-link graphs into sub-graphs with minimal cut-set and balanced density. The experiment results show that DBP-dRanking can perform faster, when compared with the other two algorithms, by eliminates the peer-meeting process. Also, because the sub-graphs are balanced, the differences of local PageRank iteration is minimal Therefore, the overall number of iteration to performs local PageRank will be minimal as well.

References

1. Andreev, K., Racke, H.: Balanced graph partitioning. Theory of Computing Systems 39(6), 929–939 (2006)
2. Brin, S., Page, L.: The anatomy of a large-scale hypertextual web search engine. Computer Networks and ISDN Systems 30(1-7), 107–117 (1998)

3. Desikan, P.K., Srivastava, J., Kumar, V., Tan, P.N.: Hyperlink analysis - techniques and applications. Technical report, Army High Performance Computing Center (2002)
4. Even, G., Naor, J.S., Rao, S., Schieber, B.: Fast approximate graph partitioning algorithms. In: Proceedings of the Eighth Annual ACM-SIAM Symposium on Discrete Algorithms, pp. 639–648. Society for Industrial and Applied Mathematics, Philadelphia (1997)
5. Fagin, R., Kumar, R., Sivakumar, D.: Comparing top k lists. In: Proceedings of the ACM-SIAM Symposium on Discrete Algorithms, pp. 28–36 (2003)
6. Feige, U., Krauthgamer, R.: A polylogarithmic approximation of the minimum bisection. SIAM J. Comput. 31(4), 1090–1118 (2002)
7. Kleinberg, J.M.: Authoritative sources in a hyperlinked environment. Journal of the ACM 46(5), 604–632 (1999)
8. Langville, A.N., Meyer, C.D.: Deeper inside PageRank. Internet Mathematics 1(1), 1–42 (2004)
9. Page, L., Brin, S., Motwani, R., Winograd, T.: The pagerank citation ranking: Bringing order to the web. Technical report, Stanford University (1998)
10. Parreira, J.X., Weikum, G.: JXP global authority scores in a P2P network. In: Proceedings of the Eight International Workshop on the Web and Databases (WebDB 2005), Baltimore, Maryland, USA, pp. 31–36 (2005)
11. Sangamuang, S., Boonma, P., Natwichai, J.: A P2P-based incremental web ranking algorithm. In: Proceedings of the 2011 International Conference on P2P, Parallel, Grid, Cloud and Internet Computing, pp. 123–127 (October 2011)
12. Sanjeev Arora, S.R., Vazirani, U.: Expander flows, geometric embeddings, and graph partitionings. In: Proceedings of the 36th ACM Symposium on Theory of Computing (STOC), Chicago, IL, USA, pp. 222–231 (2004)
13. Sankaralingam, K., Sethumadhavan, S., Browne, J.C.: Distributed pagerank for p2p systems. In: Proceedings of the 12th IEEE International Symposium on High Performance Distributed Computing, pp. 58–68. IEEE Computer Society (2003)
14. Shi, S., Yu, J., Yang, G., Wang, D.: Distributed page ranking in structured p2p networks. In: Proceedings of the 2003 International Conference on Parallel Processing, pp. 179–186 (2003)

Tor's API on iOS

Suniti Wittayasatiankul, Sittichai Chumjai, and Nuengwong Tuaycharoen

Computer and Telecommunication Engineering, Faculty of Engineering,
Dhurakij Pundit University
indynostep@hotmail.com, meed96@gmail.com,
nuengwong.tun@dpu.ac.th

Abstract. Currently, Personal Privacy Violation is aggravated via regular internet activities, such as social networks, chats, or browsing websites. The providers of those services may have no user privacy policy. This may lead an intruder leaks the user personal data and violate the user's privacy. Even worse, the intruder may cause an identity theft. Therefore, the development of an API for Tor on a Smartphone to prevent the privacy violation is crucial. Additionally, the highest usage-value Smartphone Operating System is iOS. As a result, we have developed an API to facilitate an application requiring privacy in communication with its server. The proposed API can send data without exposing the sender information via Tor Network. The experimental results show that the API can send data through Tor Network successfully, and it gives programmer convenience in the implementation process.

Keywords: Tor, API, Anonymous Communication, iOS.

1 Introduction

Currently, our internet security and privacy are very important since many intruders make a great effort to steal our identity for their own benefits. The intruders may monitor our internet activities, and learn our internet usage habits. They may redirect us to a fraud website and trick us to provide passwords with financial information. This may lead to identity thefts. Or, they can simply sell our personal information to an online marketing organization, causing endless disturbing online marketing. One method to increase our internet security is to prevent revealing our identity through the internet.

Tor Network was created to allow users to share information over the Internet without compromising their privacy. Tor also supports the application development to make communication privately, including a library, a web browser, and a Tor for Android devices. However, Tor has no API for iOS yet. Therefore, iOS developers require calling Tor with their own great efforts.

In this article, we proposed a Tor's API for iOS. The API can facilitate an iOS developer to call Tor's service easily when the application requiring communication between the applications and their servers with privacy. The experimental results

H. Unger et al. (eds.), *Recent Advances in Information and Communication Technology 2015*,
Advances in Intelligent Systems and Computing 361, DOI: 10.1007/978-3-319-19024-2_27

show that the API can send data through Tor Network successfully, and it gives programmer convenience in the implementation process.

2 Overview of Tor

Tor[8] is a distributed overlay network, designing to serve an anonymized TCP-based applications, for example, web browsing and secure shell. Tor employs a concept of onion routing, which is based on Chaum's original Mix-Net [6] for viewing World Wide Web (WWW) without revealing the user's identity. Tor process starts from a client requesting a list of available Tor routers from one of Tor's directory servers. When the client receives the list, Tor network creates a route by sending a message with Diffie-Hellman key exchange [7] to the first Tor router, called the Entry Guard. The session key is generated between the client and the Entry Guard. The route is continuously expanded with the same method hop by hop. In each expansion, Tor creates a session key for the previous route. This technique is called 'telescoping'. When it reaches 3 hops, the route is completed and ready to be used. The 'core' messages from the Tor Client, such as HTTP GET requests, are encrypted with the Internet Protocol (IP) header for the next hop. So, in each encryption layer, Tor creates multi-layered Tor 'onion'. After that, the onion is forwarded via SOCKS proxy (starting at the localhost), and each relay in the route sends the data in form of data-streams by multiplexed TCP connections between Tor routers. Then, the onion is 'un-peeled' in every passing hop, whose process reveals the next inside layer until the core message is revealed at the last hop (i.e. Exit Router). The last hop reaches the destination, such as a web server. Tor design and using 3-hop route and ephemeral (short-lived) session keys protect the user's identity from 'perfect forward secrecy' [3].

2.1 Security on Tor

Fig.1 shows the security process in Tor. The client sends a *CELL_CREATE* cell, encrypted with TSL, to the entry router. Then, the entry router replies to the client with a *CELL CREATED* cell. This process communicates with Diffie-Hellman (DH) handshake protocol. The objective is to exchange keys between the client and the entry router (base key *K1* = gxy). Note that *H(K1)* is a hash value of *K1*. This key exchange process creates keys for both parties. As a result, the first hop of the route is created. Then, the client extends the route, including the second hop and the third hop, in the same manners.

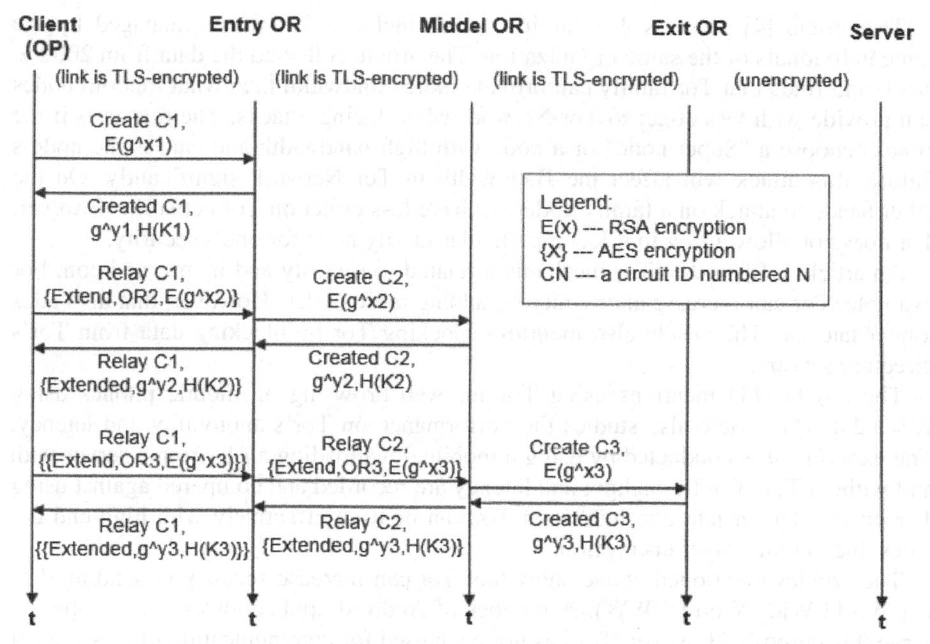

Fig. 1. Security on Tor[2]

3 Related Work

The current version of Tor is the Second Generation. The Second-Generation Onion Router [1] improves the 1st generation of Tor with perfect forward secrecy, congestion control, directory servers, exit policies, End-to-end integrity checking, Rendezvous points and hidden services.

The article [2] explains Tor hidden services, Tor nodes, and servers hiding in Tor network. With its discovery process, the article identifies real IPs of the servers hiding in Tor Network by collecting Tor cell and finding their timing correlation to identify the hidden server's real IP address. The experiment is divided into 3 phases. Phase I presumably identifies the hidden server by having the client to continue to create circuits to the hidden servers until one of their entry routers sees a special combination of cells of different types, i.e., protocol-level features. Phases II verifies the hidden servers by their rendezvous point, which deals with data cells and generates decryption errors for sending to the hidden servers. When an error occurs, the hidden server will send a destroy cell to destroy the circuit. The destroy cell can be recognized by their entry router if the hidden server uses that entry router. Phase III is to conclude the server discovery by time correlation. In this phase, the central server analyzes correlation data to identify the hidden server's real IP address, assuming that their entry router is selected and then identifies the hidden server.

The article [3] presents the performance of Tor on Mobility [9]. The authors test Tor on mobile devices (3G) while loading data with different travelling speed. Then, the authors collect the changing download rate. The experimental results show that when the travelling speed is over 10m/s, Tor performance is aggravated or unusable.

The article [4] presents Tor family, which includes Tor nodes managed by the same individuals or the same organization. The article collected the data from 2009 to 2011, and finds that Tor family can provide more bandwidth than what random nodes can provide with less effect to Tor Network when facing attacks. The reason is if the attacks choose a "Super node" or a node with high bandwidth and causes the node's failure, this attack will affect the Bandwidth in Tor Network significantly. On the other hand, an attack on a family node will have less effect on Tor network. However, Tor does not allow a user to select a particular family node for one's security.

An article [10] studies Tor mechanism related to security and its pro and con. For example, Tor can increase anonymity by adding more nodes. However, that also adds longer latency. The article also mentions blocking Tor by blocking data from Tor's directory servers.

The article [11] mentions using Tor for web browsing on mobile phones using WAP 2.0. The article also studies the performance on Tor's anonymity and latency. The experiment is conducted by using a mobile downloading a file from a server with and without Tor. The throughput and latency are recorded and compared against using Tor on PC. The article concludes that Tor can be used effectively with high-end devices due to multi-layer encryption.

The articles mentioned above show that Tor can increase security in sending data via World Wide Web (WWW). A number of Android applications are developed to serve this purpose. However, Tor has not been used for communication in iOS, except with a web browser, called Onion Browser [12]. Therefore, this article presents the development of Tor's API for iOS to serve a different purpose, which is for data communication in iOS native applications. Therefore, we develop an API to facilitate the iOS application developers to be able to send data via Tor Network securely and easily.

4 Methodology

We analyzed the program code to call Tor for sending data via its network, and found that the required algorithm to call Tor includes different multiple steps, which are:

1. Create dependencies of OpenSSL, libevent and Tor.
2. Write the program code to connect with Tor.
3. Write the program code to encrypt the sending data.
4. Write the program code to limit the mobile device speed when using the application.

The above algorithm might account for more than 20 lines of codes, including calling several third party libraries. Therefore, we implement the step 2 to step 4 above in our API and have the developer call our API instead. Any applications using our API require only 2 simple steps as the followings:

1. Create dependencies for OpenSSL, libevent, and Tor.
2. Add our API's files to your Project.

An implementation code with an iOS application is demonstrated in Section 5.2. Additionally, we also conduct an experiment to test our API performance against the Onion Browser [12], as described in Section 5 below.

5 Experimental Results

We conduct two experiments. The first experiment is to ensure that our API can send data via Tor by measuring the response time. The last experiment is to demonstrate how easy the API is to be used in an iOS application development.

5.1 Tor Response Time

To compare our API performance, we measure the respond time for 3 settings: (1) sending data via normal 3G network, (2) sending data with our API via Tor Network, and (3) sending data via Onion Browser [12]. The data include 6 strings, ranging from 8-25 letters, including 3 strings in English and other 3 in Thai. Each string is sent 5 times. We send that data via each network for 30 times with an iPhone5 device with iOS 7.1.2. The device is not moving while sending data. The response times of the three settings are shown in Table 1-3.

The 'IP' column in Table 1-3 is the IP addresses of the last node before the destination server receives the data. Table 1 shows that the IP addresses when sending data via normal network are the same. On the other hand, in Table 2, the IP addresses from our API via Tor are different because Tor assigns new route for every submission. When we look up the location of the IP with http://www.geoiptool.com/, we find that the IP addresses are from different countries, which causes by Tor preventing back-tracking to the sender. Therefore, we conclude that our API calls Tor service successfully.

Table 1. IP addresses of the last node before the destination server in Normal Network (3G)

Str	*Normal Network (3G)*		
#	*String length*	*IP*	*From*
1	8	49.230.73.16	Thailand
2	21	49.230.73.16	Thailand
3	24	49.230.73.16	Thailand
4	19	49.230.73.16	Thailand
5	15	49.230.73.16	Thailand
6	25	49.230.73.16	Thailand

Table 2. IP addresses of the last node before the destination server in our API using Tor Network

Str	*Our API using Tor Network*		
#	*String length*	*IP*	*From*
1	8	77.247.181.163	Netherlands
2	21	194.150.168.95	Germany
3	24	5.135.85.23	France
4	19	31.31.76.64	Czech Republic
5	15	204.8.156.142	USA
6	25	197.231.221.211	Liberia

Table 3. IP addresses of the last node before the destination server in Onion Browser

Str	*Onion Browser*		
#	*String length*	*IP*	*From*
1	8	77.247.181.163	Netherlands
2	21	197.231.221.211	Liberia
3	24	46.182.106.190	Netherlands
4	19	62.210.74.137	France
5	15	77.247.181.163	Netherlands
6	25	5.135.158.101	France

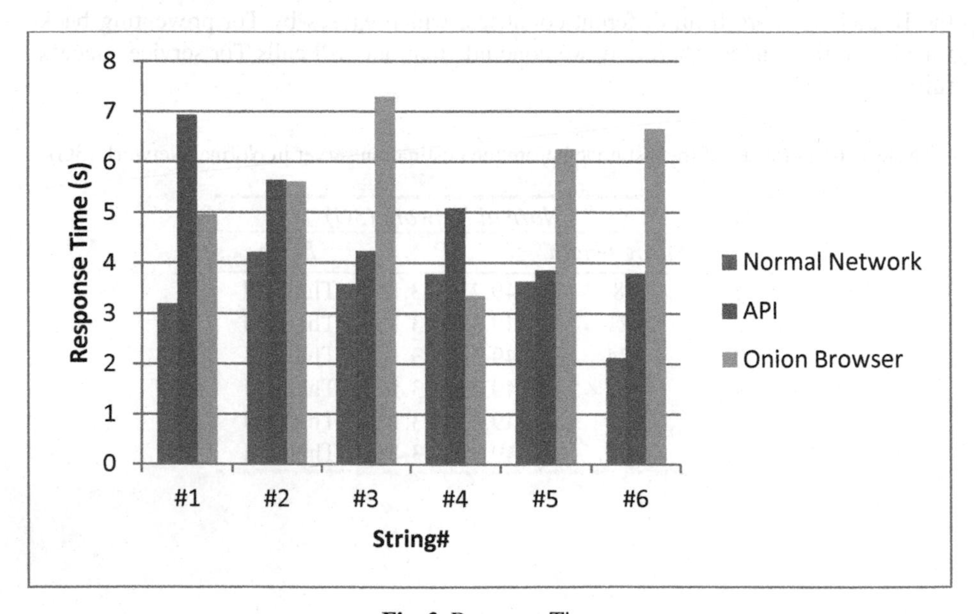

Fig. 2. Response Time

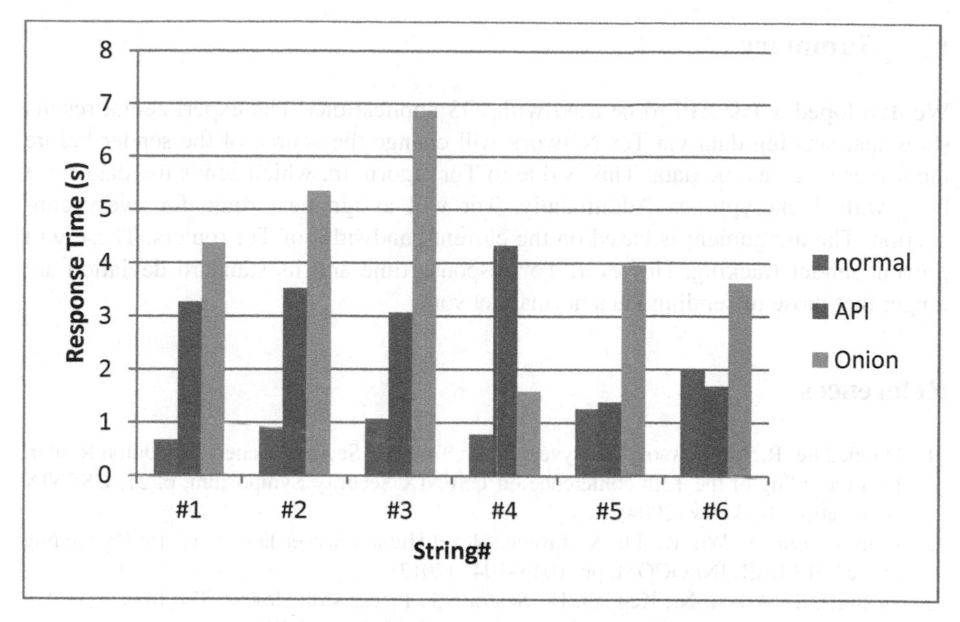

Fig. 3. Standard Deviation of the Response Time

However, in Table 3, Onion Browser has longer response time than our API. The reason is Onion Browser randomly creates a new route for every connection, while our API will create a new route only when the application is opened. Therefore, Onion Browser causes longer response time and more deviated.

As shown in Figure 2, the average response time when sending the data via Tor, either with our API or with Onion Browser, is slower than the time of sending via the normal network. The reason is Tor requires three encryptions while sending data via 3-hop Tor network. In Figure 3, Tor network also causes higher standard deviation than those of the normal network. This is because the unpredictable nature of Tor network, which decides the next node from the Tor router's current bandwidth. Therefore, the assigned route might not always be the same route, and the response time can be more deviated.

5.2 Application Development

We also modify an iOS version of an application [5] to send data via Tor network with our API. The implementation to use our API is very simple as shown below:

```
NSString *text = textForSend.text;
NSString *k = @"myvar=";
NSString *sendM = @"POST";
NSString *link = @"http://URL.com/file.php";

[TorController sendIt:link:sendM:k:text];
```

6 Summary

We developed a Tor API to be used with iOS applications. The experimental results show that sending data via Tor Network will change the source of the sender before the server receives the data. This is due to Tor algorithm, which sends the data for 3 hops with 3 encryptions. Additionally, Tor will assign new route for every connection. The assignment is based on the current bandwidth of Tor routers. These will prevent sender tracking. However, Tor response time and its standard deviation are longer that those of sending via a normal network.

References

1. Dingledine, R., Mathewson, N., Syverson, P.: Tor: The Second- Generation Onion Router. In: Proceeding of the 13th conference on USENIX Security Symposium, p. 21. USENIX Association, Berkeley (2004)
2. Ling, Z., Luo, J., Wu, K., Fu, X.: Protocol-level Hidden Server Discovery. In: Proceeding of the 2013 IEEE INFOCOM, pp. 1046–1047 (2013)
3. Doswell, S., Aslam, N., Kendall, D., Sexton, G.: Please slow down!: The Impact on Tor Performance from Mobility. In: Proceeding of the 3rd ACM Workshop on Security and Privacy in Smartphones & Mobile Devices, pp. 87–92. ACM Press, New York (2013)
4. Wang, X., Shi, J., Fang, B., Guo, L.: An Empirical Analysis of Family in the Tor Network. In: Proceeding of the IEEE International Conference on Communication, pp. 1995–2000 (2013)
5. Chimbunlang, Y., Tuaycharoen, N.: A Web-based Information System for Traffic Police Bribery Reporting via Android Smart Phone (in Thai). In: Proceeding of the ECTI-CARD 2014 (2014)
6. Chaum, D.L.: Untraceable Electronic Mail, Return Addresses, and Digital Pseudonyms. Communications of the ACM, 84–90 (1981)
7. Diffie, W., Hellman, M.: New Directions in Cryptography. Information Theory. IEEE Transactions on Information Theory 22(6), 644–654 (1976)
8. Dingledine, R., Mathewson, N., Syverson, P.: Tor: The Second-Generation Onion Router. In: Proceeding of th13th USENIX Security Symposium, p. 21. USENIX Association, Berkeley (2004)
9. Google Play, Orbot: Proxy with Tor (2014), https://play.google.com/store/apps/details?id=org.torproject.android
10. Haraty, R.A., Zantout, B.: The TOR Data Communication System. J. Communications and Networks 16(4), 415–420 (2014)
11. Andersson, C., Panchenko, A.: Practical Anonymous Communication on the Mobile Internet using Tor. In: Proceeding of the 3rd conference on Security and Privacy in Communications Networks and the Workshops, pp. 39–48. IEEE Press, Nice (2007)
12. Tigas, M.: Onion Browser. Available via iTunes Store (2014), https://itunes.apple.com/th/app/onion-browser/id519296448?mt=8

Version Management of Hierarchical Data in Relational Database

Chawarnwan Jomphrom and Kriengkrai Porkaew

School of Information Technology,
King Mongkut's University of Technology Thonburi,
Bangkok, Thailand
54442016@st.sit.kmutt.ac.th,
porkaew@sit.kmutt.ac.th

Abstract. Hierarchical data structure is organized into a tree-like structure represented by parent- child relationship. The parent can have many children but each child has only one parent. It is also known as one-to-many relationship. There are many types of data can be represented by hierarchical data structure such as organization structures and programs in academies. In some applications, there is necessary to keep historical data or version that need to be used. Temporal data management is used to handle historical data but cause high data space usage by storing every version data which decrease database efficiency. In this paper, we propose logical design to manage versions of hierarchical data in relational database that that may change overtime but historical data is still needed by reusing duplicated records. This conceptual design can avoid data redundancy and increase database efficiency

Keywords: hierarchical data structure, relational database, version management, temporal data management.

1 Introduction

Hierarchical data is not only organized structurally, but also recursive and can be unlimited depth. Most common example for hierarchy is an organizational structure showing relationship among departments and employees. The other example is programs in academies that mostly group by prescribed course, technical course and free elective course and each course contains subjects required for the program. Moreover, this kind of data can be changed overtime and both versions, old and new data, are simultaneously used. For example, organization can be reorganized that division is moved to be subordinate to other department but staffs in division need to refer that they used to be member with an old department. This situation causes application storing old structure and new structure in relational database because both need to be used. The other example is academic programs that would be closed but still need to be kept because academy has to use the programs to release former students' transcript. In this research, we propose solution to manage hierarchical and time-varying data as we just gave the examples in relational database.

H. Unger et al. (eds.), *Recent Advances in Information and Communication Technology 2015*,
Advances in Intelligent Systems and Computing 361, DOI: 10.1007/978-3-319-19024-2_28

2 Related Work

In relational database, there are several models to represent hierarchical data described in books [1, 7] such as adjacency list model, path enumeration model, nested sets model and closure table model. The most common model is adjacency list model which determines linking column that refer to parent node. Constraints between and within tables to enforce tree properties are still needed by coding procedural language to SQL or coding in application's side. To navigate data, cursors and procedural code are used by following the chain of parent value in a loop that makes going down the tree fairly simple; however, aggregation of subtrees for reporting is very slow for large trees. Self-joins method is also used for tree traversal but limited to a known depth of tree. There is recursive query in SQL: 1999 [2] that can handle unlimited depth which is also used for tree traversal.

As we know that tree retrieving in adjacency list model is expensive. This problem is solved by storing the string of ancestors instead of storing parent id as an attribute of each node called path enumeration. However, it is hard to maintain path form that depends on application and path verifying is costly.

An alternative model by storing information pertaining to be the set of its descendants rather than the node's ancestors called nested sets model. In this model, each node is given left and right number following depth-first search of the tree which is used to find ancestors and descendants of any given node. Nevertheless, node insertion and update in this model are very complex that need to recalculate all the left and right values greater than the left value of the node.

Closure table model is a simple method to store hierarchical data. An additional table is used to keep every path from each node to each of its descendants even connect to itself which requires $O(n^2)$ rows for closure table. This design could allow node belong to multiple trees but lot of rows in additional table increasing space consumption as a trade-off for reducing computing.

To store and to manage hierarchical data into relational database, we combined adjacency list model and closure table model together by storing only pairs of parent and child for each node in link table and recursive query is used for tree traversal because information stored in link table similar to information stored in adjacency list model' table. However, this model can be not managed historical data. To handle time-varying data, temporal data management such as temporal extension to ER model and temporal normalization concepts [3, 5, 9] have been proposed, including temporal dependencies, keys, and normal forms. Concept of time representation in temporal database can be captured in number of data types such as instant, period, periods and interval as describe in [4]. The practical time data types also present in SQL: 1992 as date, time and timestamp. In semantic of time domain, there are two factors associated with temporal data management such as valid time which is the possibly time spanning the past, present, and future being used to define historical data or version and transaction time which is time period during row committed in database. Most importantly, the valid time cause integrity constraints between and within tables have to be modified [6]. An identification and valid time are defined as temporal primary key. Foreign key in conventional model is using column or columns to refer to primary key in other table or within table and its value must be equal to primary key value, but foreign key in temporal data management which consists of valid time does not point to exactly matching temporal primary key. Temporal referential integrity with temporal foreign key that enforce parent valid time cover child valid time. To manipulate temporal data, temporal query languages have been defined and recently extend to SQL: 2011 [8] that valid time support is provided by tables containing application-time period and transaction time support is provided by system-versioned table holding system-time period and referential constraints and check constraints are managed by DBMS.

3 Background

Hierarchical data structure consists of nodes, links and references. Each node contains its identification and detail as follows, NODE(ID, DETAIL). Link represents parent-child relationship between nodes thus link consist of parent node identification and child node identification as follows, LINK(PARENT ID, CHILD ID). To guarantee that this structure is a tree, there is no link sharing the same child id. Reference mentions external data that refer to the node or node detail refer to external data. For example, an organizational chart is shown in fig. 1. Each unit in an organization A is represented by node that contains an internal identification and detail such as name, address, phone number and manager that refer to staff who is unit's manager as follows, NODE(ID, NAME, ADDRESS, PHONE NUMBER, MANAGER). Link between B and D show relationship that unit D is unit B's subordinate as follows, LINK(B, D). Staffs of unit D are an external data refer to unit D.

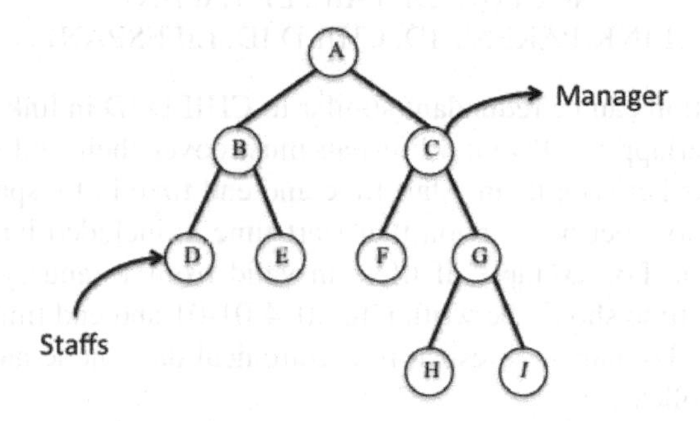

Fig. 1. Structure of organization A

Hierarchical data modification is node and link operations as follow:

- Node update is operation that changes node's detail thus there is no impact to data structure.
- Node insertion is node creation that node consists of identification and its detail without parent and child.

- Node deletion is exist node removing and reference form external node to deleted node must be informed. There are different operations depend on node types as follow:
  - If deleted node is an independent node, that node can be removed without any condition.
  - If deleted node is leaf, link has to be removed then leaf will be an independent node, after that using independent node operation.
  - If deleted node is root, every link between root and its every child has to be removed then root will be an independent node, after that using independent node operation.
  - If deleted node is an internal node, links around the node have to be removed then internal node will be an independent node, after that using independent node operation.
- Link update is operation that changes PARENT ID or CHILD ID or both. If the operation changes CHILD ID, integrity check that the data structure is a tree must be concerned that there is no repeated CHILD ID in any link.
- Link insertion is relationship specification between two nodes and CHILD ID in the link can be not duplicated.
- Link deletion is relationship removing that cause subtrees occur.

Data is always changed all the time and there is necessary to keep historical data or version that need to be used in application. To operate historical data, lifespan that contains start time and end time is added to node in hierarchical data whereas end time can be opened which specify current state. Node and link in hierarchical data can be written as follow:

NODE(ID, DETAIL, LIFESPAN)
LINK(PARENT ID, CHILD ID, LIFESPAN)

Node identification can be redundant similar to CHILD ID in link if lifespan is different and non-overlapping. Parent's lifespan must cover their children's lifespan in both node and links between them. Start time and end time in lifespan are defined by an inclusive-exclusive period notation that start time is included but end time is not included in lifespan. For example, if lifespan valid from 1 January 2014 to 31 December 2014, start time should be written to 2014-01-01 and end time should be written to 2015-01-01. To manage lifespan in hierarchical data, node and link have their own operation as follow:

- Node insertion is node creation that identification can be redundant but lifespan must be different and non-overlapping. Parent's lifespan coverage must be concerned if insertion node is not root.
- Node deletion is lifespan expiration that defines end time to deleted node and confirms that lifespan is non-overlapping for similar node deification. Node deletion affects to links around deleted node that must be considered to expire then nodes that connect to links must be expired until every node and link's lifespan in tree follow lifespan coverage rule.

- Node update cause lifespan changed. There are two operations for node update. Lifespan expiration is defining end time using node deletion operation then insert new node using node insertion operation which might be repeated follow by number of node deletion operation, after that insert link to define relationship new node and its parent.
- Link insert is relationship specification between two nodes and CHILD ID in the link can be duplicated if lifespan is different, non-overlapping and covered by parent's lifespan.
- Link deletion is link expiration that end time in lifespan is specified for deleted node relationship.
- Link update is link expiration then creates a new link to determine new relationship among nodes.

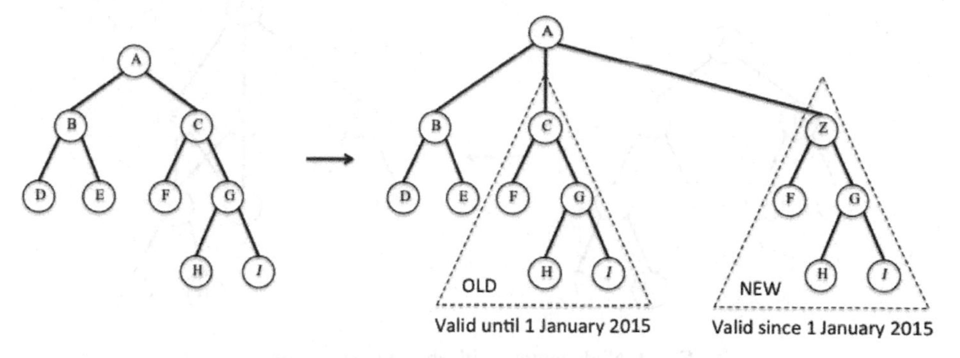

Fig. 2. Node update with historical data

In application such as an organization, when unit C has been changed to be unit Z since 1 January 2015, node deletion operation cause links around unit C and unit C's subordinates are expired then insert unit Z, unit C's former subordinates and links to define relationship among new unit nodes which valid since 1 January 2015 as shown in fig. 2.

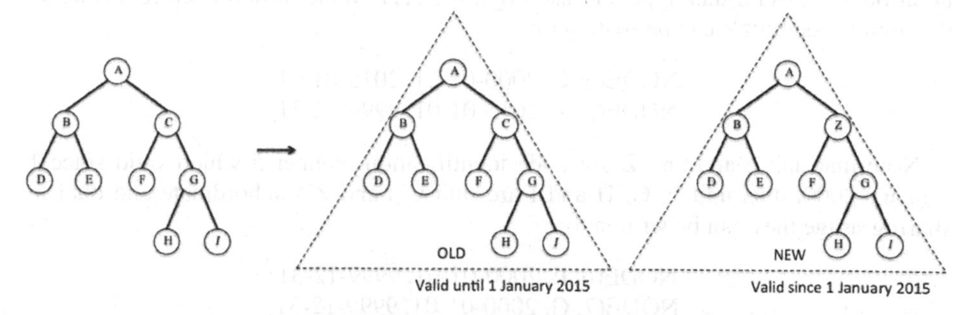

Fig. 3. Node update with historical data by duplicated the whole tree

An alternative method that mostly used is an organization A would be expired then duplicate the whole organizational structure to a new tree that unit Z's subtree valid since 1 January 2015 as shown in fig. 3. In this method, node and link identifications in new tree can be the same identifications or new identifications. However, if identifications have been changed, relationship between old tree and new tree can be not inherited thus historical tracking is difficult to handle.

4 Sharing Data Usage Model

Hierarchical data management we just expanded causes high data space usage. The models keep redundant data cause by data change. To reduce redundant and data space, we considered to reuse duplicated data. For example, subtree in dash line in data structure in fig. 2 can be sharing used as shown in fig. 4.

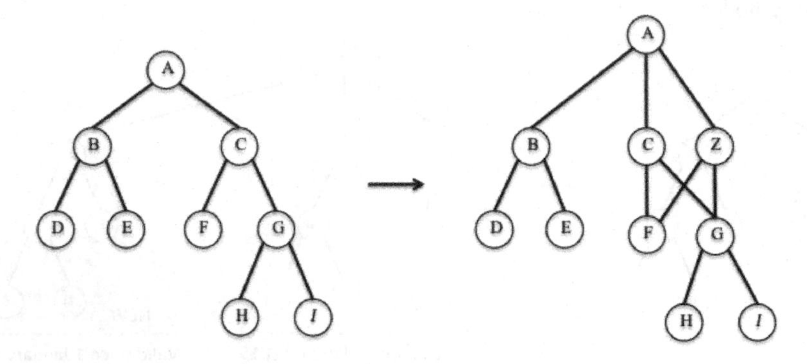

Fig. 4. Node update with sharing data usage

Historical data management in this concept is how to manage sharing data and it version. Unit C valid since organization A has been established which is 1 January 2000. On 1 January 2015, unit C has been changed their name to unit Z then lifespan of node C is started from 1 January 2000 to 31 December 2015 and lifespan of unit Z is started since 1 January 2015. We replace lifespan with inclusive-exclusive period notation and DATE data type and use highest DATE value to define current state so that unit C and unit Z can be written to:

NODE(5, C, 2000-01-01, 2015-01-01)
NODE(5, Z, 2015-01-01, 9999-12-31)

Note that unit c and unit Z are node identification number 5 which valid since 1 January 2000, thus unit F, G, H and I are unit C's and Z's subordinate and dual to sharing usage they can be written to:

NODE(6, F, 2000-01-01, 9999-12-31)
NODE(7, G, 2000-01-01, 9999-12-31)
NODE(8, H, 2000-01-01, 9999-12-31)
NODE(9, I, 2000-01-01, 9999-12-31)

Because data modification follow lifespan modification, all links except links of AC, CF and CG have to be expired then insert links of AZ, ZF and ZG as follow:

LINK(A, C, 2000-01-01, 2015-01-01)
LINK(C, F, 2000-01-01, 2015-01-01)
LINK(C, G, 2000-01-01, 2015-01-01)
LINK(A, Z, 2015-01-01, 9999-12-31)
LINK(Z, F, 2015-01-01, 9999-12-31)
LINK(Z, G, 2015-01-01, 9999-12-31)

Implementation in relational database, data integrity must be maintained through constraints assuring accuracy and consistency. Primary key in node table is defined by temporal primary key which consists of identification and lifespan. Similarly, primary key in link table is defined by parent, child and lifespan. Foreign key is defined by temporal foreign key that parent and lifespan refer to parent node and child and lifespan refer to child node and check constraints is defined by lifespan modification as we just explained.

5 Experimental Evaluation

Our experiments were performed on a virtual machine that has single core Intel Core i7-4700HQ 2.40 GHz CPU, 2 GB of RAM running Windows XP. We used DB2 10.1 as database and implemented application in PHP 5.4.31. To handle version management of hierarchical data in relational database, five different academic programs are used as sample datasets that program table contain list academic programs' nodes, faculty table contain reference that programs refer to and link table contain relationship among them as shown as ER Diagram in fig. 5.

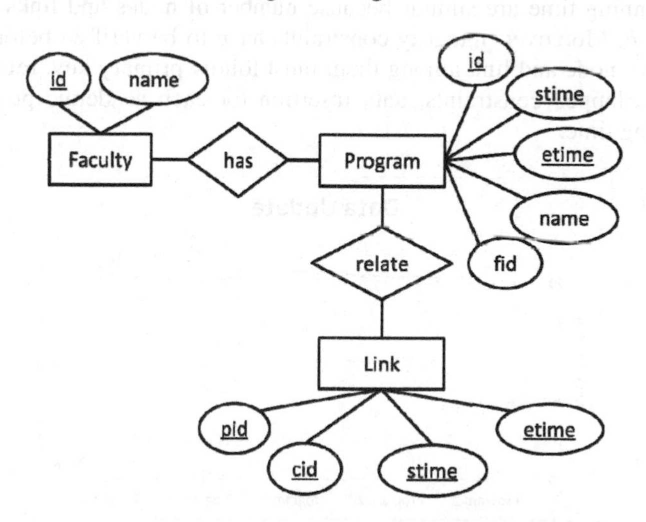

Fig. 5. Academic programs' ER Diagram

Data insertion in our experiments, we inserted academic programs represented by hierarchical data into relational database following lifespan modification constraints. Number of nodes and links in each academic program that we used in experiments are shown in table 1.

Table 1. Number of nodes and links in each academic program

Program	*Number of nodes*	*Number of links*
Program 1	932	931
Program 2	974	973
Program 3	966	965
Program 4	941	940
Program 5	963	962

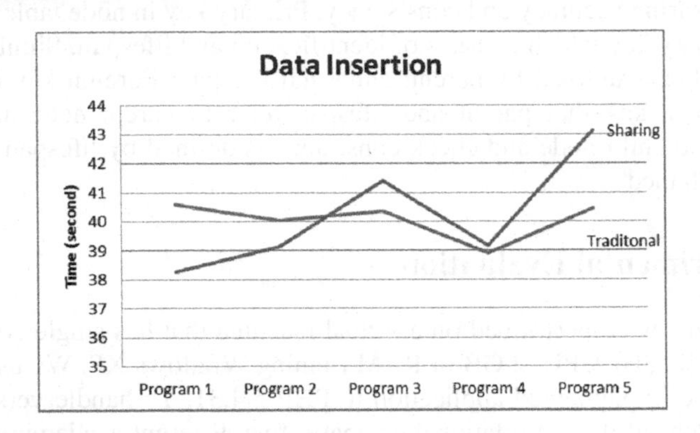

Fig. 6. Data insertion running time

The data insertion result show that both traditional model and our sharing data usage model running time are similar because number of nodes and links are equal as shown in fig. 6. Moreover, integrity constraints have to be verified before insert into database, every node and link among them must follow primary key, referential, coverage and overlapped constraints, data insertion for each academic program highly used processing time.

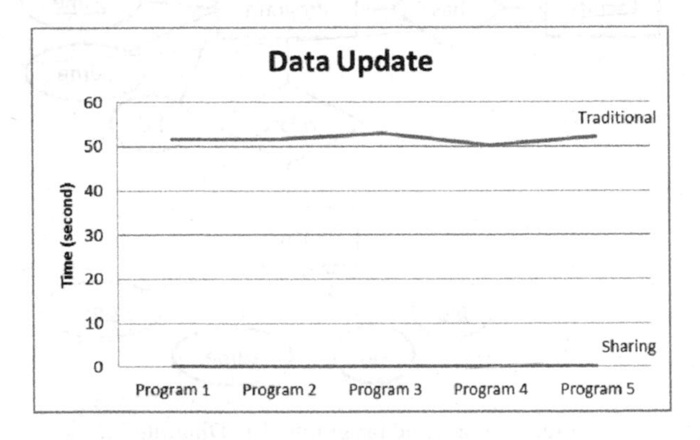

Fig. 7. Data update running time

We updated internal node which is general subject in each academic program affect to data tree in database that has to be duplicated when this operation occurred in traditional model. Even though update on leaf whole data tree also has to be duplicated that causes highly used data space. Differently, we updated only changed nodes in sharing data usage model. Data update running time compared between traditional model and our model is shown in fig. 7 which traditional model highly used running because of nodes insertion. Data space usage in term of recorded number in database using our method that recorded number of nodes is 4,781 and recorded number of links is 4,776 records less than traditional method that rapidly grows up to 9,552 records in node table and 9,542 in link table after update on 5 programs.

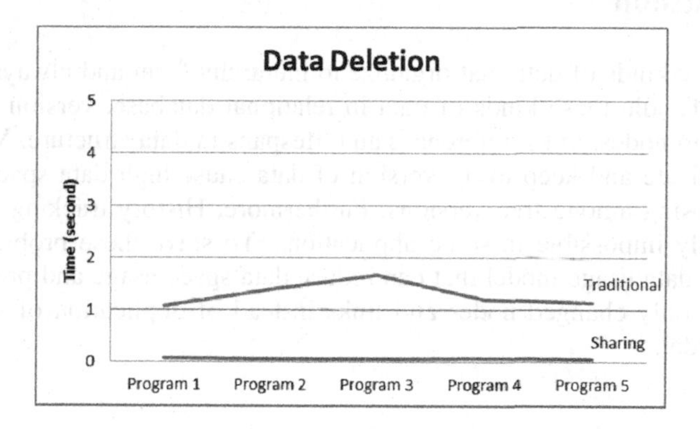

Fig. 8. Data deletion running time

Node deletion experiments also deal with internal nodes that similar to node update. Selected nodes have to be expired and cause subtrees have to be expired either. However, following sharing data usage concept, the subtree need not to be updated in our model. Data deletion running times which traditional model is slightly higher than sharing data usage model as shown in fig. 8.

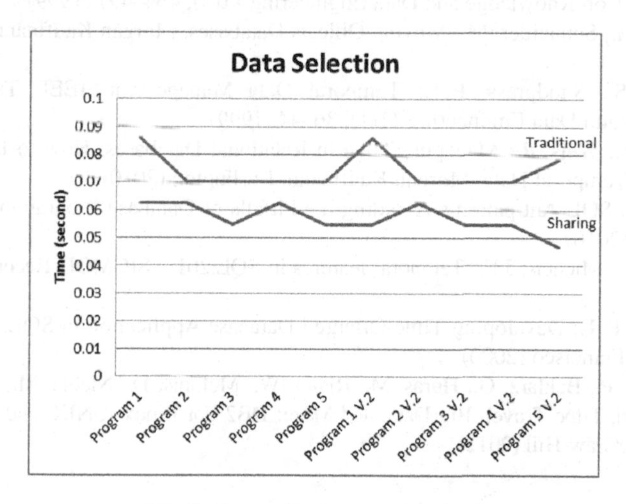

Fig. 9. Data selection running time

We walked through each tree in our data selection that returns academic program structure as a result. Tree traversals result in both models using recursive query on link table that traditional model used running time greater than sharing data usage model because number of rows in both node table and link table are greater than sharing data usage model as shown in fig. 9.

6 Conclusion

There are many kinds of data that organize to hierarchy form and always be changed overtime. To handle these kinds of data in relational database, version management has to deal with nodes, links, references and lifespans in data structure. Version snapshot that duplicate and keep every version of data cause high data space usage sing relationship losing among tree versions. Furthermore, History tracking is very difficult and mostly impossible in some applications. To solve these problems, we proposed sharing data usage model that can reduce data space usage and processing time by modifying only changed nodes and links instead of duplication of whole tree in traditional model.

References

1. Celko, J.: Joe Celko's Tree and Hierarchies in SQL for Smarties. Morgan Kaufmann, San Francisco (2004)
2. Eisenberg, A., Melton, J.: SQL:1999, formerly known as SQL3. SIGMOD Record 28(1), 131–138 (1999)
3. Gregersen, H., Jensen, C.S.: Temporal Entity-Relationship Models—A Survey. IEEE Transactions on Knowledge and Data Engineering 11(3), 464–497 (1999)
4. Güting, R.H., Schneider, M.: Moving Objects Databases. Morgan Kaufmann, San Francisco (2005)
5. Jensen, C.S., Snodgrass, R.T.: Temporal Data Management. IEEE Transactions on Knowledge and Data Engineering 11(1), 36–44 (1999)
6. Johnston, T., Weis, R.: Managing Time in Relational Databases: How to Design, Update and Query Temporal Data. Morgan Kaufmann, Burlington (2010)
7. Karwin, B.: SQL Antipatterns Avoiding the Pitfalls of Database Programming. Pragmatic Bookshelf (2010)
8. Kulkarni, K., Michels, J.E.: Temporal features in SQL:2011. SIGMOD Record 41(3), 34–43 (2012)
9. Snodgrass, R.T.: Developing Time-Oriented Database Application in SQL. Morgan Kaufmann, San Francisco (2000)
10. Zikopoulos, P., Baklarz, G., Huras, M., Rjaibi, W., McInnis, D., Nicola, M., Katsnelson, L.: Warp Speed, Time Travel, Big Data, and More; DB2 For Linux, UNIX, and Windows New Feature. McGraw-Hill (2012)

Deadlock Detection for Resource Allocation in Heterogeneous Distributed Platforms

Ha Huy Cuong Nguyen[1], Hung Vi Dang[2], Nguyen Minh Nhut Pham[2], Van Son Le[2], and Thanh Thuy Nguyen[3]

[1] Department of Information Technology, Quangnam University, Viet Nam
nguyenhahuycuong@gmail.com

[2] Danang University of Education, Danang University, Viet Nam
dhungvi@ued.vn, minhnhutvh@gmail.com, levansupham2004@yahoo.com

[3] VNU University of Engineering and Technology, Viet Nam
nguyenthanh.nt@gmail.com

Abstract. In this paper, we study the resource allocation at the infrastructure level, instead of studying how to map the physical resources to virtual resources for better resource utilization in a cloud computing environment. We propose a new algorithm to resource allocation for infrastructure that dynamically allocate the virtual machines among the cloud computing applications based on approach algorithm deadlock detection and can use the threshold method to optimize the decision of resource reallocation. We have implemented and performed our algorithm proposed by using CloudSim simulator. The experiment results show that our algorithm can quickly detect deadlock and then resolve the situation of approximately orders of magnitude in practical cases.

Keywords: Cloud Computing, Resource Allocation, Heterogeneous Platforms, Deadlock Detection.

1 Introduction

"Recently, there has been a dramatic increase in the popularity of cloud computing systems that rent computing resources on-demand, bill on a pay-as-you-go basis, and multiply many users on the same physical infrastructure. These cloud computing environments provide an illusion of infinite computing resources to cloud users that they can increase or decrease their resources. In many cases, the need for these resources only exists in a very short period of time" [1], [2], [3].

Since them system of information and communication technology (ICT) was introduced, and has played a significant role in the lives of smart cities, the role of information technology infrastructure virtualization has contributed significantly to the solution of the major problem of the succession system of distributed computing, grid computing and parallel computing. In particular, tend to use cloud computing as a key is distributing virtual servers.

H. Unger et al. (eds.), *Recent Advances in Information and Communication Technology 2015*,
Advances in Intelligent Systems and Computing 361, DOI: 10.1007/978-3-319-19024-2_29

The increasing use of virtual machine technology in data centers distributed on a global scale, to provide cloud computing services with competitive low cost and solve the problem rapidly, spurred researchers to propose solutions to solve the complex problem thoroughly. Developing algorithms to provide automatic resources to prevent incidents which frequently occur, we found that most of these studies have acknowledged a platform including identical nodes connected by a cluster. However, we need to apply an algorithm to solve the problem for heterogeneous platforms.

In this work, we propose resource allocation in heterogeneous platforms, we approach a deadlock detection algorithm, to detect deadlock in providing resources for the virtualization heterogeneous platforms. More specifically, our contributions are as follows:

- We provide an algorithmic approach to detect deadlock and resource allocation issues in the virtualization of heterogeneous platform. This algorithm is, in fact, more generally, even for heterogeneous platforms, and only allows allocating minimal resources to meet QoS arbitrary force.
- Using this algorithm, we extend previously proposed algorithms for the heterogeneous case. We evaluate these algorithms via extensive simulation experiments, using statistical distributions of application resource requirements based on a real world dataset provided by CloudSim.
- Most resource allocation algorithms rely on estimates regarding the resource needed for virtual machine instances, and do not refer to the issue of detecting and preventing deadlocks. We studied the impact of estimation errors and propose different approaches to mitigate these errors, and identify a strategy that works well empirically.

The work is organized in the following way: in section 2, we introduce the related works; in section 3, we introduce existing models; in section 4, we present approaches for improving the parallel deadlock detection algorithm; in section 5, we present simulation results and analysis; in section 6, we present our conclusions and suggestions for future work.

2 Related Works

Resource allocation in cloud computing has attracted the attention of the research community in the last few years. In [4], Srikantaiah et al studied the problem of request scheduling for multi-tiered web applications in virtualized heterogeneous systems in order to minimize energy consumption while meeting performance requirements. They proposed a heuristic for a multidimensional packing problem as an algorithm for workload consolidation. Garg et al proposed near optimal scheduling policies that consider a number of energy efficiency factors, which change across different data centers depending on their location, architectural design, and management system [5]. Warner et al discussed the challenges and opportunities for efficient parallel data processing in a cloud environment and presented a data processing framework to exploit the dynamic resource provisioning offered by IaaS clouds in [6]. In [7], Wu et al proposes a resource allocation for SaaS providers who want to minimize infrastructure cost and SLA violations. Addis et al in [8] proposed resource allocation policies for the management of multi-tier virtualized cloud systems with the aim to maximize the profits associated with multiple class SLAs. A heuristic solution based on a local search that also provides availability, guarantees that running applications have developed. Abdelsalem et al. created a mathematical model for power management in a cloud computing environment that primarily serves clients with interactive applications such as web services. The mathematical model computes the optimal number of servers and the frequencies at which they should run in [9]. A new approach for dynamic autonomous resource management in computing clouds introduced by Yazir et al [10].

Their approach consists of a distributed architecture of news that perform resource configurations using MCDA with the PROMETHEE method. Our previous works mainly dealt with resource allocation, QoS optimization in the cloud computing environment. There are more general types of resource allocation problems than those we consider here. For instance:

- We consider the possibility that users might be willing to accept alternative combinations of resources. For example, a user might request elementary capacity CPU, RAM, HDD rather than a specific.
- We consider the possibility that resources might be shared. In this case, some sharing is typically permitted; for example, two transactions that need only to read an object can be allowed concurrent access to the object.
- We begin by defining our generalized resource allocation problem, including the deadlock problem as an interesting special case. We then give several typical solutions.

3 Problem Formulation

In the past, grid computing and batch scheduling have both been commonly used for large scale computation. Cloud computing presents a different resource allocation paradigm than either grids or batch schedulers. In particular, Amazon C2 is equipped to, handle may smaller computer resource allocations, rather than a few, large request as is normally the case with grid computing [5]. The introduction of heterogeneity allows clouds to be competitive with traditional distributed computing systems, which often consist of various types of architecture as well. Heterogeneous processor pool can be represented as a weighted undirected graph H = (P, E), which we refer to as the system graph. It consists of a set of vertices $P=\{p_1,p_2,\ldots,p_n\}$, denoting processors, and a set of edges $E = \{(p_i ,q_j) \mid p_i ,q_j \in P\}$, representing communication links between processors.

We consider a service hosting platform composed of H heterogeneous hosts, or nodes. Each node comprises D types of different resource(*such as CPUs, network cards, hard drives, or system memory*). Each type of resource under consideration a node may have one or more distinct resource elements (*a single real CPU, hard drive, or memory bank*) [11], [12], [13].

The service is initiated within the same virtual machine provides virtual elements. For some resources, such as system memory or hard drive space, it is relatively easy for by the different elements together in the hypervisor or the operating system to organize the virtual machine efficiency can interact with only a single major factor. For other types of resources, such as CPU cores, the situation becomes more complicated [2].These resources can be partitioned arbitrarily among virtual elements, but we cannot be effectively pooled together to provide a single virtual element with a greater resource capacity than that of a physical element. For these types of resources, it is necessary to consider the maximum capacity allocated to individual virtual elements, as well as the aggregate allocation to all vital elements of the same type.

The allocation of resources to a virtual machine determines the maximum number of each individual element of each type of resource that will be used, as well as the aggregate amount of each resource of each type. Typically, such resource allocation, represented by two vectors, vector elementary level and a maximum allocation vector synthesis. Note that in a distributed valid it is not necessarily the case in which each value of the second vector a multiple of the corresponding values in the first vector, because the demand for resources may be unevenly distributed on the virtual resources.

Figure 1 illustrates an example with two nodes and one service. Node A, B are comprised of 4 cores and a large memory. Its resource capacity vectors show that each core has elementary capacity 0.8 for an aggregate capacity of 3.2. Its memory has a capacity of 1.0, with no difference between elementary and aggregate values because the memory, unlike cores, can be partitioned arbitrarily. No single virtual CPU can run at the 0.9 CPU capacity on this node. The figure shows two resource allocations one on each node. On both nodes, the service can be allocated for memory it requires.

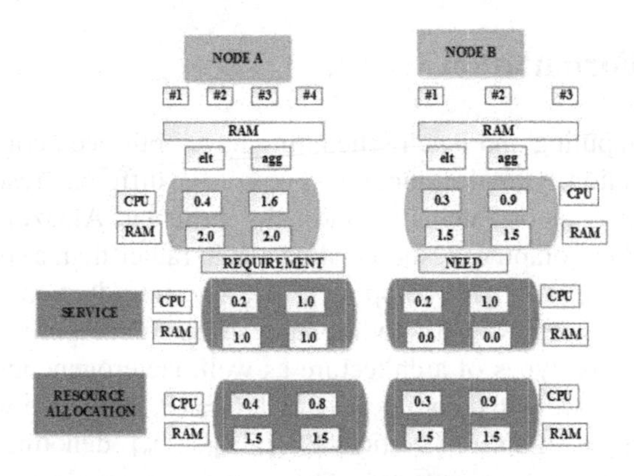

Fig. 1. Resource allocations example

In any large IaaS system, a request for *r* VMs will have a large number of possible resource allocation candidates. If n servers are available to host at most one VM, the total number of possible combinations is (*n*,*r*). Given that $n \geq r$, exhaustively searching through all possible candidates for an optimal solution is not feasible in a computationally short period of time.

Figure 2 shows such a system having two nodes, a VM_1 and VM_2, and two resources, S_1 and S_2. Each processor (VM_1 or VM_2) has to use both resources exclusively to complete its processing of the streaming data. The case shown in Figure 2 (b) VM_1 holds resource S_1 while VM_2 holds resource S_2. Further, VM_1 requests S_2, and VM_2 requests S_1. When VM_2 requests S_1, the system will have a deadlock since neither VM_1 nor VM_2 gives up or releases the resources they currently hold; instead, they wait for their requests to be fulfilled.

Informally speaking, a deadlock is a system state where requests are waiting for resources held by other requesters which, in turn, are also waiting for some resources held by the previous requests.

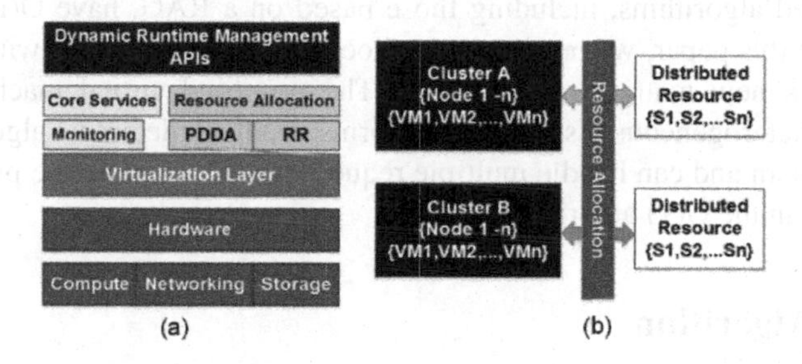

Fig. 2. Deadlock example

In this paper, we only consider the case where requests are processors on virtual machine resource allocation on heterogeneous distributed platforms. A deadlock situation results in permanently blocking a set of processors from doing any useful work.

There are four necessary conditions which allow a system to deadlock[3]:

(a) *Non – Preemptive*: resources can only be released by the holding processor;
(b) *Mutual Exclusion*: resources can only be accessed by one processor at a time;
(c) *Blocked Waiting*: a processor is blocked until the resource becomes available;
(d) *Hold – and – Wait*: a processor is using resources and making new requests for other resources that the same time, without releasing held resources until some time after the new requests are granted.

Deadlock detection can be represented by a Resource Allocation Graph (RAG), commonly used in operating systems and distributed systems. A RAG is defined as a graph (V,E) where V is a set of nodes and E is a set of ordered pairs or edges (v_i,v_j) such that $v_i,v_j \in V$. *V* is further divided into two disjoint subsets: $P = \{p_0, p_1, p_2, \dots, p_m\}$ where P is a set of processor nodes shown as circles in Figure 1; and $Q = \{q_0, q_1, q_2, \dots, q_n\}$ where *Q* is a set of resource nodes shown as boxes in Figure 1. A RAG is a graph bipartite in the *P* and *Q* sets. An edge $e_{ij}=(p_i,q_j)$ is a racist edge if and only if $p_i \in P$, $q_j \in Q$. The maximum number of edges in a RAG is m $\times$ n. A node is a sink when a resource (processor) has only incoming edge(s) from processor(s) (resource(s)). A node is source when a resource (processor) has only outgoing edge(s) to processor(s) (resource(s)).

A path is a sequence of edges $\varepsilon = \{(p_{i1}, q_{j1}), (q_{j1}, p_{i2}), ..., (p_{ik}, q_{jk+1}), (q_{js}, p_{is+1})$ where $\varepsilon \in E$. If a path starts from and ends at the same node, then it is a cycle. A cycle does not contain any sink or source nodes.

The focus of this paper is deadlock detection. For our virtual machine resource allocation for heterogeneous distributed platform deadlock detection implementation, we make three assumptions. First, each resource type has one unit. Thus, a cycle is a sufficient condition for deadlock. Second, satisfies request will be granted immediately, making the overall system expedient [3]. Thus, a processor is blocked only if it cannot obtain the requests at the same time.

All proposed algorithms, including those based on a RAG, have O(m×n) for the worst case.. In this paper, we propose a deadlock detection algorithm with O (min (m, n)) based on a new matrix representation. The proposed virtual machine resource allocation on heterogeneous distributed platforms deadlock detection algorithm makes use of parallelism and can handle multiple requests/grants, making the proposed algorithm faster than the O(1) algorithm[8],[9].

4 Our Algorithm

On the use of graphs representing RAG matrix is presented, we approach me to propose a deadlock detection algorithm in heterogeneous platforms. The basic idea of this algorithm is reported to reduce the matrix by removing the corresponding columns or rows. This is continued until the matrix can not be reduced any more columns and rows. At this time, if the matrix still contains row(s) or column(s), which may also consider other factors, not anymore, then consider a cycle exists, it cannot declare published at least one deadlock n the system. If not, there is no deadlock. The description of this algorithm shows in the algorithm 1.

Table 1. The description of notations

Notations	Meanings
$x_i^{j(CPU)}$	*CPU required by a* VM_i *from the IaaS provider j*
$x_i^{j(\mathrm{RAM})}$	*RAM required by a* VM_i *from the IaaS provider j*
C_j^{CPU}	*The maximum capacity of CPU of IaaS provider j*
C_j^{RAM}	*The maximum capacity of RAM of IaaS provider j*

The following example illustrates how the algorithm works. In each iteration of this parallel algorithm, at least one reduction can be performed if the matrix is reducible. Hence, it takes at most min(*m*,*n*) iterations to complete the deadlock detection. This example has two processors: VM_1 and VM_2, as p_1 and p_2 respectively. The devices are S_1, S_2, and S_3, as q_1,q_2 and q_3 respectively as shown in Fig 3.

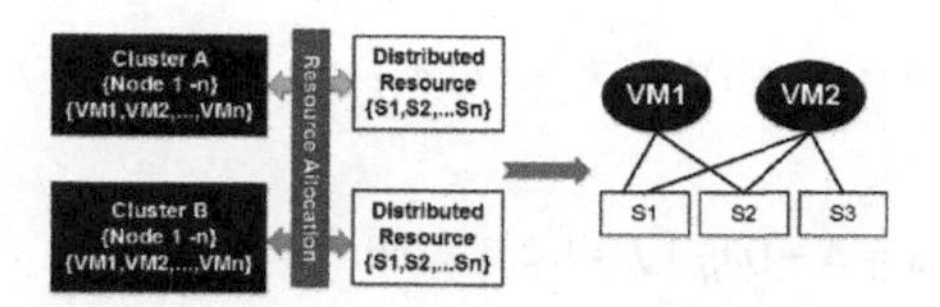

Fig. 3. Resource allocation on heterogeneous platform

Algorithm 1 *Parallel Deadlock Detection Algorithm (PDDA Improved)*

Input: $P_i^{j(CPU)^*}$; $P_i^{j(RAM)^*}$ from **IaaS** provider i;

Output: new resource $r_j^{CPU^{(n+1)}}$; $r_j^{RAM^{(n+1)}}$;

BEGIN

Calculate optimal resource allocation to provide VM:

$x_i^{j(CPU)^*}, x_i^{j(RAM)^*} = Max\{U_{IaaS}\}$;

Computes new resource:

If $C_j^{CPU} \geq \sum_i x_i^{j(CPU)}, C_j^{RAM} \geq \sum_i x_i^{j(RAM)}$

{

$r_j^{CPU^{(n+1)}} = \max\{\varepsilon, r_j^{CPU^{(n)}} + n(\sum_i x_i^{j(CPU)} - C_j^{CPU})\}$

$r_j^{RAM^{(n+1)}} = \max\{\varepsilon, r_j^{RAM^{(n)}} + n(\sum_i x_i^{j(RAM)} - C_j^{RAM})\}$

Return new resource $r_j^{CPU^{(n+1)}}$; $r_j^{RAM^{(n+1)}}$

}

Else

{

$$M = [m_{ij}]^{m \times n}, \text{ where } m_{ij} = \begin{cases} r & if \quad \exists (p_i, q_j) \in E. \\ g & if \quad \exists (p_i, q_j) \in E., (i=1,...,m; j=1,...,n) \\ 0 & otherwise \end{cases}$$

$\Lambda = \{m_{ij} \mid m_{ij} \in M, m_{ij} \neq 0\}$;

DO

Reducible = 0;

For each column:

$if\ (\exists m_{ij} \in \forall k, k \neq i, m_{kj} \in \{m_{ij}, 0\})$

{

$\Lambda_{column} = \Lambda - \{m_{ij} \mid j = 1, 2, 3, ..., m\}$,

$reducible = 1$;

};

For each row:

$$if\ (\exists m_{ij} \in \forall k, k \neq i, m_{kj} \in \{m_{ij}, 0\})$$

{

$$\Lambda_{row} = \Lambda - \{m_{ij} \mid j = 1, 2, 3, ..., m\},$$

$reducible = 1;$

};

$$\Lambda = \Lambda_{column} \cap \Lambda_{row};$$

$UNTIL(reducible = 0);$

}
Detect Deadlock
If ($\Lambda \neq 0$),
Deadlock
Else
No deadlock;
END;

The matrix representation of this example is shown in Table 1. In this matrix, the first and second column contains both g and r, and hence is not reducible. However, the third column contains only g. Thus m_{12}=g can be reduced. At the same time, each row is also examined, however, there is no reduction possible. Since there is one reduction, the next iteration will be carried out. In the second iteration, the first and second columns still contain both g and r, and hence are not reducible. At the same time, each row is also checked, but no reduction is possible for any row. Since there are no more reductions, a conclusion is drawn. In this case, hardware deadlock detection takes two iterations and finds a deadlock.

Table 2. Example with 2 Processes and 3 Resources

	Q_1(S3)	Q_2(S3)	Q_3(S3)
P_1 (VM1)	g	R	0
P_2 (VM2)	r	R	g

Let us remove the edge (p_2,q_2) in this case and consider it again. The matrix is shown in Table 2. In this matrix, the first column cannot be reduced, because of the existence of both g and r, while the second and third columns can be reduced, because the second column has only one r and no g's, and the third column has only one g and no r's. At the same time, the first and second rows cannot be reduced, because of the existence of both g and r in each row. Since this iteration has a reduction, Step 1 will be re-executed by the second and third columns having been removed. During the second iteration, the first column is not reduced, because there are both r and g in this column. However, the first row can be reduced because on r is in this row. Then Step 1 is executed again in what is now a third iteration of the Parallel Deadlock Detection Algorithm. There are no more reductions, because the matrix now is empty. Step 2 concludes that there is no deadlock. In this case, three iterations are taken to complete detections.

Table 3. Example without Deadlock

	Q_1(S1)	Q_2(S2)	Q_3(S3)
P_1 (VM1)	g	R	0
P_2 (VM2)	r	0	g

5 Experiments and Results

In this paper, cloud computing resource allocation method based on improving PDDA has been validated on CloudSim, the platform is an open source platform, we use the Java language to program algorithm implementation class [12]. The experiments give 9 tasks, by CloudSim's own optimization method and improved algorithm PDDA to run the 9 tasks, experiment data as follows Table 4[13]:

Table 4. Comparison the optimal time of our algorithm to PDDA algorithm

			PDDA			PDDA Improved			
Cloudlet ID	Data center ID	VM ID	Start	End	Time	Start	End	Time	Improved (%)
0	2	0	0.1	100.1	100	0.1	90.1	90	10.00%
5	2	0	0.1	110.1	110	0.1	100.1	100	9.09%
1	2	1	0.1	132.1	132	0.1	110.1	110	16.67%
4	2	1	0.1	145.1	145	0.1	160.1	160	-10.34%
6	2	2	0.1	200.1	200	0.1	165.1	165	17.50%
9	2	2	0.1	220.1	220	0.1	172.1	172	21.82%
2	2	4	0.1	235.1	235	0.1	175.1	175	25.53%
3	2	4	0.1	248.1	248	0.1	180.1	180	27.42%
8	2	3	0.1	260.1	260	0.1	182.1	182	30.00%
7	2	3	0.1	290.1	290	0.1	185.1	185	36.21%

The comparative analysis of experimental result can bc seen in many times, after task execution, although there were individual time improved PDDA algorithm response time was not significantly less than an optimal time algorithm, in most cases, improved algorithm is better than the optimal time algorithm, thus validated the correctness and effectiveness.

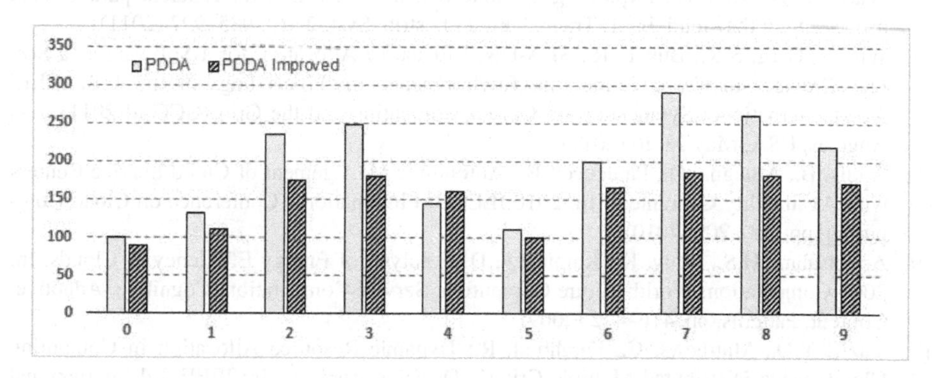

Fig. 4. Comparison the optimal time of algorithms

6 Conclusion and Future Works

A deadlock detection algorithm is implemented for resource allocation on heterogeneous distributed platforms. The deadlock detection algorithm has O(min(*m*,*n*) time complexity, an improvement of approximately orders of magnitude in practical cases. In this way, programmers can quickly detect deadlock and then resolve the situation, e.g., by releasing held resources.

Our main approach focuses on applying deadlock detection algorithms for each type of lease contracts and applying the proposed algorithm in resource allocation on heterogeneous distributed platform.

Through this research, we found that the application of appropriate scheduling algorithms would give optimal performance to distributed resources of virtual server systems.

References

1. Armbrust, M., Fox, A., Griffith, R., Joseph, A., Katz, R., Konwinski, A., Lee, G., Patterson, D., Rabkin, A., Stoica, I., Zaharia, M.: A view of cloud computing. Commune. ACM 53(4), 50–58 (2010)
2. Stillwell, M., Vivien, F., Casanova, H.: Virtual Machine Resource Allocation for Service Hosting on Heterogeneous Distributed Platforms. In: IPDPS 2012, Proceedings of the 2012 IEEE 26th International Parallel and Distributed Processing Symposium, pp. 786–797 (2012)
3. Lee, G.: Resource Allocation and Scheduling in Heterogeneous Cloud Environments. EECS Department, University of California, Berkeley (2012)
4. Srikantaiah, S., Kansal, A., Zhao, F.: Energy aware consolidation for cloud computing. Cluster Comput. 12, 1–15 (2009)
5. Garg, S.K., Yeo, C.S., Anandasivam, A., Buyya, R.: Environment-conscious scheduling of HPC applications on distributed cloud-oriented data centers. J. Parallel Distrib. Comput. (2011)
6. Warneke, D., Kao, O.: Exploiting dynamic resource allocation for efficient parallel data processing in the cloud. IEEE Trans. Parallel Distrib. Syst. 22(6), 985–997 (2011)
7. Wu, L., Garg, S.K., Buyya, R.: SLA-based Resource Allocation for a Software as a Service Provider in Cloud Computing Environments. In: Proceedings of the 11th IEEE/ACM International Symposium on Cluster Computing and the Grid (CCGrid 2011), Los Angeles, USA, May 23-26 (2011)
8. Addis, B., Ardagna, D., Panicucci, B.: Autonomic Management of Cloud Service Centers with Availability Guarantees. In: 2010 IEEE 3rd International Conference on Cloud Computing, pp. 220–207 (2010)
9. Abdelsalam, H.S., Maly, K., Kaminsky, D.: Analysis of Energy Efficiency in Clouds. In: 2009 Computation, World: Future Computing, Service Computation, Cognitive, Adaptive, Content, Patterns, pp. 416–422 (2009)
10. Yazir, Y.O., Matthews, C., Farahbod, R.: Dynamic Resource Allocation in Computing Clouds using Distributed Multiple Criteria Decision Analysis. In: IEEE 3rd International Conference on Cloud Computing, pp. 91–98 (2010)

11. Benin, S., Bucur, A.I., Epema, D.H.: A measurement-based simulation study of processor co-allocation in multi-cluster systems. In: JSSPP, pp. 184–204 (2003)
12. Calheiros, R.N., Ranjan, R., Beloglazov, A., Rose, C.A.F.D., Buyya, R.: CloudSim: A toolkit for modeling and simulation of cloud computing environments and evaluation of resource provisioning algorithms
13. http://www.cloudbus.org/cloudsim/

Formal Verification of Multi-agent System Based on JADE: A Semi-runtime Approach

Chittra Roungroongsom and Denduang Pradubsuwun

Department of Computer Science, Faculty of Science and Technology,
Thammasat University, Thailand
gravypunk@hotmail.com, denduang@cs.tu.ac.th

Abstract. Since a multi-agent system (MAS) is a system composed of multiple interacting agents, verifying multi-agent interaction is acquiring increasing importance. Several researches have been proposed to verify the multi-agent interaction, so we present our proposed method as an alternative way to verify how MAS meets their specification in term of messaging among agents. This paper describes an ongoing effort for a formal verification of MAS based on Java Agent Development Framework (JADE) in semi-runtime approach. The timed trace theoretic verification is applied to detect time constraint failures in such a system. We use a well-known book trading case as an example to illustrate our proposed method.

Keywords: Formal Verification, JADE, Multi-agent System, Semi-runtime Verification, Time Petri Net.

1 Introduction

Multi-agent systems are systems composed of multiple interacting computing elements, known as agents. Agents are computer systems with two important capabilities. Firstly, they are capable of autonomous action to decide what they need to do in order to achieve their design purposes. Secondly, they are capable of interacting with other agents to perform some kinds of social activities like, cooperation, coordination, and negotiation [1].

Carrying out of such activity needs to ensure the correctness of interacting among agents that communicate via messaging in a platform. Invalid scenarios, such as agents cannot reach the success state according to the given specification, should not be permitted. In addition, in some case that involved with time constraint, such as, e-auction, performing tasks in a given time is crucial and time violation is not allowed to occur. It is desirable to be verified the execution of messaging in MAS satisfies its specification. The scope of this paper is that we only focus to verify the events of messaging that may lead some time constraint failures.

Formal verification is a systematic process that uses mathematical reasoning to verify that specification is preserved in implementation [2]. Verification can take place at design time (static verification) or at runtime (dynamic verification) [3]. Runtime verification is a testing/debugging technique that combines monitoring and

H. Unger et al. (eds.), *Recent Advances in Information and Communication Technology 2015*,
Advances in Intelligent Systems and Computing 361, DOI: 10.1007/978-3-319-19024-2_30

formal verification during the program execution to checks the run of program against properties, unlike traditional testing techniques such as unit testing which are ad hoc and informal [4]. This paper will focus on semi-runtime verification. We run the program and generate logs of messaging events between agents. The log file is converted to be in form of Time Petri Net (TPN) which is the input of verification tool [5] that we are going to use in this paper. Then we run the verification tool to check how the result of the program running satisfies the given specification.

JADE [6], an Application Programming Interface (API) implemented in Java, is the state-of-art tool for developing MAS and has been widely used for many real industrial applications. So we propose the approach to verify programs implemented by JADE. As we trace back into the Time Petri Net that will be modeled a run of a JADE program, Petri Net is a formal model to represent the flow of activities with explicit concurrency. It is powerful due to simplicity and generality. Since it is support for describing asynchronous event, we will use Petri Net with time constrain or Time Petri Net to analyzes the communication and messaging between agents.

In summary, the contributions of this work are following,

- This paper proposes a formal verification from [5], which based on timed trace theory, for detecting time constraint failures that do not satisfy with time specification.
- The verification is focused on failures during the interaction of agents in MAS which is implemented by JADE.
- The proposed method does in semi-runtime way by keep logging of the messaging events and propose an algorithm for converting the logs to Time Petri Net as runtime program model.

The remainder of this paper is organized as follows. Section 2 will discuss the other works related to this paper. Section 3 presents an overview of timed trace theory and Verification. Section 4 shows the proposed method of this paper. Section 5 describes a worked example. Finally, Section 6 presents our conclusions and points to directions for future work.

2 Related Works

Several researches of formal verification use formal methods, such as model checking and theorem proving, to detect bugs in distributed multi-agent programs [7, 8]. However, in traditional testing, formal methods are usually applied on an abstract model of the program. Consequently, even if a program has been formally verified, we cannot guarantee the dependability of a particular implementation. In this scope, we concern the JADE implementation in which is mostly concerned the verification at runtime.

A few JADE researches focus on the testing of program, such as, [9] presented the use of mock agents for unit testing. Mock agent is a dummy implementation of a single agent role to interact with the agent being tested. The test is under successful and exceptional scenarios of the testing role so it will not cover the integration process of the whole system. Later, the same developer team presented a JAT framework [10] for building and running MASs test which develop from mock agent approach for monitoring agents during test and do as integration test. [11] did the mutation testing by proposing mutation operators for JADE Mobile Agent System. The tester needs to modify the program source code to find test inputs that will cause the modified programs to behave differently from the original one. [12] only proposed a platform design for performance testing based on JADE which is not related to our work.

Few researches take the verification at runtime. [3] proposed a framework of runtime verification of JADE together with Jason to monitor and verify the sequence of events. The protocol representation is expressed in AIP global types form and project to SWI Prolog to verify the compliant of agent interaction. [4] proposed a runtime verifier, implemented in AspectJ. It accepts a user-specified Pomset automaton and the associated atom predicates for a given JADE source program to monitor and debug both ordering requirement and atom requirements. [13] proposed a framework that extends JADE allowing programmers to monitor global states, to detect the occurrence of certain events and to react to these events at runtime. The framework is located as the layer between the JADE and the application. So the programmers who want to use this framework need to modify the source code to extend the class of proposed framework instead of the JADE.

In this paper, our main contribution is that we use the timed trace theory, an approach of formal verification provided by [5] which was used in formal verification of asynchronous circuits and has never been used for MAS, to present an alternative ways of verification based on JADE. Our method achieves other JADE researches in term of simplicity. Since it is only use the fundamental service of JADE (mentioned later in section 3) to capture messaging events as a runtime verifier, the tester does not need to deal with other extensions connected to JADE or to interfere the existing source code.

3 Timed Trace Theory and Verification

3.1 Time Petri Net

Time Petri Net is a directed bipartite graph, represented by a 6-tuple N = (P, T, F, lb, ub ,μ_0) where P = {p_1, p_2, …, p_m} is a finite set of places, T = {t_1, t_2, …, t_n} is a finite set of transitions in which $P \cup T \neq \emptyset$ and $P \cap T = \emptyset$, $F \subseteq (P \times T) \cup (T \times P)$ is set of flow relations, lb and ub: $T \rightarrow R^+ \cup \{\infty\}$ are functions for earliest and latest firing time of transitions, satisfying $lb(t) \leq ub(t)$ for $t \in T$, and μ_0 is initial marking of the net. Each transition t must be enables and fires within the time bound lb(t) and ub(t).

3.2 Timed Trace Theory

The implementation and specification are modeled by module and semi-module respectively. A module is a tuple M = (I, O, N), where I is a set of input transition, O is a set of output transition, and N is a Time Petri net. A semi-module is the same as a module but the definition of timed trace structure is defined differently.

A timed trace structure of a module M is denoted by 4-tuple T(M) = (I, O, S, F) where S is called success trace set and F is called failure trace set. The timed trace theory uses the mirror property to check the conformance between modules. So that, the semi-module is considered as a mirror of module M denoted by M^{sm} = (O, I, N).

In order to check the correctness between a module M_1 and a module M_2, we use notion $T_1 = (I_1, O_1, S_1, F_1)$ and $T_2 = (I_2, O_2, S_2, F_2)$ such that $I_1 \cup O_1 = I_2 \cup O_2$. Intersection of T_1 and T_2, denoted by $T_1 \cap T_2$, is shown as $(I_1 \cap I_2, O_1 \cup O_2, S_1 \cap S_2, (P_1 \cap F_2) \cup (F_1 \cap P_2))$. If $(P1 \cap F2) \cup (F1 \cap P2) = \emptyset$, then the module M_1 conform to the module M_2.

Thus, the implementation module conforms to its specification module if the composition of $T(M_1)$ and $T_s(M_2^{sm})$, written by $T(M_1) \parallel T_s(M_2^{sm})$, is failure-free. Intuitively the conformation relation implies that the implementation module behaves similarly to the specification module with any environment with respect to failure-freeness. Conformation checking can be inherited by Theorem 1 called "Hierarchical verification" which is shown as follows.

Theorem 1. $\{M_1,\ldots,M_{k-1},M_{k_1},\ldots,M_{k_m},M_{k+1},\ldots,M_n\}$ conforms to M_s, if $\{M_{k_1},\ldots,M_{k_m}\}$ conforms to M_k, and $\{M_1,\ldots, M_{k-1}, M_k, M_{k+1}, \ldots , M_n\}$ conforms to M_s.

Note that the implementation of a system may consist of several modules including $M_1, \ldots, M_{k-1}, M_{k_1}, \ldots, M_{k_m}, M_{k+1}, \ldots, M_n$ such that $M_k = \{M_{k_1},\ldots,M_{k_m}\}$ and its specification is represented by module M_s.

3.3 Time Constraint Failures

By the means of timed trace theory to check the conformance of modules, it is simply said to be failure-free if and only if there does not exist failures, involving with the firing time of transitions in this case, called safety and timing failure.

Consider the example in Fig 1, assume that M1 is the implementation module, M2 is the specification module, and t0 is a transition with the bounded time that we are going to check the conformance.

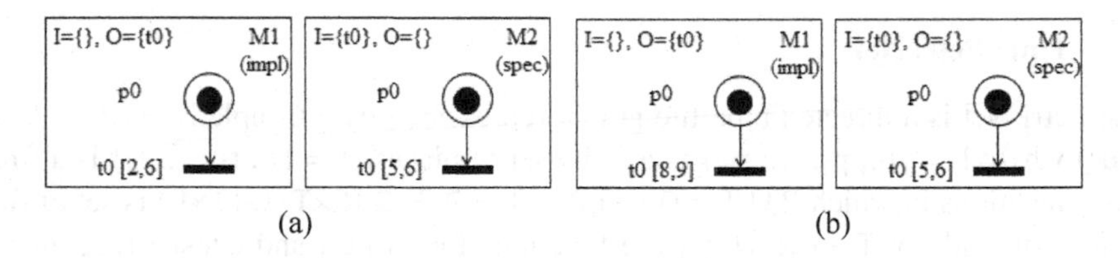

Fig. 1. Examples of time constraint failure cases (a) safety failure (b) timing failure

Fig. 1 (a) shows that t0 in M2 will enable during [5, 6], but in M1, the firing is allowed to occur during [2, 6]. So that, if a token in M1 fires at the time unit 3 while t0 still does not enable, a failure occurs which we call "Safety Failure".

Fig. 1 (b) shows that t0 in M2 will enable during [5, 6], but in M1, the firing is allowed to occur during [8, 9]. So that, if a token in M1 fires at the time unit 8 while t0 has disabled already, a failure occurs which we call "Timing Failure".

4 The Proposed Method

The perspective of a proposed approach is shown in the Fig. 2. In order to verify the JADE program at semi-runtime by using Timed Trace theory, both the specification and the implementation are modeled as the time Petri net.

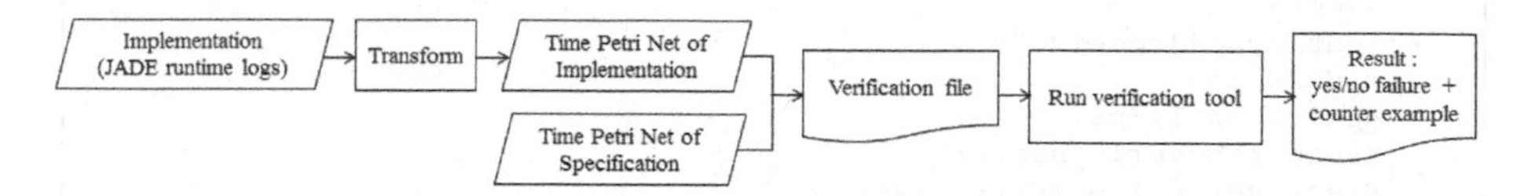

Fig. 2. Workflow of the proposed method

As we mention earlier, in this paper, we focus on events of messaging occurrence to detect the time constraint failures. We do not consider the content inside the messages. The events of messaging in this scope, including, sending and receiving messages at runtime and we omitted the details of other operations, i.e. message preparation, agent selection, etc., by presenting a processing event. We modelled those events in Time Petri Net forms which are shown in Fig. 3. We modelled the events based on the "Pair-wise negotiation process model" provided by [14] because of the simplicity and clear representation.

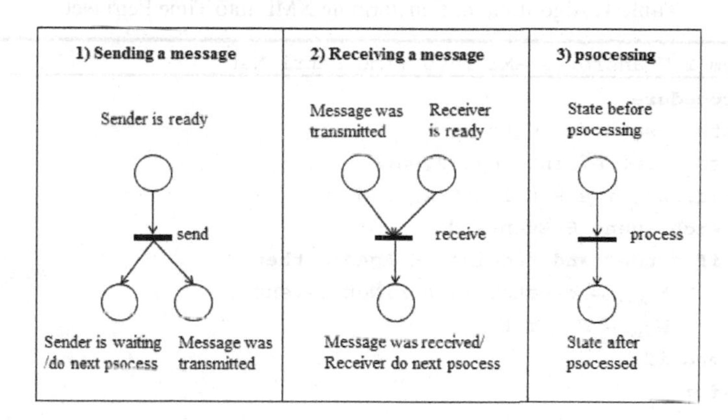

Fig. 3. Time Petri Net of messaging events

In order to generate logs while the program is running, we need to implement additional source codes to keep logging of events. We implement the "distributed coordinated filter" which is the common service available in JADE to detect the incoming and outgoing message events. It is running on the top of the program so we do not need to intervene with the existing source codes. The detail of source codes with explanation can be found in [15].

For the log format, we create the messaging event in EXtensible Markup Language (XML) for the description of the model from messaging event which inspired from [16] since it is the easy to understand and can be useful for the future development. Fig. 4 is an example of the description by XML.

```
<events>
  <event type="send">
    <id>1</id>
    <pair>cfp_1</pair>
    <sender>buyer1</sender>
    <receiver>seller1</receiver>
    <performative>cfp</performative>
    <timestamp>5</timestamp>
  </event>
  <event type="receive">
    <id>2</id>
    <pair>cfp_1</pair>
    <sender>buyer1</sender>
    <receiver>seller1</receiver>
    <performative>cfp</performative>
    <timestamp>10</timestamp>
  </event>
  ...
</events>
```

Fig. 4. Partial codes of XML description of messaging events

After getting the log file, we propose an algorithm to transform the XML description into Time Petri Net which shown as follows in Table 1.

Table 1. Algorithm for transforming XML into Time Petri Net

```
Algorithm 1 Transforms XML into Time Petri Net
begin procedure
    Events ← set of events
    Agents ← set of interest agents
    initialize Time Petri Net N_main, N_subnet
    for each event ∈ Events do
        if sender and receiver ∈ Agents then
            N_subnet ← createTimePetriNet (event)
            N_main ← N_main ∪ N_subnet
        end if
    end for
end procedure

createTimePetriNet (event)
begin sub-procedure
    define states according to the event type
    map event type to transition with states and default time bound
end sub-procedure
```

Because an event of running program occur at a point of time (timestamp), for example a sent event occurs at time 5, therefore we set the earliest and latest with the same value of timestamp like [5,5] for Time Petri Net of implementation module.

5 The Worked Example

We demonstrate an application of book-trading for our verification approach. The system consists of agents that play a BookSeller and BookBuyer roles. The system was implemented follow to the Contract Net Protocol (CNP) [17], a task-sharing protocol in multi-agent systems provided by The Foundation for Intelligent Physical Agents (FIPA). The basic idea of the protocol describes the case of one agent (the Initiator) that wishes to have some task performed by one or more other agents (the Participants).

In this application, BookBuyer roles as an initiator require to buy a book from other participants which role as BookSeller. We assume that there exists at least one BookSeller in the system and BookBuyer knew the list of BookSeller agents to trade with as the pre-conditions in this example.

In such the system, BookSeller agents are ready and wait for incoming message. When a BookBuyer agent joins the system and knows the list of BookSeller, it sends a "cfp" (call for proposal) message with a book title to all BookSeller agents. A BookSeller agent that receives the cfp message will look for the specified book title in its catalogue. Then the BookSeller agent sends a "propose" message with book price back to the BookBuyer agent if the book is available. Otherwise, it sends a "refuse" message to tell BookBuyer that it does not sell that book. Once the BookBuyer agent receives all propose/refuse messages from BookSeller agents, it considers all proposals of book prices and select the lowest price from the BookSeller agent who owns that price. After that, it sends an "accept-proposal" message with the book title to the selected BookSeller agent. When the BookSeller agent receives an accept-proposal message, it deletes the specified book from its catalogue and sends an "inform" message to tell the BookBuyer that the book purchasing was successful. In some case that the BookSeller is unable to sell the book, such as, the book was sold out before the trade was finished. The BookSeller agent will send a "failure" message instead.

Fig. 5 (a) below shows the Time Petri Net from the log of running program in success scenario with the time bound [5, 5] of transition tb1 which means the agent "Buyer" sends a CFP message during the interval [5, 5].

For the specification of Fig. 5, Fig. 5 (b) we set the time bound of tb1 to [0, 5]. The result of firing occurs in [5, 5] interval. So the result of verification says that there is no failure for this model. Fig. 5 (c) we set the time bound of tb1 to [7, 10]. The result of firing occurs in [5, 5] interval. So the result of verification says that there exists a safety failure at transition tb1. Another case in Fig. 5 (d) which the time bound for the same transition tb1 is set to [0, 2]. So the result of verification tells the existing of timing failure at transition tb1.

In addition, the result of transition trace from running the verification tool can be adapted to check the reachability of the Time Petri Net model to fulfill some requirement specifications as the following example,

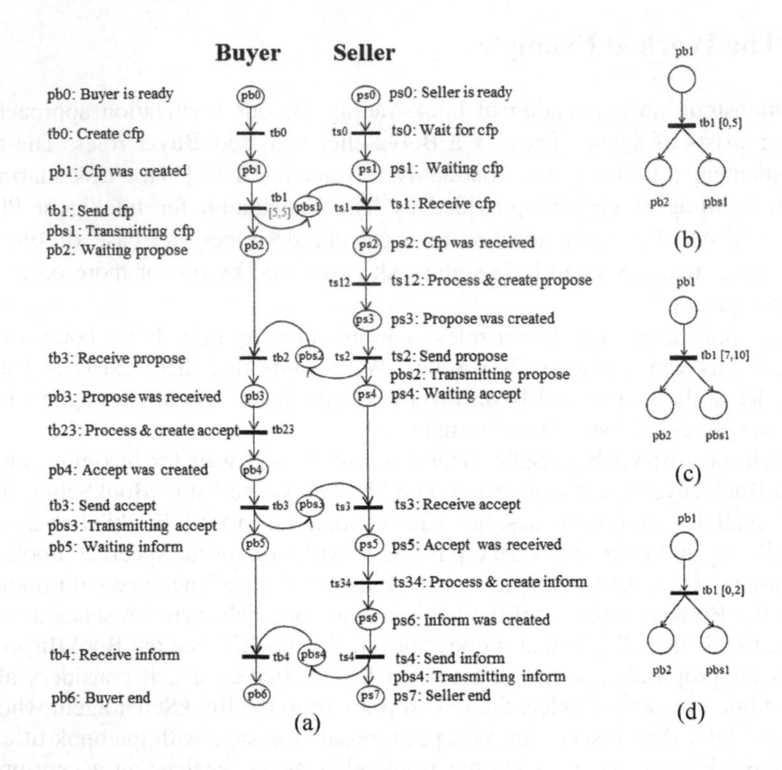

Fig. 5. Time Petri Net for implementation and specification (a) implementation module of book trading in success scenario (b, c, and d) specification modules cases that specified different time bound at transition tb1

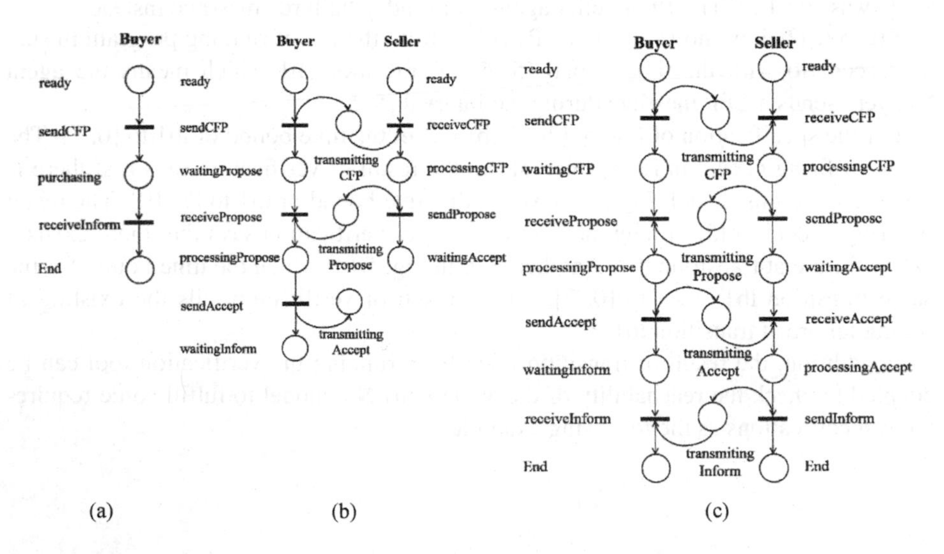

Fig. 6. Some examples of Time Petri Net of specification modules to check reachability

Fig. 6 (a) *Buyer can purchase the book successfully.* This means that from the beginning state, there exists at least a sequence of firing that the buyer is able to reach the successful state of purchasing.

Fig. 6 (b) *If at least one agent sells that book, then the buyer will eventually accept to purchase it.* This means that, once the transition of receiving propose has been fired, the transition of accepting proposal will be enabled to fire also.

Fig. 6 (c) *The ordering of messaging follows the Contract Net Protocol.* This means that the sequence of firing transition must satisfy the workflow of performatives according to the Contract Net Protocol.

6 Conclusion and Future Work

In this paper, we have proposed a methodology for formal verification of MAS based on JADE by keeping logs of running program at runtime and convert into the input form of verification tool. The verification process use timed trace theory to detect time constraint failures, including safety and timing failures. Since we focus on the work of program at runtime in this paper, we plan to focus on the verification of system model at design level which has more variety of agent behaviors than messaging events as future work.

References

1. Wooldridge, M.J.: An introduction to Multiagent systems, 2nd edn., p. xi. John Wiley, Chichester (2002)
2. Sanghavi, A.: What is formal verification? EE Times_Asia (2010)
3. Briola, D., Mascardi, V., Ancona, D.: Distributed Runtime Verification of JADE and Jason Multiagent Systems with Prolog. In: Proceeding of the 29th Italian Conference on Computational Logic, Torino, pp. 319–323 (2014)
4. Liu, Y.: Design and implementation of a Pomset automaton based runtime verifier for distributed Jade programs. Dissertation, Concordia University (2008)
5. Pradubsuwun, D., Yoneda, T., Myers, C.: Partial Order Reduction for Detecting Safety and Timing Failures of Timed Circuits. In: Wang, F. (ed.) ATVA 2004. LNCS, vol. 3299, pp. 339–353. Springer, Heidelberg (2004)
6. JADE: Java Agent DEvelopment Framework, `http://jade.tilab.com/` (accessed January 10, 2015)
7. Bourahla, M., Benmohamed, M.: Formal Specification and Verification of Multi-Agent Systems. Electronic Notes in Theoretical Computer Science, ENTCS, vol. 123, pp. 5–17. Elsevier, Amsterdam (2005)
8. Wooldridge, M., Fisher, M., Huget, M., Parsons, S.: Model Checking Multi-Agent Systems with MABLE. In: Proceeding of the 1st International Joint Conference on Autonomous Agents and Multiagent Systems, pp. 952–959. ACM, New York (2002)
9. Coelho, R., Kulesza, U., Staa, A.V., Lucena, C.: Unit Testing in Multi-agent Systems using Mock Agents and Aspects. In: Proceeding of the International Workshop on Software Engineering for Large-scale Multi-agent Systems, pp. 83–90. ACM, New York (2006)

10. Coelho, R., Cirilol, E., Kuleszal, U., Staa, A.V., Rashid, A., Lucena, C.: JAT: A Test Automation Framework for Multi-Agent Systems. In: Proceeding of the 23rd IEEE International Conference on Software Maintenance, pp. 425–434 (2007)
11. Saifan, A., Wahsheh, H.A.: Mutation Operators for JADE Mobile Agent Systems. In: Proceeding of the 3rd International Conference on Information and Communication Systems. ACM, New York (2006)
12. Baiquan, X.: Design of Platform for Performance Testing Based on JADE. In: Proceeding of the 6th International Conference on Measuring Technology and Mechatronics Automation, pp. 251–254 (2014)
13. Maghayreh, E.A., Samarah, S., Alkhateeb, F., Doush, I.A., Alsmadi, I., Saifan, A.: A Framework for Monitoring the Execution of Distributed Multi-agent Programs. International Journal of Advanced Science and Technology, IJAST 38, 53–66 (2012)
14. Chen, Y., Peng, Y., Finin, T., Labrou, Y., Cost, S., et al.: A negotiation-based Multi-agent System for Supply Chain Management. In: Proceeding of the Agents 1999 Workshop on Agent Based Decision-Support for Managing the Internet-Enabled Supply-Chain (1999)
15. Bellifemine, F., Caire, G., Greenwood, D.: JADE Internal Architecture. In: Wooldridge, M. (ed.) Developing Multi-agent Systems with JADE, pp. 131–136. John Wiley, New York (2007)
16. Taibi, M., Ioualalen, M., Abdmeziem, R.: An Automatic Petri-net Generator for Modeling Multi-agent Systems. In: Proceeding of the 8th International Conference on Software Engineering Advances, pp. 128–133 (2013)
17. FIPA Contract Net Interaction Protocol Specification. Foundation for Intelligent Physical Agents, http://www.fipa.org/specs/fipa00029/SC00029H.html

Author Index

Printed in the United States
By Bookmasters

Printed in the United States
By Bookmasters